To Grasp the Essence of Life

# To Grasp the Essence of Life

## A History of Molecular Biology

by

**Rudolf Hausmann**

*Institut für Biologie III,*
*University of Freiburg, Germany*

Translated by

**Celma and Rudolf Hausmann**

KLUWER ACADEMIC PUBLISHERS
DORDRECHT / BOSTON / LONDON

A C.I.P. Catalogue record for this book is available from the Library of Congress.

ISBN 1-4020-1092-3

Published by Kluwer Academic Publishers,
P.O. Box 17, 3300 AA Dordrecht, The Netherlands.

Sold and distributed in North, Central and South America
by Kluwer Academic Publishers,
101 Philip Drive, Norwell, MA 02061, U.S.A.

In all other countries, sold and distributed
by Kluwer Academic Publishers,
P.O. Box 322, 3300 AH Dordrecht, The Netherlands.

*Printed on acid-free paper*

There is no Majesty and there is no Might,
save in Her, Nature and Her Laws.
We all shall perish miserably
and none will know of us.

After: The Third Voyage of Sindbad
(see also: Yčas, 1959).

# CONTENTS

x

# PREFACE

History: is its study a science? Most historiographers certainly would assert that it is. To a layman, though, doubts about that will often arise: he is convinced that one of the basic principles of science should be its legendary "objectivity". Notwithstanding, he witnesses the spread of subjective biases, disguised and made unassailable under the cover of the supposed scientific objectivity. Actually, history appears to me to be always intrinsically subjective, arbitrarily selective. To someone dedicating himself to some aspects of history, it would be advantageous to immediately put aside the ideal of objectivity, and to reveal at least the critera for his partial, personal choice of themes. He also should explain what goal he aims to achieve.

My aim is somehow didactical. I am often surprised how the data in textbooks are presented such, that student readers will come to accept them, consciously or not, as dogmas; and this indoctrination occurs without any explanation of the procedures by which results were obtained, not to mention the questions from which the research initially originated. Also rarely considered are the problems relevant at the time before the "ultimate truth" was brought to light. This is, one is told, the sin of the "whig interpretation of history" (Harrison, 1987) which judges the past according to its relevance to the present; on the other hand, it is basically impossible to consider the past the way it "really was", and trying to do so is probably an act of self-deception.

But openly admitting one's personal point of view on a subject does not imply it was not derived from real facts. And if I allow myself to describe here how I see the development of molecular biology in general terms, without claiming a historiographer's professionalism, I do that specially for two reasons: first, because I personally participated in a good part of this development (although, mainly, by sitting on the fence) during 5 decades of teaching and doing research in universities and scientific institutions in Brazil and the U.S.A., as well as in Germany; second, because the professional view of the history of science, in fact, also propagates nothing less than a distorted view of reality. For almost 20 years I commenced my genetics course at the University of Freiburg pointing out an article – it must have impressed me, indeed! – from the University of Maryland's physicist and historiographer, Stephen Brush, "Should the history of science be rated X?" (Brush, 1974). It describes how the official history of science was distorted, aiming, supposedly, at the preservation of the scientific establishment's good image. The article, referring especially to the history of physics, aroused my curiosity so that I looked for parallels in biology. It was not difficult to find them: not only in classical festschrifts but also in the key studies on the history of molecular biology. Nevertheless, many such publications, containing minutely described personal interviews with scientists directly involved, in addition to pure scientific literature, contributed crucially to the elaboration of this text. An overview of these sources is found as an appendix to the literature list at the end of this book.

I hope this work will be considered in the way that I see it: as a documentation of the "inner reality" of someone who followed the development of molecular biology with the greatest interest.

The present English version is an updated and slightly expanded translation of the German edition: "... und Wollten Versuchen, das Leben zu Verstehen ... Betrachtungen zur Geschichte der Molekularbiologie" Wissenschaftliche Buchgesellschaft Darmstadt; ISBN 3-534-11575-9

# CHAPTER 1

## ENZYMES

Louis Pasteur, in 1885, investigated the fermentation of sugar: yeast cells, in the absence of oxygen from the air, transformed this substance into alcohol. Pasteur believed that this reaction could only be brought about by living cells: a mysterious force should reside in them, unfathomable forever perhaps, the "élan vital", quintessential to life. This force would be able to drive all complex life phenomena, unknown and following inscrutable natural laws. And it is said that many decades later Einstein expressed the thought that the known natural laws of physics would not suffice to explain all phenomena of life. Not to mention whole generations of classical biologists who revolted against such ideas as, for example, that the metamorphosis of an ugly caterpillar into a beautiful butterfly could be simply explained by chemistry and physics.

For many of these so-called vitalists, it came as a huge surprise – a disappointment even – when Eduard Buchner in Munich, in 1896, demonstrated that alcoholic fermentation could also occur in cell-free extracts of yeasts (ground with sand). Apparently, the typically biological phenomenon was driven not by that quintessential "vis vitalis", but by a mere material factor – a ferment, as it was then called – present in the cell-free yeast extract. By the way, Buchner's was a chance discovery: he had added sugar to his yeast cell extract, expecting thus to preserve the mixture over the weekend.

In 1906, Arthur Harden and W. J. Young, at the Lister Institute in London, described a method for separating such cell extracts into a filtrable sediment and a supernatant; in their case, again, chance was at work: they simply had let their yeast extract stay untouched for longer than planned (Leloir, 1983). Each fraction was, for itself, unable to ferment the sugar; mixing the two separated parts, though, restored the original capability. This was the beginning of the still today crucially important technique of cell fractionation (Excursus 1-1). Transformation of sugar into alcohol, apparently, depended not only on one "ferment", but on the conjoint activity of separate individual factors, each one acting on a specific step.

Figure 1.1: Memento. Pasteur's mausoleum at the Institut Pasteur in Paris. Louis Pasteur (1822-1895) in Paris, and Robert Koch (1843-1910), first a country doctor near Breslau (today: Wroclaw) and later in Berlin, were the main protagonists of a revolution in biology. They recognized specific microorganisms as causative agents of infectious diseases, feats which, regarding their consequences for mankind, overshadowed all twentieth century achievements. Tuberculosis, a scourge claiming at that time more than 300,000 victims per year in Europe, posed an insoluble challenge. Upon the discovery of the tuberculosis bacillus by Koch, this problem could be tackled rationally. For this finding, Koch was laureated with the Nobel Prize in 1905. But Pasteur was already dead at the beginning of the century, when Nobel Prizes started being conferred – to living personalities only. After Robert Koch's death, a mausoleum was also built for him, at the Robert Koch Institute in Berlin.

It was soon established that a multitude of "ferments" existed, each one directing a different reaction within the complex cell metabolism. But metabolic complexity was such as to evade all detailed understanding.

Nevertheless, the key to the mystery of cell metabolism had already been found some time ago; it lay in fields other than biology. In 1836, the Swedish chemist Jöns Berzelius had described the phenomenon of catalysis: a substance, the so-called catalyst, could, simply by its presence, accelerate a certain chemical reaction or even allow it to occur; and, at the end of the reaction, the catalyst remained unaltered. Berzelius imagined thousands of different metabolic reactions occurring, each one being directed and maintained by its special catalyst. "Ferments" were the catalysts – biocatalysts! Or enzymes, as they are called nowadays. The word enzyme is derived from the ancient greek εν ζυμη, meaning simply "in the yeast".

The nature and mode of action of enzymes, though, still had to be elucidated. Emil Fischer (Fig. 1.8) was already dedicating himself to this task before the end of the 19th century. He characterized the structural formulas of different sugars which were chemically similar. One enzyme then already known, invertase*, specifically cleaved sucrose (cane sugar) into glucose and fructose. To investigate this enzymatic activity, Fischer synthesized a series of so-called sucrose-analoga – substances with chemical formulas similar to sucrose – and tested them for their ability to be recognized and cleaved by invertase. For example, by substituting a methyl group for the fructose group of sucrose, he obtained methyl-glucose, which could be cleaved by invertase into glucose and methanol. Thus, the enzyme recognized not the substrate molecule as a whole but only a partial structure, which remained unaltered in methyl-glucose. But by modifying the spatial configuration of the substrate molecule – for example by synthesizing a methyl-glucose with the methyl group in β-configuration instead of the α-configuration (Fig. 1.2) – he obtained a product unresponsive to enzymatic activity. Fischer concluded, with a logic that even today could not be more poignant: in order to react with a substrate molecule, an enzyme must fit to it like a key to its lock (Fig. 1.3).

All this refers to the enzymatic mode of action ... but what about the material characteristics of enzymes? By means of biochemical methods one could separate various enzymatic activities present in cell extracts from other cell components, like lipids and carbohydrates (Excursus 1-1). Some examples, worked out at the beginning of the 20th century, showed that enzymatic activity could be enriched together with the cellular protein fraction; this led to the suspicion that enzymes themselves had a proteic nature. [Proteins were those macromolecular compounds containing nitrogen and displaying a chemical constitution indistinguishable from that of egg white (thus, the German word for protein: Eiweiss).] Ideally, one could fractionate cell extracts until a stage was reached when one fraction catalyzed only one single biochemical reaction, the corresponding biocatalyst was then considered to be in a chemically purified form.

_____
* The suffix „ase" is used to denote enzymes

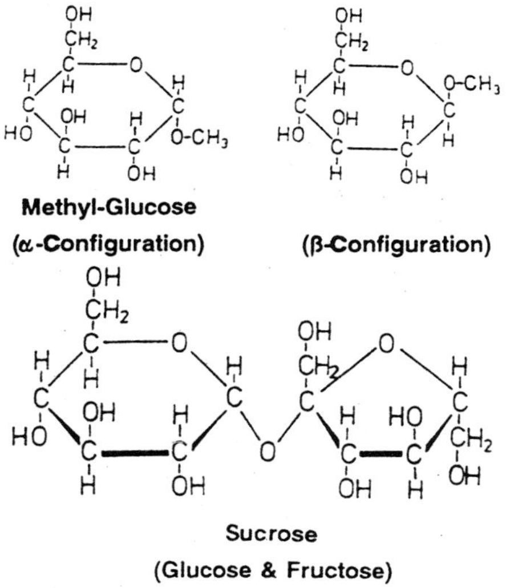

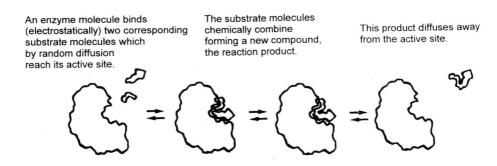

**Methyl-Glucose**

**(α-Configuration)**        **(β-Configuration)**

**Sucrose**

**(Glucose & Fructose)**

Figure 1.2: Substrate molecules for invertase must display the α-configuration.

In early 1920s Munich, a preparation endowed with such properties led to a fatal conclusion: After many purification steps and although the presence of proteins could no longer be detected, even utilizing the most sensitive methods then available, Richard Willstätter's preparations still had catalytic activity (invertase activity). That proved – proclaimed, self-assured, the Nobel laureate Willstätter – that whatever they were, enzymes were not proteins! Even when James Sumner, in 1926 at Cornell University, crystallized the enzyme urease, which displayed an obvious proteic

An enzyme molecule binds (electrostatically) two corresponding substrate molecules which by random diffusion reach its active site.

The substrate molecules chemically combine forming a new compound, the reaction product.

This product diffuses away from the active site.

Figure 1.3: Substrate-specific enzymatic mode of action. The reverse reaction is also possible: a substrate molecule reaches the active site and is cleaved, yielding two reaction produts.

composition, this was not enough to contradict and refute Willstätter; Sumner's crystals contained many water molecules, and who knows what traces of small organic molecules could be hidden in them? The last reasonable doubts would only be put aside in the 1930s by Moses Kunitz (who, fleeing the czar's police, had found refuge in America) and John Northrop, working at the Rockefeller Institute in Princeton. Employing methods of electrophoresis and centrifugation for analyzing purified enzymes, they were able to demonstrate their identity with the protein fraction. Willstätter had made a gross mistake. Enzymes were apparently so active that even the smallest traces of the corresponding protein – not detectable by the methods then available – still displayed easily observable catalytic activity.

How could one derive the infinitely diverse specific activities of different enzymes from the characteristics common to all proteins? For many decades this question prompted biochemists to investigate the molecular composition of proteins. First of all, it was essential to discover that proteins were assembled from amino acid building blocks. These were bound together by peptide bonds, forming long polypeptide chains (Fig. 1.4 and 1.5). This was the hypothesis simultaneously proposed in 1902 by Franz Hofmeister in Strasbourg and Emil Fischer in Berlin (Excursus 1-2). The general validity of their proposition was repeatedly challenged

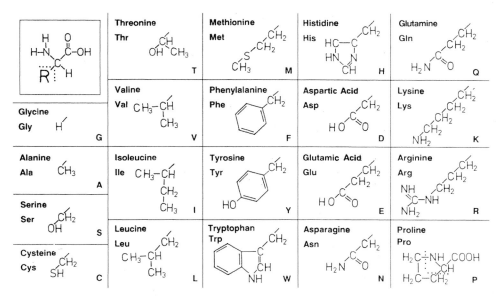

Figure 1.4: The 20 amino acids, building blocks of proteins. Above left, the general formula of an α-amino acid; inside boxes, the distinct side chains (R) with their respective 3-letter and 1-letter symbols. In the case of proline, the complete formula is shown because its side chain undergoes a covalent bond with the α-amino group (forming a ring).

for years. But never refuted! According to Frederick Sanger in 1952, this represented the most convincing argument for its acceptance. In that year, Sanger and his co-workers, in Cambridge, U.K., after eight grinding years, had determined for the first time the amino acid sequence of a protein, namely insulin, which contained 51 amino acids (Ryle et al.,1955). Only then were the most obstinate sceptics really convinced. For this feat, Sanger was laureated with the 1958 Nobel Prize. This also

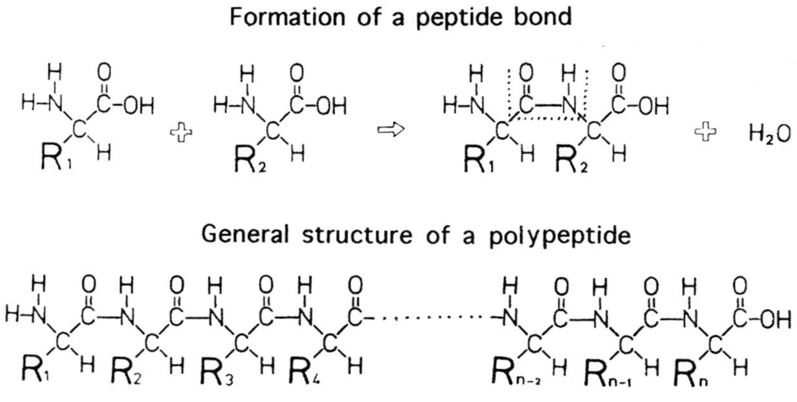

### Formation of a peptide bond

### General structure of a polypeptide

Figure 1.5: Proteins, as chain-like molecules derived from peptide bonding of the amino acid building blocks. The peptide bond is framed by dots in the top right scheme.

provided an impressive documentation of the concept, still not fully accepted, that each distinct protein had a different, well defined and highly specific chemical formula. The amino acid sequence of the polypeptide chain (with an amino and a carboxyl terminus) fully described such formulas, since the amino acids were invariably bound by peptide bonds. Nevertheless, knowledge of the amino acid sequence, the so-called primary structure (Fig. 1.6), did not provide any insight into its metabolic action – a sobering disillusion.

Well, were not many proteins crystallizable? This property suggested that the thousands of atoms composing the protein macromolecule displayed an organized spatial structure. This is tantamount to the assertion that polypeptide chains could not possibly be folded up by chance, at random. If this were the case, amorphous masses would result, but not ordered crystalline aggregates. The next logical step for clarifying this point was to gather information on the specific folding of polypetides. Hopefully, knowledge of the specific spatial assemblages of the building blocks would throw a glimmer of understanding on their functioning. Nevertheless, the methods of organic chemistry were not suited to analyze spatial structures of molecules, the so-called secondary and tertiary structures (Excursus 1-3). This is so, because the forces that determine and preserve spatial structures are not like the strong chemical bonds

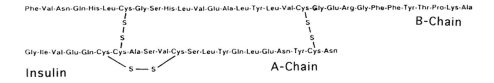

B-Chain

Insulin    A-Chain

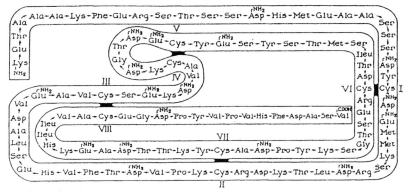

RNAase from bovine pancreas

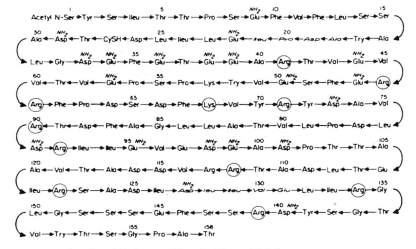

Coat protein subunit of Tobacco Mosaic Virus (TMV)

Figure 1.6: Primary structure of the three proteins first sequenced (amino end at left): insulin (Ryle, Sanger, Smith & Kitai, 1955); RNAase (Spackman, Stein & Moore, 1960; see also Hirs, Moore & Stein, 1960), TMV subunit (Tsugita, Gish, Young, Fraenkel-Conrat, Knight & Stanley, 1960).

that bind together the atoms within a molecule (covalent bonds); they only correspond to weak electrostatic attractions between different atoms within an amino acid chain (for example: hydrogen bonds, ionic bonds) and hydrophobic interactions between amino acid residues.

Promising tools for this research were X-ray diffraction analysis ( Excursus 1-4), as well as empiric model-building (see Chapter 5).

An understanding of the mode of action of enzymes – even if only in general terms – was brought about by their structural analysis. Over a thousand metabolic pathways, leading to the synthesis of a multitude of small organic molecules, could then be regarded as a consequence of particular interactions between substrate molecules and the so-called active sites of corresponding enzymes. Active sites of enzymes were thought of as being exactly defined cavities located on the surface of the folded polypeptide structure. The suiting substrate molecule, with its appropriate shape, would fit into this cavity and be kept there for a brief lapse of time, allowing the catalytic reaction to occur.

Understanding the chemical principles of the entire cell metabolism (Fig 1.7) became feasible. The astonishment shown by vitalists in face of the cell's huge synthesizing capabilities turned out to be pointless. [One should remember that, for example, a coli bacterium can, in a matter of minutes, synthesize an immense variety of organic substances from sugar and a few mineral salts, substances which do not react with each other spontaneously (Excursus 2-1).]

However, gaining an insight into cell metabolism did not help to grasp life's mystery. The basic problem was still unsolved: the enigma of the living cell's self-replication. One has to consider that whenever such a sophisticated living form as, for example, a coli cell divides in two, after about one hour, in a medium containing sugar and salts, it has not only achieved the enzymatic synthesis of innumerable small organic molecules but also the synthesis of a complete set of cell enzymes.

How these newly synthesized enzymes made their appearance became the next crucial question. This problem had remained repressed from the professional consciousness of biochemists for many decades, as a veritable taboo theme: since it could not be tackled by the usual biochemical methods, even to think about it was seen as frivolous. This attitude remained pervasive till after the Second World War.

Of course, one could postulate an enzymatic function able to catalyze peptide bonding of amino acids. But this did not solve the problem, which lay in specifying the precise amino acid sequence of different enzymes. Even for small proteins with, say, only 100 amino acids, $20^{100}$ possible different primary structures are to be considered, since each one of the 20 amino acids can, theoretically, occupy any one site in the polypeptide chain (see also Excursus 1-5). Obviously, accepting the idea of an almost infinite number of enzymes, each one adding a certain amino acid to a certain polypeptide, sums up to sheer fantasy; it would become the paradox of the enzyme which makes an enzyme, which makes an enzyme... Vitalism, thrown out of the window by biochemists, was again creeping in through the back door.

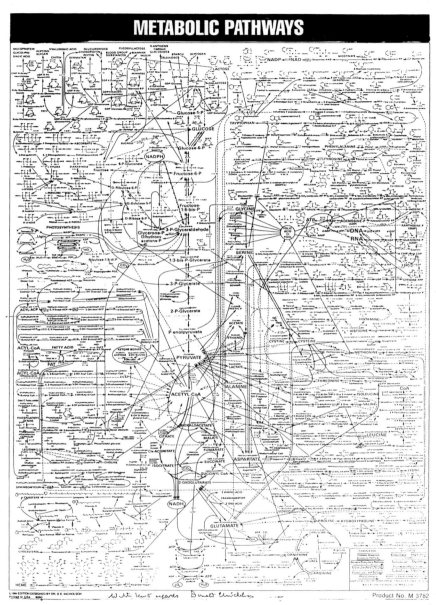

Figure 1.7: Cellular metabolism. This synoptic overview confers an idea of the complexity of the network of biochemical reactions as they have been understood for some decades now (© Donald E. Nicholson, Univ. Leeds).

**Excursus 1-1**
**CELL FRACTIONATION**

Preparing cell-free extracts, or homogenates, is the first step for obtaining sub-cellular fractions. These, as well defined as possible, can then be used for further analyses. In the 19th century, cells were already being ground with sand and subsequently filtered through paper, gauze or even through gelatine. In 1893 in Munich, Eduard Buchner (1860-1917; Nobel Prize for Chemistry, 1907) tried to patent this procedure, only to get his application turned down with the argument that the method had already been utilized before (Kohler, 1971). Nowadays, such homogenates are obtained by osmotic shock, ultrasound, blade or glass homogenizers or by using the so-called French Press (sudden strong pressure variation), depending on the characteristics of the cell walls, or their absence. Centrifugation of the resulting homogenate (diluted with buffer solution) frees it from larger debris and still unbroken cells. The supernatant can then be subjected to different techniques of choice (Deutscher, 1990). A first and important method of protein separation, precipitation by salts, was already known by the end of the nineteenth century. Hofmeister then refined this method: inorganic salts, especially ammonium sulfate, are gradually added, in increasing concentrations, to cell homogenates; different protein fractions, according to their solubility in different ionic concentrations, precipitate from the suspension as fine flakes, being subsequently collected by filtration or centrifugation. Protein fractions can also be separated by other methods, like isoelectric precipitation (which selectively utilizes different pH values) or solubilization in various special organic solvents. As early as 1833, alcohol was used to achieve fractionation of biological substances. Material obtained by such procedures can be further subfractioned by, for example, electrophoretic or chromatographic techniques or ultracentrifugation – techniques developed and improved after the Second World War. In many cases, after various fractionation steps, extensive purification of a specific protein may allow it to crystallize. (Obtaining crystals under appropriate conditions is a challenge in itself.) Some proteins can be purified by the very step of crystallization; such examples are hemoglobin, which, already in 1870, was obtained in pure crystallized state directly from rough erythrocyte homogenates; and crystals of ovalbumin were obtained by Hofmeister, in 1889, from a partially purified fraction of egg white. For good preservation, it is required that protein preparations are processed at low temperatures, around the freezing point, but in the case of temperature resistant proteins, heating can be used as an enrichment step.

**Excursus 1-2**
**EMIL FISCHER, FRANZ HOFMEISTER, AND THE PEPTIDE BOND**

In 1902 and by chance on the same day, Emil Fischer (Fig. 1.8) from Berlin and Franz Hofmeister from Strasbourg presented at a scientific congress their hypotheses of the peptide bond, which represented the connexion between amino acids in a protein. Fischer was an organic chemist who had dedicated himself to the study of pigments, sugars and purines (which brought him the honor of the Nobel Prize in 1902), but also of proteins. While he considered proteins simply as another interesting organic chemical substance, the physician and biologist Hofmeister (1850-1922) was specially interested in their biological characteristics. Both in temperament and style the two scientists could not have been more dissimilar. Fischer was a domineering personality, although not totally without wit. He was talented in making human contacts, which guaranteed him social and professional recognition. Even for his era, he was an extreme autocrat who did not accept any contradictions. In most of his over 600 scientific publications, he did not even bother to cite his assistants who actually performed the work. Sometimes he thanked them, but he never cited them as co-workers. Hofmeister, communicative in small circles and liberal in his laboratory – out of more than 300 publications from his Institute in Strasbourg, only about a dozen show his name as author –, preferred solitude, attending congresses only seldom and viewing human society with scepticism, even contempt. These characteristics led consequently to his scientific contributions being underestimated. Thus, it seems appropriate to emphasize that it was Hofmeister who first recognized proteins as diverse and precisely defined chemical substances and who drew attention to their possible function as enzymes. Each of the biochemical reactions of the cell was controlled by a specific enzyme protein, he declared prophetically (Dressler & Potter, 1991; Fruton, 1990).

**Excursus 1-3**
**SPATIAL STRUCTURES OF POLYPEPTIDES**

Hsien Wu, living in post-imperial China, pondered the possible physico-chemical basis of protein folding. He postulated that native proteins were subject to special folding patterns, intrinsic to the corresponding polypeptide chain, stabilized by concerted actions of non-covalent weak forces. Denaturation of these proteins – by heating, for example – would undo their peculiar folding, leading to disorganization and random entanglement of the chains (Wu, 1931). Alas, Wu did not go into details on the nature of these weak forces.

In the 1930s, Linus Pauling (Fig. 1.9), from Caltech (California Institute of Technology, in Pasadena), also gave some thought to the mechanisms directing the folding of polypeptides. He concluded that, especially, hydrogen bonds – in other

Figure 1.8: Emil Fischer (1852-1919).

words, weak electrostatic forces – were holding protein molecules in their characteristic forms (Mirsky and Pauling, 1936). For years, he and his collaborators analyzed the detailed atomic coordinates of the simplest peptides, by means of X-ray diffraction. In the 1940s, he was certain that the peptide bond was flat and stiff; its atoms could neither gyrate relative to each other, nor alter their binding angles. Nevertheless, X-ray diffraction patterns from his peptides and other, more complex ones, like hair α-keratin, clearly showed a spiralization of the monotonous backbone. He decided to build a model encompassing both facts. Bound to bed by flu, Linus Pauling began tentatively to discern some models using paper stripes with peptide backbones sketched on them – amino acid side chains were not taken into account. In this way he discovered the α-helix (Fig. 1.10): the peptide backbone can form a stable screw, in which hydrogen bonding between the carboxyl oxygen atom from one amino acid and the hydrogen atom linked to the amid-nitrogen from the 4th next amino acid takes place (amino acid n° 1 bonds to n° 5, n° 2 to n° 6, n° $n$ to n° $n+4$, starting from the amino terminal). Besides the α-helix (Pauling & Corey, 1950),

Figure 1.9: Linus Pauling (1901-1994). Pauling studied and worked for decades at the California Institute of Technology (Caltech), Pasadena. He described chemical bonds as quantum-mechanical electrostatic phenomena, thus transforming the then current concepts of molecular structures. His classical work "The nature of the chemical bond" (1939) revolutionized chemistry. The discovery of the α-helix and the identification of sickle-cell anemia as a molecular disease marked his triumphs in biochemistry. After that, he was caught in years-long controversies on the origin of cancer, which, according to him, could be avoided and halted by means of megadoses of vitamin C (Pauling, suffering from prostate cancer after he was 90, then asserted that it would have happened 20 years before, were it not for his vitamin C regimen). During the post-war era, Pauling was actively propagating pacifist politics of disarmament, against the development of the hydrogen-bomb. This attitude brought him innumerable frictions not only with McCarthy era officials, but also with his colleagues. The 80-years-old Pauling was still passionately involved in the peace movement; this brought him the distinction of being the only laureate to be accorded two unshared Nobel Prizes: one for chemistry in 1954, one for peace in 1963.

Pauling and his collaborators (Pauling & Corey, 1951; Marsh, Corey & Pauling, 1955) described other so-called polypeptide secondary structures, which are structures based on interactions between atoms of the backbone: parallel and anti-parallel pleated sheets. These are, as later shown, together with the α-helix, the crucial structural elements of most proteins.

However, secondary structures could not, by themselves, be responsible for the peculiar and precise polypeptide chain folding. For example, if the α-helix were the only determining spatial element of polypeptides (as was the case for the monotonous polyalanine studied by Pauling), then all polypeptides would have similar forms: longer or shorter rods with amino acid side chains sticking out in all directions as bigger or smaller lumps or protuberances. In addition to the oxygen and hydrogen atoms of the backbone, the side chains of the amino acids – as Pauling already suspected – can also interact with each other, for example, through the mutual attraction of opposing weak electrostatic charges. Non-polar groups are also involved: let us consider that water molecules establish and maintain direct contact, preferentially, with each other or with other polar molecules, because their respective opposite positive and negative

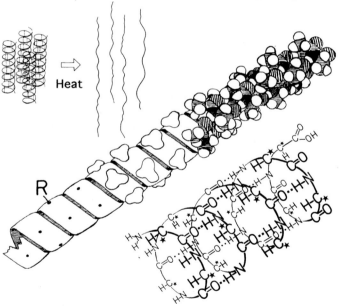

Figure 1.10: The α-helix as a secondary structure of the backbone of polypeptides. Top left: scheme of bundled parallel α-helices, as occurring in hair; heat dissociates the hydrogen bonds, allowing the hair thread to be stretched almost twice its original length. Middle: scheme of polyalanine, structured as an α-helix. Bottom right: the corresponding chemical bonds (hydrogen bonds symbolized by dots and amino acid side chains by stars).

charges will then neutralize each other. This leads to the repulsion by water of non-polar molecules or side chains, consequently denominated as hydrophobic. Hydrophobic amino acid side chains, like those of phenylalanine or valine, once repulsed by water, tend to come into contact with each other. Considering a polypeptide, this means that many non-polar amino acid side chains, located at different sites of the primary structure, aggregate, forcing the α-helix to bend at different points. A very typical folding of the polypeptide chain results – the so-called tertiary structure (Fig. 1.11 and 1.12) – because, despite the many imaginable interactions between amino acid side chains, only one combination is actually accomplished. Since proline does not fit sterically into an α-helix structure, sites occupied by this amino acid turn out to be preferential points for bending.

One could assume that those very special tertiary structures, derived from peculiar foldings, are thermodynamically the most stable. Consequently, one would also expect that polypeptide chains tended to restore their original forms spontaneously, once acted upon by disrupting external factors such as heat, acids, or high concentrations of urea. Truly, this is the case documented by some exceptional examples:

Christian Anfinsen (Nobel laureate in 1972) and his co-workers at the National Institutes of Health (Bethesda, Maryland), experimented, at the end of the 1950s, with ribonuclease from bovine pancreas. They observed that its tertiary structure totally unfolded in 8 molar urea, after chemical reduction of its disulfide bridges (whose formation greatly contributes to stabilize already formed tertiary structures – one of the few spontaneously occurring metabolic reactions). The resulting loose peptide chains simultaneously lost their ribonuclease activity. Upon chromatographic separation of the peptide chains from the urea, the original enzymatic activity would be restored to 80% of its initial rate (Anfinsen et al.,1961). The disulfide-bridges were reestablished, apparently in the very same combination as in the original structure, although 105 different combinations of the eight cysteine residues were possible. This case was used to reiterate the bold assertion made by Francis Crick (1958) that the spatial structures of polypeptides were exclusively determined by their primary structure; specific folding, then, would necessarily result from the many non-covalent interactions between individual atoms from the backbone of the polypeptide and the amino acid side chains bound to it; this, occurring spontaneously, should not pose any problem. According to Crick, the really fundamental question of molecular biology was: how is the primary structure determined? On this point, Crick was right: establishing the primary structure was crucial. But, is it not an exaggeration to take some special examples (Epstein et al., 1963) as proofs of a general principle? In the case of hundreds of other proteins, the reestablishment of the original tertiary structures after previous denaturation was not achieved. (It is surprising that Anfinsen's elegant experiment still is fashioned by many textbooks as proof of Crick's idea.)

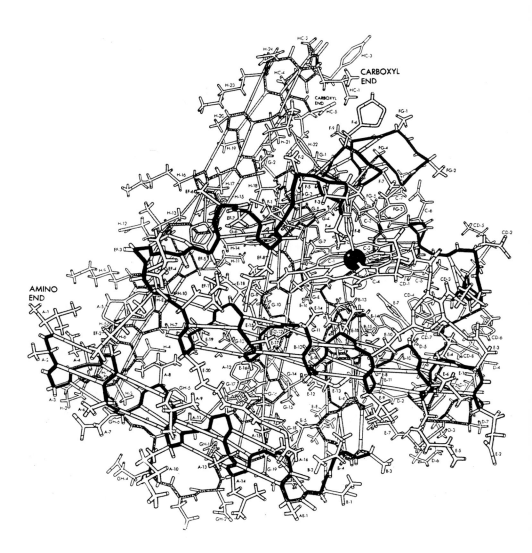

Figure 1.11: Tertiary structures of proteins, exemplified by myoglobin: in this model, hydrogen bonds and atomic radii are represented by bars. The peptide backbone is emphasized in black, α-helix regions being well recognizable. The black sphere represents the iron atom of the heme group (built into the peptide structure only after its completion, and responsible for binding oxygen). Left: amino end. Middle right, almost at the top: carboxyl end (© Sir John C. Kendrew, see also Kendrew, 1961).

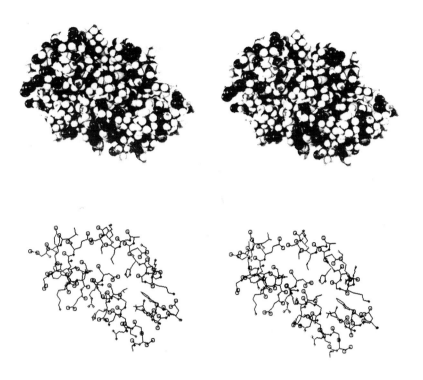

Figure 1.12: Stereo images, consisting of a sphere model (top) and a simplified computer diagram of myoglobin (most inner atoms deleted). One should stare at the stereo image with relaxed eyes, at a distance of about 30 cm; after three images emerge, one concentrates on the one in the middle, which then evoques a spatial illusion (Yankeelov & Coggins, 1971).

Could it be that native tertiary structures are progressively formed as the synthesis of the corresponding polypeptide chain proceeds? The first foldings occurring without interference from amino acids to be attached next to the growing chain? Yes, but not only that. Experiments set up to investigate something totally different (Georgopoulos, 1992), gradually made it clear that Crick's spontaneous folding was not always occurring. Can one still call this process spontaneous, if special adjuvant proteins, dubbed chaperones, must be present in order that the specific folding proceeds correctly? Only these chaperone proteins can, in many cases, create the appropriate conditions for the correct folding to occur during peptide synthesis (see, for instance, Hartl & Hayer-Hartl, 2002). In defense of Crick: he never actually asserted that a spontaneous process excluded the necessity for special conditions. A mere detail of semantics may be at issue.

The same principle of mutual interaction due to electrostatic forces operates at still another level: two or more of the same, or different, polypeptide structures, each one composed of one amino acid chain, may aggregate to form larger complexes. The condition for this to occur is a steric and electrostatic fit between the molecules in question. Only by complementary electrostatic charges and an appropriate, complementing configuration can the many weak bonds sum up to create a relatively stable aggregate. Under appropriate biological conditions, two or more of such fitting polypeptides bind together non-covalently, forming stable complexes, the so-called quaternary structures. Examples of quaternary structures are multi-enzyme complexes, virus coats, and, of course, hemoglobin (Fig. 1.13; Excursus 1-4).

Enzymatic activity and many other properties of proteins in general derive automatically from polypeptide tertiary or eventual quaternary structures. For their part, these tridimensional structures are ultimately determined by the sequence of amino acids of the polypeptide chain – whether they can be produced only *in statu nascendi*, with or without the help of chaperones, is irrelevant in considering the general context.

Understanding that each protein, each enzyme, displays a particular tertiary or quaternary configuration led to the notion that enzymatic activity depends directly on the spatial form of the corresponding polypeptide. This concept is convincingly supported by some examples; it was shown that in some enzymes, amino acids localized at totally different sites on the primary structure came to occupy neighboring positions at the concavity of the active site of the enzyme (Fischer's key) as a result of folding. The methods of organic chemistry do not allow us to analyze the spatial structures of proteins. Practically, the only method for elucidating tertiary and quaternary configurations is X-ray diffraction analysis ( Excursus 1-4).

**Excursus 1-4**
**X-RAY DIFFRACTION ANALYSIS**

After the discovery of X-rays in 1895 by Wilhelm Röntgen, in Würzburg, Max von Laue in 1912, in Zürich, pondered the meaning of images created when these radiations penetrated and crossed matter. These were primarily diffraction patterns similar to those already known in optics. These patterns revealed the presence of lattices corresponding to the disposition of atoms in the matter hit by the X-rays. (By the way, this observation proved the wave nature of X-rays.) Basically, it was possible to predict the diffraction pattern if the spatial atomic lattice structure and the wavelength of the radiation were known. The inverse prediction was however not so easy, since different atom arrangements may originate similar or even identical patterns of X-ray diffraction. In the relatively simple case involving the crystal lattices of mineral salts, one could overcome this difficulty by postulating a model, to be then confirmed and refined by data from X-ray diffraction. This methodology was

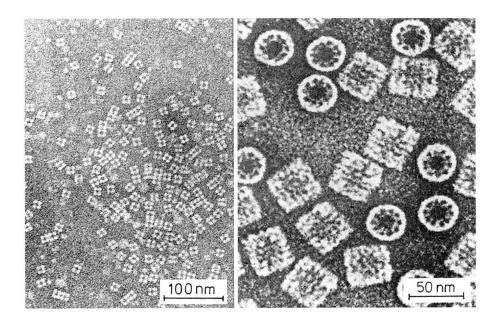

Figure 1.13: Quaternary structures of proteins as evidenced by electron microscopy. Each clear spot in a structure is the image of a polypeptide subunit, that is, of a tertiary structure. Left: pyruvate carboxylase from chicken liver (Cohen et al., 1979); right: hemocyanin (oxygen transporter of snails), in upright or lateral position (Fernández-Morán et al., 1966).

systematically and intensively explored in Cambridge, U.K., in the 1920s by William Henry and William Laurence Bragg, a father and son team. A fine X-ray beam was directed at a small crystal; the diffraction pattern obtained (table salt produced the simplest pattern) was fixed on a photographic plate for subsequent analysis. In this way, the Braggs clarified several crystal structures, first of simple salts, and later of more complex minerals, such as beryl and other silicates.

As one can imagine, scientific ambition and challenges offered by gradually more complex molecules contributed to the development of a self-supporting dynamic inside the newly established field of X-ray crystallography, complete with its own specialists: the X-ray crystallographers. Desmond Bernal, for example, one of Bragg's students, a universal genius in physics but, that aside, simple-minded in politics (he remained convinced of and dedicated to communism all his life, despite Stalin), was one of the first to experiment with biological material. In particular, he was the first to obtain X-ray diffraction patterns from the crystals of a protein (pepsin). And William Astbury, also from Bragg's school, reached the pinnacle of sophistication for the time before the Second World War with his diffraction patterns of hair, horn and, yes, DNA. But the ultimate challenge was protein structure. Crystals of relatively simple proteins, like hemoglobin, produced hundreds of diffraction spots of different intensities in an apparently chaotic arrangement (Fig. 1.14). A life-time task!

At least that is how Max Perutz, an Austrian emigrant and graduate student of Bernal, saw it. He worked tirelessly for 17 years without any real progress. In 1950, finally, a first breakthrough was attained, though not by him. Linus Pauling was the one who, utilizing the most simple and elementary method, discovered the α-helix (Excursus 1-3). For Max Perutz, a relative consolation remained in the fact that he was able to demonstrate that regions of α-helix also existed in horse hemoglobin. A second breakthrough happened when Perutz, convinced by Francis Crick's ideas,

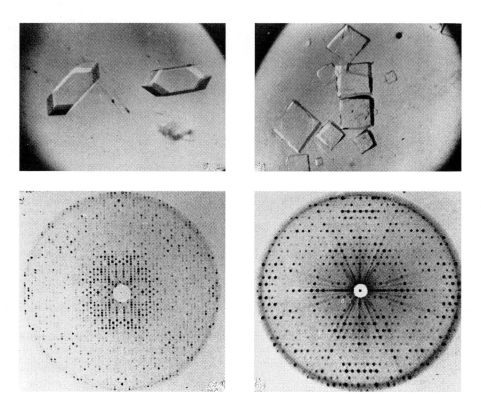

Figure 1.14: Crystals of proteins (enlargement: about 20-fold), with their corresponding X-ray diffraction patterns. Left: phosphoglucose isomerase from pig; right: phosphoglycerate kinase from yeast (Campbell et al., 1971).

marked his hemoglobin molecules with mercury atoms at certain reactive sites (the SH-groups of cysteines). These heavy atoms, surrounded by thick electron coats, altered the intensity of many diffraction spots, allowing the calculation of an important parameter, the phase of the diffracted wave (whose amplitude could be determined by the intensity of the spots). It was John Kendrew who was to achieve a third breakthrough by architecturing a rough spatial model of myoglobin. He was a younger colleague who had joined Perutz' small group (Perutz was basically a "lone fighter") some years before. Myoglobin was relatively easy to analyze, since it consisted of a single polypeptide, while hemoglobin had four subunits (2 $\alpha$ and 2 $\beta$ subunits); these subunits were structurally similar to myoglobin (not surprisingly, since we now know that the $\alpha$ and $\beta$ subunits, as well as myoglobin, are molecular descendants of a common ancestral protein). When Perutz finally presented his hemoglobin model in 1958 (Fig. 1.15), after over two decades of work, both, he and Kendrew, deserved the Nobel Prize conferred to them in 1962. While Kendrew afterwards dedicated himself to other tasks – he was for many years director of EMBL (European Molecular Biology Laboratory) in Heidelberg, Germany –, Perutz, venerably aged, continued to work in Cambridge on his hemoglobin, investigating the most subtle alterations of its form, in accordance with its function as blood oxygen transporter.

In these pioneering old days, years, or even decades, slowly passed before the thousands of calculations and measurements achieved their final aim – the building of a model corresponding to the tertiary structure of a protein. Nowadays, these time intervals are measured in days or weeks. Besides that, the resolution achieved is routinely much better: 2Å (0,2 nm), in contrast to 6Å in the first models. But any method still requires the availability of good crystals as a precondition for successful work; obtaining them is often still very laborious, laden with difficulties, and sometimes shear impossibilities.

The current trend, utilizing computer technology to perform this sort of analysis, was in fact initiated by Kendrew. The most modern techniques utilize sensors directly linked to X-ray cameras. And suddenly, by the turn of the millennium, protein structure determination became a commercial commodity, being carried out at an industrial scale by small upshot biotechnology companies founded in the wake of the proteomics craze (see Chapter 23). Hundreds of protein tertiary structures are to be determined within a period of months and the data sold to customers, like pharmaceuticals companies, which hope to use this information for drug search and drug design (see Fletcher, 2000; Harris, 2000).

Figure 1.15: Max Perutz (1914-2002) and Sir John Kendrew (1917-1997) analyze a model of a hemoglobin molecule.

**Excursus 1-5**
**PASCUAL JORDAN'S MISTAKE**

During the post-war era, quantum physicist Pascual Jordan (see also Chapter 5) – like many of his colleagues in the field – pondered the biological riddles. He imagined having found the scientific proof of God's existence as creator of life (Jordan, 1970). He considered the following: the probability of a chance occurrence of the correct sequence of amino acids in a polypeptide, like the 135 amino acids of the $\alpha$-subunit of hemoglobin, was $1 : 20^{135}$. If the entire matter of the universe consisted solely of amino acids, and these, since the beginning of time, had tried a thousand new combinations every second, then only an infinitesimal fraction of all possibilities would have been tested by now; $20^{135}$ was practically the same as an infinite number of possible combinations; thus, the probability of getting the winning sequence by chance was practically zero. This, according to Jordan, was tantamount to requiring the action of conscient creation. But his proposition was based on false premises; his question posed wrongly. It should have been: what is the probability of getting, during millions of years of evolution, a protein capable of transporting oxygen? And there is an infinite number of imaginable proteins possessing this characteristic – infinite divided by infinite: indeterminate!

# CHAPTER 2

## GENES

𝒜 new approach had to be considered.

Would life not be easier to be understood if one thought of cells, living beings, as machines, self-replicating units, automata, that continuously reproduce themselves, yielding descendant automata identical to themselves?

The concept of a self-replicating machine was elaborated by John von Neumann (Fig. 2.1) in Princeton at the beginning of the 1950s. Von Neumann proved mathematically that constructing such self-replicating machines would be feasible, at least theoretically. For this purpose three conditions had to be fulfilled:

- Building blocks should be available.
- A source of energy should be on tap.
- A blueprint and a mechanism for the correct assembly of building blocks should be at hand.

In our case, the first two points are trivial (Excursus 2-1 and 2-2); the third stipulation was the actual problem – at least that was the way some of the first molecular biologists saw it. Implicit in the concept of a self-replicating machine was not only the availability of information for the correct assembly of building blocks, but also the ability to replicate this information and propagate it to descendant machines.

In fact, von Neumann could have saved the effort spent trying to demonstrate his point, because self-replicating machines have existed for about 3 billion years on our planet (Excursus 2-1). Nevertheless, accepting this obvious truth was only a later insight – von Neumann never acknowledged living cells as examples of his wonder machine. [Man-made self-replicating automata, or robots, do not yet exist, despite media reports having for many years suggested that steps in this direction were already being taken. Notwithstanding, such robots build robots that make cars, or whatever – but not identical robots (Fig. 2.2).]

The concept of information, currently so crucial in molecular biology, permeated only slowly and belately, through tortuous paths, the minds of geneticists and biochemists. We will try to analyze this process from a practical, experimental point

of view. [Although every working scientist is familiar – to his or her own satisfaction – with the concept of information, this concept is extremely difficult do cope
with theoretically. For a masterly treatment of this aspect in the context of the history
of molecular biology, see Kay (2000).]

Figure 2.1: John von Neumann, "the most intelligent person on earth" (Monk, 1993) with his
wife, Klara. Johann (later John) von Neumann was born in Budapest and raised in Germany.
He studied in Switzerland and emigrated in 1930 to the U.S.A. (Princeton). On the trail of
mathematical logic and theoretical physics, he found himself in the role of a scientific counselor
for issues involving cold war politics and nuclear détente. Among his varied bright
contributions was the development of a mathematical theory of games which revolutionized the
understanding of many facets of social phenomena, like economy, military planning, but also
biology [evolutionary theory, sociobiology (Maynard Smith, 1982)]. (Life itself can be viewed
in the light of game theory: the individual plays his personality against an unpersonal universe.)
Von Neumann also tackled the theoretical problem of self-replication, delivering a
mathematical proof for the possibility of self-replicating machines (Neumann, 1966).

650                              New Scientist   28 August 1980

# Technology

## Japanese robots are prepared to self-multiply . . .

Peter Marsh, Tokyo

One of Japan's biggest electronics firms is building a factory where robots will make other robots. Fujitsu Fanuc will convert science fiction into fact at a cost of £20 million.

Thirty robots and one hundred and fifty people will work in the factory at the foot of Mount Fuji near Tokyo. At full capacity, the plant will make 350 robots per month, in addition to about 500 other types of electronic machines, such as computer-controlled machine tools.

Fujitsu Fanuc's current robot production at its main Tokyo factory is thirty per month. But the new factory is opening because orders are fast expanding. The company says that it has doubled the output from its Tokyo factory in the last year. But because production there is also highly automated it did not need to increase its workforce of 850 people.

The new plant, which opens in January 1981, will include flats for the families of 30 workers. There will also be 60 dormitory rooms for employees as well as 20 guest rooms.

The plant will operate for 24 hours a day, says the company. Yet people will work there for only eight hours a day. The rest of the time the robots and other automated equipment will work away by themselves.

The job of the robots will mainly be to load metal parts, which eventually will form the basis of new robots, into automatic machine tools for cutting and shaping. TV cameras will watch the robots while nobody is on the shop floor to make sure nothing goes wrong. If something does happen, a solitary maintenance worker looking at a TV console will shut the plant down.

Fujitsu Fanuc says it will also install special robot monitors which will automatically stop the robots from working when certain types of faults occur. At this stage, however, the firm will not reveal details of how this monitor operates.

Most of the one hundred and fifty employees in the new factory will be trained engineers. Because of its policy of running highly automated factories the company says it does not need to recruit very often.                     □

Figure 2.2: Technology still has not been able to construct a self-replicating machine – despite a long list of suggestive media reports.

Already in 1865, Gregor Mendel demonstrated, through experimental crosses with peas, that individual "elements" – as he dubbed them – existed in the living cell and were responsible for the appearance of specific hereditary characteristics (Excursus 2-3). Later, in 1906, these "elements", or hereditary factors, were given a new name by the Danish geneticist Johannsen: genes. Genes were endowed with the same quality of the information-carrying element of von Neumann's machines: they possessed two fundamental characteristics, essential for the maintenance of life. That is, due to their very presence, they had the ability of eliciting the appearance of specific characteristics (such as red petals or dwarfism in pea plants), and in addition to that, they could replicate and propagate themselves from one generation to the next.

The totally mysterious genes were, apparently, located inside the cell nucleus – or, more precisely, in the chromosomes. This conclusion was reached in the beginning of the 20th century not only by William Sutton at Columbia University, N. Y., but also by Theodor Boveri in Würzburg, Germany, and – mostly forgotten – by Nettie Stevens (Fig. 2.3), and later confirmed in much detail by the pioneering work of the Morgan school. The Morgan school, also located at Columbia, with its "fly room" (Fig. 2.4), made its impact especially during the decade 1910-1920, and represented the coronation of classical genetics. Detailed chromosomal maps of the fruit fly *Drosophila melanogaster* were elaborated (Excursus 2-4), and crucial insights on the correlations between chromosome structure and gene expression were reached. As representative examples of such achievements, one could cite the observation of position effects by which genes, depending on their location on the chromosome, actively express themselves or not, and the correlation between puffs

Figure 2.3: Nettie Stevens, working in the laboratory of E.B. Wilson at Columbia University in 1905, was the first to observe that the sex (of many insects) was determined by special chromosomes, thus contributing crucially to the chromosome theory of heredity (Booth, 1989; Dunn, 1965).

(protrusions of the giant chromosomes in fly larvae) and metabolic activity of specific genes at particular stages of development. Nevertheless, the actual mode of action of genes – the biochemical description of their metabolic activity – remained absolutely cryptical. Many a classical geneticist made a virtue out of this state of affairs, considering the enigmatic hereditary factor – the gene – as imbued with the mathematical beauty of the abstract. Genes could be demonstrated, if not directly by the visible characteristics of the individual – the phenotype – then by analysis of the progeny.

This state of mind in face of the enigmatic genes was not universal, though. Some scientists pondered their material status, as for example Hermann Muller did in 1926. The maverick Muller, who originally belonged to the Morgan school of thought, worked out a physico-chemical approach utilizing X-radiation to induce genetic mutations (Muller, 1927). Thus, the material nature of genes, as opposed to the assumption of an incorporeal creative force, became an indisputable fact. However, the question of the physical basis of the genetic material remained the object of vague speculations. The scientific consensus then accepted exclusively proteins as candidates, since only these substances, in their multiple forms – diverse enzymes, keratin, colagen, etc... – seemed to parallel the enormous diversity of gene action. It was known that chromosomes were composed of about equal parts of

Figure 2.4: The "fly room" at Columbia University. From left to right: Calvin B. Bridges (1889-1983), who had a special talent for discerning new mutants; Alfred H. Sturtevant (1891-1970); Thomas H. Morgan (1866-1944), who received the Nobel Prize in 1933 and (in the background) Hermann J. Muller (1890-1967) laureated in 1946 with the Nobel Prize.

proteins and nucleic acids – deoxyribonucleic acid (DNA) and ribonucleic acid (RNA) (Excursus 2-5). Moreover, it was generally accepted that nucleic acids were molecules as invariable and uninteresting as, for instance, starch or cellulose, whose basic structures did not offer any possibility of diversification. A hypothesis raised in the 1920s by Phoebus Levene (Rockefeller Institute, New York) proposed that the four different building blocks of DNA were bound together, forming tetranucleotides (Excursus 2-5), which by polymerization originated long threads. These threads within the chromosomes were supposed to, somehow, hold the genes together – maybe comparable to fine lingeries hanging on a line. This tetranucleotide hypothesis – admired and often cited in textbooks – influenced a new scientific generation, acquiring little by little the status of a proven fact. As a consequence, genes were assumed to have no other than a proteinaceous – or a complex "colloidal" (whatever that meant) – nature.

Nothing challenged this widely accepted and well established concept – much to the contrary! Experiments performed by George Beadle and Edward Tatum in the 1940s indicated that enzymes acted as bridges between genes and their corresponding phenotype. How did they come to this conclusion?

**Excursus 2-1**
***ESCHERICHIA COLI*: AN EXAMPLE OF A SELF-REPLICATING MACHINE**

In 1885, a young physician in Munich – later councillor of the court in Vienna, Prof. Dr. Theodor Escherich (Fig. 2.5) – described the bacterium *Escherichia coli* under the name of *Bacterium coli commune*. These germs were present in the diapers of all babies – except newborns (Meyer & Arber, 1986). These mostly innocuous inhabitants of the bowels of mammals came to be the main research subject for pioneering molecular biologists – a fact that Escherich could not have dreamt about. Coli*, a single-celled, rod-shaped microorganism about 2 $\mu$m in length and 1 $\mu$m in diameter, has a very frugal way of living. At 37 °C, in simple media with mineral salts and a source of organic carbon and energy (see table), it divides in two about every hour; in rich culture media with organic nutrients (amino acids, vitamins, etc.) the generation time shrinks to 20 minutes.

Figure 2.5: The councillor of the Viennese court, Prof. Dr. Theodor Escherich (1857-1911).

---

\* *Escherichia coli*, *E. coli*, coli bacterium, coli cells, or simply coli

Let us consider *E. coli* as a black box, a mysterious machine whose function is self-replication, resulting in exponential growth – a fact to be easily monitored (Fig. 2.6). Nevertheless, neither microscopic nor electron-microscopic observations offer any clues for clarifying this phenomenon (Fig. 2.7). However, it becomes obvious that in this von Neumann's replication machine, simple building-blocks, that is nutrient molecules from the medium, are assembled so as to build new identical machines.

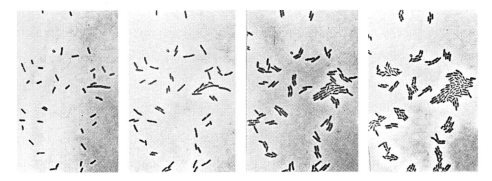

Figure 2.6: Coli cells as "black boxes". A droplet from a culture of coli cells was spread over solid medium and incubated at 37 °C. The immobilized cells were monitored through microscopic photos taken every 20 minutes (enlargement: approximately 750-fold).

| Culture Media for *Escherichia coli*: | | |
|---|---|---|
| **Minimal Medium:** | | **Complete Medium (Nutrient Broth):** |
| Ammonium chloride, $NH_4Cl$ | 1 g | Meat Extract (Bouillon)   10 g |
| Primary potassium phosphate, $KH_2PO_4$ | 5 g | Table Salt, NaCl   5 g |
| Secondary sodium phosphate, $Na_2HPO_4$ | 10 g | Water*   1 liter |
| Magnesium sulfate, $MgSO_4$ | 0.1 g | |
| Glucose (or other carbon and energy source) | 10 g | |
| Water* | 1 liter | |

*Since coli needs trace amounts of several metal ions, it is better to use tap water instead of distilled water. For solid media, add 15 g agar per liter.

Microorganisms able to synthesize all substances needed for growth and replication from the constituents of minimal medium are dubbed *prototrophs*. *Auxotrophs*, in contrast, need supplements (specific amino acids or vitamins) in order to grow.

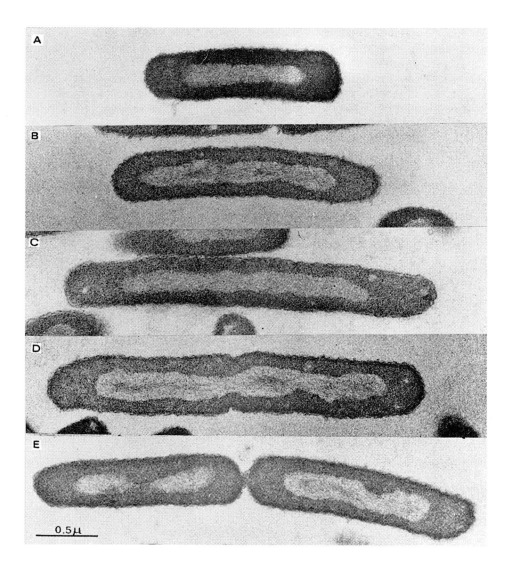

Figure 2.7: Electron microscopic images of thin slices of coli cells at different phases of growth and cell division (Woldringh, 1976). Two layers on the periphery (cell wall and cell membrane) enclose the contents of the cell. The only recognizable structural elements enclosed are the granular ribosomes and the so-called nucleoid. Nucleoids, the bacterial equivalents of cell nuclei, are bundles of DNA (Excursus 2-5) comprising the bacterial genetic material.

**Excursus 2-2**
**A BRIEF DIGRESSION: WHERE DOES THE ENERGY COME FROM?**

In 1937, during the apogee of classical biochemistry, Hans Krebs (1900-1981) at Cambridge clarified some of the most elementary biochemical reactions: the citric acid cycle, also called Krebs cycle in his honor. (Krebs, later bestowed with the title Sir, had arrived in England in 1933 as a refugee, coming from Freiburg, a university town in the Black Forest of Germany.) It was clear that oxidation of sugars somehow yielded adenosine triphosphate (ATP), the universal donor of energy for many synthesis reactions. All oxidation steps were described. However, why was the energy from these reactions not dissipated as heat? Where was the compound directly linked to ATP synthesis? Which specifically defined chemical reactions originated ATP? These questions were by no means purely rhetorical; they were rather crucial to innumerable frustrating experiments performed for decades by scores of biochemists trying to solve this riddle. The solution, as visualized by them, was never to be found; no wonder – it did not exist. None of the biochemists was able to have this insight; all were too deeply immersed in their professional bias of running after conventional reactions, and thus they were not in a position to conceive something innovative. An outsider had a better chance; he arrived in the form of a physicist who demonstrated that ATP synthesis was, basically, a physical phenomenon.

Peter Mitchell, first in Cambridge and later in Edinburgh, worked on biomembranes and transmembrane transport. Polarity was an obvious feature of cell membranes because their outsides and insides were different, not exchangeable, and transport of ions and molecules from the outside inwards or vice-versa was distinct, not to be explained by simple diffusion or osmosis. And it also was known that all enzymes of the respiratory chain were located on the membrane of aerobic bacteria, or, in higher organisms, on the inner membrane of mitochondria.

These two facts could be combined – this was Mitchell's ingenious insight: enzymes of the respiratory chain were bound to membranes in such a way as to allow them to eject the electrons and protons, yielded by the oxidative reactions, to different sides of the membranes, thus creating a cumulative electrostatic potential on these electrically isolating membranes. Bacterial cells or mitochondria (Fig. 21.2) were virtually batteries, recharging themselves as their respiratory metabolism proceeded, positive on the outside, negative on the inside (in the special case of chloroplasts with its photosynthetic activity – also recognized by Mitchell as a similar phenomenon – the outside membrane is the negative pole, the inside the positive one). And this electrostatic gradient could be tapped in order to synthesize ATP. To achieve this, though, an apparatus was needed – ATPase (a better name would be ATP synthase) – which itself was to be anchored in the membrane, serving as support for adenosine diphosphate (ADP) and inorganic phosphate. These substances, when bombarded by $OH^-$ ions from the inside and $H^+$ ions from the outside, would condense to ATP and water (Fig. 2.8).

In 1961, Mitchell was isolated with his ideas; no one took him seriously. He was even derided by some. Two years later, he himself decided to become an outsider. A generous inheritance (his father was a building contractor) allowed him to give up his position at the University of Edinburgh, and to retire, together with his collaborator of many years, Jennifer Moyle (Fig. 2.9), to a lonely estate in Cornwall. There he had bought a derelict farmhouse, restored it as a family residence and laboratory, and was then able to immerse into his experimental pursuits. His results and arguments became ever more convincing and detailed (Mitchell, 1967; Mitchell & Moyle, 1965, 1968). Still, years should pass until a new generation of scientists became persuaded by his innovative ideas (Mitchell, 1979). In 1978, Mitchell was honored with the Nobel Prize – and, in 1997, Paul Boyer (Univ. of California, Los Angeles), and John Walker (Medical Research Council, U.K.) received the Nobel Prize for a detailed elucidation of the mechanism of action of ATP synthase, along the lines proposed by Mitchell (see, for instance, Capaldi & Aggeler, 2002).

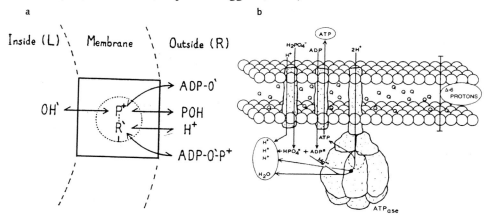

Figure 2.8: Mitchell's hypothesis for ATP synthesis. *a*) Mitchell's original notion of an "ATPase" bound to the membrane: enzymes from the respiratory chain, anchored in the membrane, yield an electrostatic and/or a pH-gradient (outside H+ ions, inside OH- ions). Adenosine diphosphate (ADP-O`) and inorganic phosphate (POH), placed within the ATPase, once acted upon by OH- ions from one side and H+ ions from the other side, lose a water molecule and condense to ATP (ADP-O`-P+) (from Mitchell, 1961). *b*) Mitchell's hypothetical ATPase is now identified as a complex quaternary structure dubbed Fo-F1-complex, composed of 16 protein subunits. The Fo component is a tube imbedded in the membrane, a veritable proton channel, a linear particle accelerator. Through this channel, H+ ions shower into the balloon-like F1-complex, located at the inside of the membrane. In this complex, ADP and phosphate condense to ATP – basically as Mitchell conceived it. In the case of aerobic bacteria, newly synthesised ATP diffuses into the cytoplasm, becoming available to metabolic reactions. In the case of mitochondria of eucytes (see also Chapter 21), the ATP has to be redirected to the outside, that is, into the cytoplasm, a fact hinted at in the scheme (Elthon & Stewart, 1983).

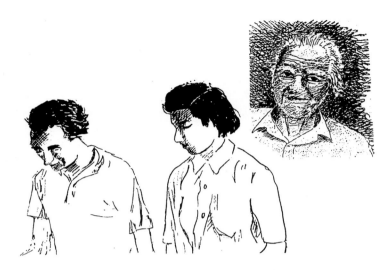

Figure 2.9: Peter Mitchell (1920-1992) and Jennifer Moyle in Cambridge at the end of the 1940s; insert right, P. Mitchell around 1990.

## Excursus 2-3
## MENDEL

For eight years, the Augustinian monk Gregor Mendel (Fig. 2.10) performed crossing experiments with pea plants in his cloister garden in Brno (Czech Republic – then Brünn, Austro-Hungarian Empire) (Fig. 2.11). He came to the conclusion that individual hereditary units, responsible for the emergence of specific characteristics, propagated immutably from generation to generation (Fig. 2.12). This novel idea went against the established notion that heredity derived from the diffuse influences of freely mixable "hereditary fluids". Mendel's inference was based on dozens of crossing experiments between pea plants distinct in characteristics like flower color (red or white), structure of the pea kernel (rough or smooth) or its color (green or yellow). For example, upon crossing hereditarily constant (nowadays one would say homozygous) red-flowered plants with white-flowered ones, all direct descendants (F1) were red-flowered; nevertheless, when these were crossed with each other, their descendants (F2) could be of two different types, namely red- or white-flowered. This means that the hereditary factor responsible for the white flower color was propagated individually and unchanged through the heterozygotic F1 generation, being able to manifest itself in the next generation. This F2 generation displayed red- and white-flowered plants, segregating in a 3:1 ratio. Because the gene (or better, the allele – see glossary) responsible for red-colored flowers was dominant over that responsible for the white-colored ones, all F2 heterozygotes (like those in the F1 generation) were red-flowered.

Figure 2.10: Gregor Mendel (1822-1884), né Johann, studied four semesters physics and biology in Vienna, and for eight years he then crossed distinct types of pea plants in his cloister garden in Brno (then Brünn). Highly appreciated by his cloister's brethren, he was elected abbot in 1868, thus – unwillingly – putting an end to his scientific career (Photo: Iconografia Mendeliana, Brno's Moravsk Museum, 1965).

Figure 2.11 (right, top): Mendel's cloister in Brno. Along the wall, Mendel's garden (see also Fig. 2.12). Above the gate, second window from left, his room.

Figure 2.12: Mendel's crucial discovery (represented as flower bed motives, in the summer of 1993, at Mendel's cloister garden which today is part of a Mendel Museum worthy of a visit): by crossing two types of true breeding pea plants (parental generation, P), differing in only one characteristic, in this case, red- or white-flowered plants,  all direct progeny (generation F1) is of the same type (in the case here represented, all F1 plants display red flowers). By crossing these F1 plants with each other, the descendants (generation F2) will show the characteristic of one or the other parental type. In the flower bed, the two front rows symbolize the results of

back-crossing the recessive parental type (white-flowered) with F2 individuals: only red-flowered plants (left) result from the cross with the F2 homozygote dominant plant; from the cross with F2 heterozygotes, red-flowered and white-flowered plants appear in equal proportions (in the middle); the back-cross with white-flowered F2 individuals produces white-flowered plants only (at right).

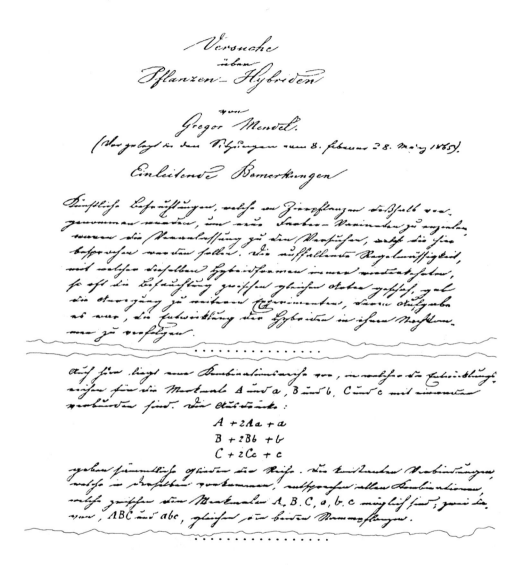

Figure 2.13: Facsimile of the first lines of Mendel's 48-page manuscript, and of the passage (lacking the relevant notations for diploidy, AA and aa, which would be applicable to the homozygotes) which fed the speculation that Mendel, actually, did not really understand the phenomenon as viewed today (Olby, 1979). At right, the printed text.

VERHANDLUNGEN DES NATURFORSCHENDEN VEREINES IN BRÜNN

Bd. IV. für das Jahr 1865, Abhandlungen, 3–47, 1866.

## VERSUCHE ÜBER PFLANZEN-HYBRIDEN.

### Von Gregor Mendel.

(Vorgelegt in den Sitzungen vom 8. Februar und 8. März 1865.)

### Einleitende Bemerkungen.

Künstliche Befruchtungen, welche an Zierpflanzen desshalb vorgenommen wurden, um neue Farben-Varianten zu erzielen, waren die Veranlassung zu den Versuchen, die hier besprochen werden sollen. Die auffallende Regelmässigkeit, mit welcher dieselben Hybridformen immer wiederkehrten, so oft die Befruchtung zwischen gleichen Arten geschah, gab die Anregung zu weiteren Experimenten, deren Aufgabe es war, die Entwicklung der Hybriden in ihren Nachkommen zu verfolgen.

. . . . . . . . . . . . . . . . . .

Auch hier liegt eine Kombinationsreihe vor, in welcher die Entwicklungsreihe für die Merkmale $A$ und $a$, $B$ und $b$, $C$ und $c$ mit einander verbunden sind. Die Ausdrücke:

$$A + 2Aa + a$$
$$B + 2Bb + b$$
$$C + 2Cc + c$$

geben sämmtliche Glieder der Reihe. Die konstanten Verbindungen, welche in derselben vorkommen, entsprechen allen Kombinationen, welche zwischen den Merkmalen $A$, $B$, $C$, $a$, $b$, $c$ möglich sind; zwei davon, $ABC$ und $abc$ gleichen den beiden Stammpflanzen.

. . . . . . . . . . . . . . . . . .

Somatic cell diploidy, as opposed to germ cell haploidy, was not wholly grasped by Mendel, a fact that led to the bizarre statement that Mendel was no "Mendelian" in the modern sense of the word (Olby, 1979; Hartl & Orel, 1992; Fig. 2.13). Actually, the terms diploidy (double set of chromosomes) and haploidy (single set of chromosomes) were only coined after Mendel's work. The observations made by Mendel remained forgotten for 30 years, being finally rediscovered in 1900. Understanding his results was only possible at the beginning of the 20th century when advances in cytology had clarified the principles of meiosis and fertilization.

**Excursus 2-4**
**CHROMOSOME MAPS**

The chromosomal theory of heredity was confirmed in detail by comparing the abstract chromosomal maps with chromosomes, the cytological structures. The merit of this feat goes to the Morgan school, specially to Alfred Sturtevant, who, as a graduate student, conceived the idea of chromosomal maps. The physical distance between two genes located on the same chromosome could be estimated from the percentage of recombinants originating from crosses between progenitors differing with regard to the characteristics determined by the genes in question. In the case of the fruit fly *Drosophila melanogaster*, made famous by Morgan's group, the mutation $w$ (white eyes) and wild type $w^+$ (red eyes) were examples of such alternative characteristics, as were the mutation $m$ (miniature wings) and wild type $m^+$. The two respective genes (those for eye color and wing form) were both located on the X chromosome (sex-linked heredity); X-linkage simplified genetic analysis, because the *Drosophila* males – as males of many other species – possess only one X chromosome. Supposing that the alleles* segregated freely (Mendelian segregation), that is, if they were located on different chromosomes, the double heterozygote descendant females would produce gametes with all four possible combinations ($w^+m^+$, $w$ $m$, $w^+m$ and $w$ $m^+$) in equivalent numbers. This would be detected by observing the next generation of males. However, this supposition did not hold true; instead of the expected 50% recombination rate, which would correspond to the free segregation hypothesis (or, alternatively, no segregation at all, if the alleles for eye color and wing form were totally indissociable) there was a segregation rate of 30%. This result indicated that the alleles $w$ and $m$ were somehow linked, not totally physically separated. Sturtevant grasped this state of affairs: the alleles were indeed linked on the same chromosome structure but not in a totally indissociable manner. They could be separated by crossing-over (material exchange between homologous chromosomes occurring during the maturation process of gametes): the larger the distance between the alleles on the chromosome, the higher was the chance that a crossing-over would occur. Consequently, alleles only seldom separated by crossing-over were located near to each other as close neighbors. During a sleepless night (Crow, 1988), Sturtevant visualized that, by knowing the distance between $A$ and $B$ and between $B$ and $C$, one could infer the distance between $A$ and $C$: if the incidence of crossing-over really reflected a physical distance, then $AC$ should approximately be $AB$ plus $BC$ or $AB$ minus $BC$ (Sturtevant, 1913). Figure 2.14 shows the first chromosome map deduced by this method, as well as maps obtained during the following few years. These results were the fruit of the obstinate – even obsessive – search for *Drosophila melanogaster* mutants and the corresponding innumerable crosses that followed.

---

* term meaning alternative forms of a gene

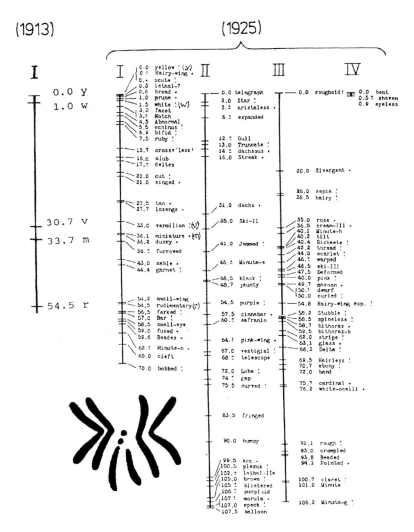

Figure 2.14: Comparison of the first chromosomal map by Sturtevant (1913), comprising 5 markers, all situated on the X-chromosome of *Drosophila melanogaster*, with the map of all four chromosomes from the same organism (Morgan, Bridges & Sturtevant, 1925). Since then, these results from 1925 have undergone only minor revisions. However, the sequencing of the *Drosophila* genome (Adams et al., 2000) in the wake of the genomics initiatives (see Chapter 23) gave a completely new quantitative and qualitative dimension to the genetic map of *Drosophila*. This map comprises now about 13,600 genes, i.e., virtually all the genes of this organism. In addition, these genes are now known not only as genetic markers, but as well defined molecular structures. Their functions, nevertheless, remain unknown in most instances.

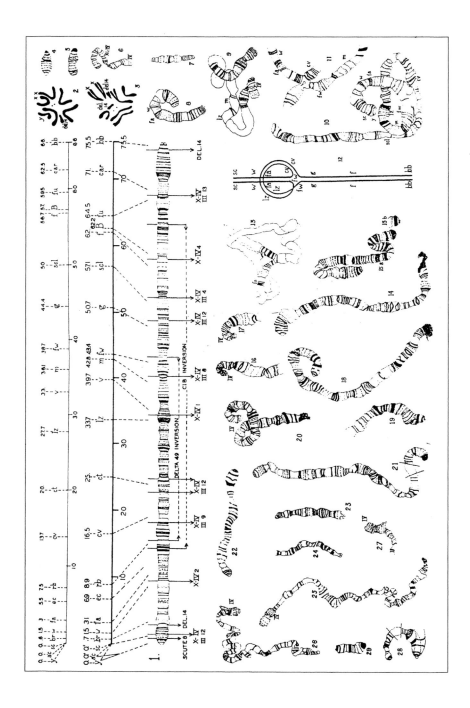

Figure 2.15 (left): Giant chromosomes. The linear arrangement of genes on chromosomes was progressively documented in detail by comparing the anomalies (deletions, inversions) of the banding patterns of the giant chromosomes in the salivary glands of *D. melanogaster* larvae with respective data obtained from genetic crosses. Here, this approach is illustrated by a representation of the X-chromosome by Painter (1935) which shows the normal banding pattern aligned with the genetic markers (top); bottom: different chromosomal anomalies, whose genetic consequences can be detected by crosses. Interesting is, for example, the pairing depicted at bottom right and its respective scheme (at its left). This pairing involves the normal X-chromosome and its homologous partner, a mutant, carrier of an inversion in the region *cv-lz-fw* and displaying a loop configuration in order to achieve correspondence of homologous regions.

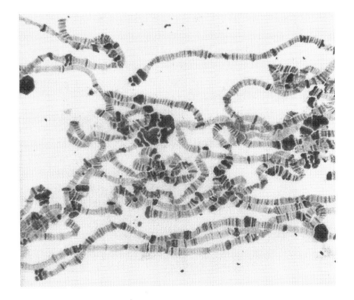

Figure 2.16: Giant chromosomes are not only found in salivary glands of diptera larvae, but also, for example, in the so-called macro-nuclei of such protozoa as the ciliate *Stylonychia mytilus*. Each band probably corresponds to a gene region. But, in contrast to *Drosophila*, none of those genes have been further characterized [Photo: D. Ammermann, Univ. Tübingen (unpublished); see also Ammermann, 1979].

Comparison of anomalies occurring on the giant chromosomes from salivary glands of *Drosophila* larvae with data obtained from crosses, gradually allowed an ever deeper comprehension about how genes were located on the chromosome structures in a strictly linear fashion (Fig. 2.15).

**Excursus 2-5**
**THE EARLY BIOCHEMISTRY OF THE CELL NUCLEUS**

In 1869, the supposedly high standard achieved by biochemistry made Felix Hoppe-
Seyler, a professor in Tübingen, feel really proud; that was the reason why he viewed
with suspicion a new organic substance containing phosphate, recently discovered by
his young collaborator Friedrich Miescher. Miescher had isolated such a substance
from cell nuclei in the pus from bandages discarded by the nearby surgical clinic.
This was a sensational finding; at that time the only known organic compound
containing phosphate was lecithin (today a frequent addition to beauty creams).
Hoppe-Seyler's scepticism made him repeat Miescher's experiments for two years
before daring to publish them in his journal (Miescher, 1871; Hoppe-Seyler, 1871). It
was undeniable: cell nuclei under mild acid treatment yielded the so-dubbed "nuclein"
– containing phosphate! In 1870 Miescher returned to his hometown, Basel,
dedicating himself for the rest of his life to the study of nuclein. The Rhine, which
runs through Basel, provided him with a plentiful and more genial new source of
material: the sperm of the Rhine salmon. After several years of research, it became
clear that Miescher's nuclein was dissociable in two fractions, a protein fraction and
another one, lightly acidic, containing phosphate, the "nucleic acid" (Altmann, 1889),
then to be fully characterized. After Miescher's death (Miescher: 1844-1895), a small
group of followers worked on this task. By the turn of the century, Albrecht Kossel,
in Heidelberg, had identified some nitrogeneous compounds: the bases cytosine,
adenine and thymine. (It should be mentioned that Kossel was actually one of the first
to recognize and emphasize that biological macromolecules were assembled from
smaller organic building blocks.) But it was not until 1930 that two fundamental
classes of nucleic acids were recognized from an organic-chemical point of view,
characterized by distinct building blocks. Sperm or thymus, for example, yielded
deoxyribonucleic acid (DNA); ribonucleic acid (RNA) could be isolated from yeast,
for example. Acid hydrolysis yielded the most simple building blocks from these
nucleic acids: a pentose sugar (2-deoxyribose from DNA, and ribose, from RNA), 5
different nitrogenated heterocyclic bases (the so-called purine bases, adenine and
guanine, and the pyrimidine bases, cytosine, thymine and uracil – thymine being
specific to DNA and uracil to RNA), and, of course, inorganic phosphate. Milder
enzymatic hydrolysis of DNA or RNA yielded more complex building blocks, dubbed
nucleotides, composed of a phosphorylated sugar and one base (Fig. 2.17). The
original subject of analysis was vertebrate DNA, where the 4 typical bases occurred
in roughly equimolar proportions. This consequently led to the assumption that
nucleotides occurred in nucleic acids as foursomes (maintained together by phosphodiester bridges); this notion resulted in the already mentioned fateful tetranucleotide
hypothesis advanced by Phoebus Levene (Levene, 1921).

Proteins from cell nuclei were also early targets for analysis. Besides a large number of diverse free proteins, it became clear that a fraction of basic proteins, the histones (protamines, in sperm), were closely associated with DNA. To give a biological meaning to all these observations (genes, after all, what are they?) was a task which remained fuzzy and full of prejudices until well into the 1950s (Mirsky, 1951).

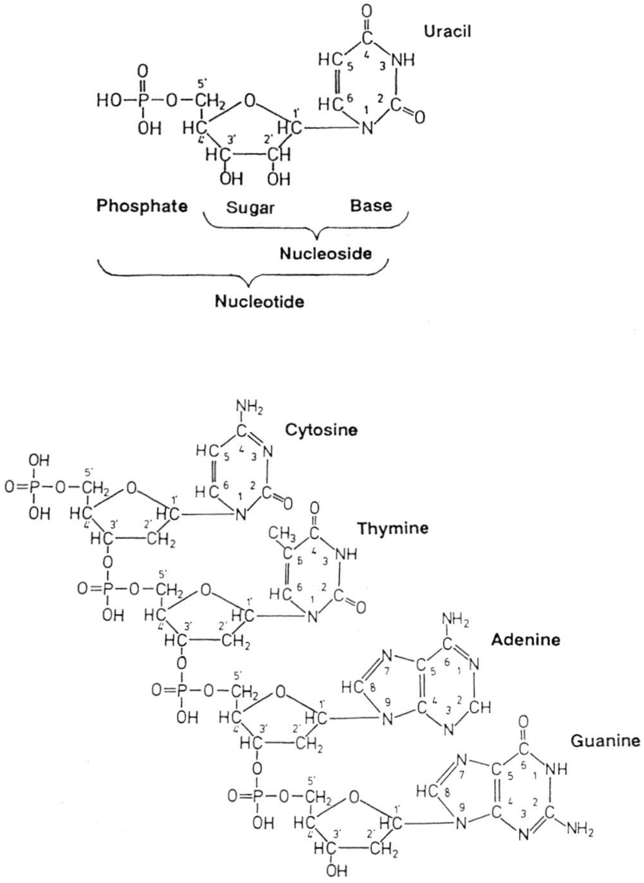

Figure 2.17: Top: the building blocks of nucleic acids: the sugar represented here is ribose and the base is uracil. Both compounds are characteristic of ribonucleic acid (RNA). Bottom: a tetranucleotide, Levene's hypothetical foursome which, according to him, represented the repeating unit of nucleic acids. The bonds between individual nucleotides are here depicted correctly; there were many hypotheses postulating erroneous chemical bondings. This scheme displays the sugar deoxyribose, characteristic of deoxyribonuclei acid (DNA).

# CHAPTER 3

## ONE GENE – ONE ENZYME

Originally, Morgan thought of himself as an embryologist in the quest of "how" – whatever that meant in chemical-physiological terms – an individual evolved out of an egg cell. His students coerced him to become a geneticist. In doing so, they encountered little resistance, especially since the discovery of the sex-linked inheritance of the white-eyed characteristic of the fruit fly *Drosophila* (Morgan, 1910) had introduced him to the field of genetics. In contrast with embryology, the results obtained in genetics led to good, palpable progress. In the back of his mind, though, he never totally abandoned the wish to grasp the real mechanism of gene action governing the development of an embryo. The hopelessness of that intent – considering the state of the art at that time – could not then be sensed; an understanding of it was only possible in retrospect, from a viewpoint still to be generated in the future. Nonetheless, concrete projects were developed to proceed towards the desired research goal.

In 1934, Boris Ephrussi, a Russian emigrant with a Rockefeller Foundation fellowship in hand, came from Paris to join Morgan's group (since 1928 at Caltech), with the intent of becoming familiar with *Drosophila* genetics. His basic interest was the mechanisms of gene action, just as Morgan's was. He imagined approaching this theme by means of cell cultures and tissue transplants and won a young assistent over to his cause, George Beadle. However, both soon realized that genetics and embryology, although theoretically closely related, actually, in laboratory practice, followed extremely divergent paths. The most probable explanation for this state of affairs could be found in the fact that the fruit fly *Drosophila* – the pet of geneticists – was of no use for embryology as worked on at the time, while sea urchins and frogs – the beloved subjects of study of embryologists – were not accessible to genetic analysis. When Ephrussi's fellowship came to an end, Morgan arranged for Beadle to accompany Ephrussi back to Paris (according to Beadle's suspicions, financed out of Morgan's own pocket) in order to continue their work, now back in the old world. In 1935, in Paris, utilizing quite an intricate procedure consisting of two microscopes to be used simultaneously, they succeeded in transplanting tissues of *Drosophila* larvae.

Among many different experiments, they transplanted, for example, larval tissues corresponding to the adult eye region from *vermilion* into *cinnabar* larvae. *Vermilion* flies are pink- and *cinnabar* flies orange-eyed; *Cinnabar* flies with transplanted *vermilion* eye tissue evolved into adult flies displaying dark-red eyes, typical for normal wild type flies. Another eye color mutation, *claret*, behaved relative to *vermilion* as *vermilion* to *cinnabar*. If one assumed that different diffusible substances were synthesized by the wild type genes of normal flies, then these results could be explained. One of these genes would elicit the synthesis of "substance 1", another would be responsible for transforming "substance 1" into "substance 2" (Beadle & Ephrussi, 1936). Considering that biochemical reactions are controlled by enzymes, nothing was more logical than to postulate the existence of 2 enzymes: the first, "enzyme 1", responsible for the synthesis of "substance 1" (from a precursor substance), and then, "enzyme 2", determining the transformation of "substance 1" into "substance 2". Apparently, the enzyme, with its specificity determined by the corresponding gene, was the instrument that this gene utilized in expressing its respective phenotype. The seed of the "one gene – one enzyme" hypothesis was sowed in Beadle and Ephrussi's field of thoughts. Still, ten years would go by before this hypothesis was actually formulated in these terms. [However, the emphasis on the enzymatic or metabolic aspect of gene action had already been instigated, not only by such earlier research as that on the genetics of pigmentation of the flour moth *Ephestia* (Caspari, 1933) but also by still much older observations on the synthesis of the plant pigment anthocyanine.]

An important next step was the identification of the precursor substance which originates the redish-brown eye pigment of *Drosophila* flies. Ephrussi in Paris, Beadle now in Stanford, and their collaborators tried exactly that. They would certainly have succeeded if a research group in Berlin had not first identified the *vermilion* substance as kynurenine, a derivative of tryptophan (Butenandt et al., 1942). (Even during the war years in Germany, there was still some basic research going on.)

But, if – as Beadle and his younger collaborator, Edward Tatum, who joined the Stanford group in 1937, believed – reactions catalyzed enzymatically were basically controlled by specific corresponding genes, then there should exist an easier way to clarify this point. They figured that mutant organisms defective in a straightforward and already known enzyme reaction should exist; the *vermilion* story seemed too complex. Their choice fell on *Neurospora crassa*, the bread mould, and this for two reasons. First, the life cycle of this haploid fungus had been outlined some years ago: meiosis occurred. That would permit a Mendelian approach of crossing analysis. Second, culturing techniques for fungi in strictly defined synthetic media had recently been developed. For example, it turned out that *Neurospora*, besides some mineral salts and glucose, needed only the vitamin biotin as a nutritional supplement (Excursus 1-1).

Cultures of *Neurospora* were treated with ultraviolet or X-rays (let us remember here that, in 1926, Hermann Muller had shown the mutagenic effect of X-rays on

*Drosophila*). After this procedure, the ability of mycelia developed from individual spores to grow in minimal medium was tested. The researchers succeeded already with their 299th spore (Beadle and Tatum had decided to check at least 5000 spores before giving up). Mycilia derived from spore number 299 would grow normally in complete medium – but not in minimal medium! As a next step, the quest for the putative lacking vitamin, amino acid or other molecular building-block followed. A full battery of substances was tested before it was decided that vitamin B6 (pyridoxine) was the important ingredient (full growth was only achieved by addition of vitamin B6). Back-crossing the mutant strain with the wild type originated asci with 4 wild type and 4 vitamin B6-dependent spores: vitamin B6 dependency was inherited in a mono-factorial, Mendelian way. Beadle and Tatum were in possession of a mutant gene whose wild type allele was responsible for the synthesis of vitamin B6.

Figure 3.1: George Beadle (1903-1989) and Edward Tatum (1909-1975).

The discovery of other mutant types followed in quick succession; altogether they isolated mutants dependent on a series of different vitamins or amino acids, but in each case exclusively one substance was affected, as compared to normal metabolism.

During their involvement with *Neurospora* in 1942, Beadle and Tatum came across the fact that – as a huge surprise – something similar had already been described for human metabolism and had even been similarly interpreted, and this at the beginning of the century! (Beadle later: "On learning of this long-neglected work it was immediately clear to us that in principle we had merely rediscovered what Garrod had so clearly shown forty years before.")

It was in London that Archibald Garrod, not long after Mendel's rediscovery, had shown that alkaptonuria – a relatively innocuous anomaly noticeable because the urine of affected persons turns dark-colored upon exposure to air – is inherited according to Mendel's rules as a recessive characteristic. Dark-coloring of a patient's urine is explained by the presence of homogentisic acid (2,5-dihydroxy-phenylacetate or alkaptone), which is not broken down further in such patients and therefore is excreted by the kidneys. With exposure to air, alkaptone undergoes spontaneous oxidation, turning dark. Garrod also observed other genetic anomalies, like albinism (a patient's failure to synthesize the skin pigment melanin), cysteinuria and pentosuria (excretion of, respectively, cysteine or pentose in urine) – mostly occurring in the progeny of consanguineous parents. He correctly recognized that these observations could be interpreted as enzymatic defects (the enzymes themselves being still unknown) caused by recessive mutations of the respective genes.

Garrod published his findings in 1909 in a well-selling book ("Inborn Errors of Metabolism"), and in first-class medical journals. He was a respected, even famous physician, knighted "Sir" in recognition of his scientific accomplishments. The whole scientific establishment used to attend his presentations at the Royal Society. Nevertheless, apparently nobody understood him: neither the geneticists (who knew nothing about medicine and biochemistry) nor the biochemists (who knew nothing about medicine and genetics) nor the physicians (who knew nothing...).

Figure 3.2: Sir Archibald Garrod (1858-1936).

Beadle and Tatum produced many further examples of enzymatic defects originated from gene mutations. Contrary to humans, *Neurospora* – but also other ascomycetes, like *Penicillium* (Bonner, 1946) – was an excellent subject for experimental analysis. In a matter of a few years of work, Beadle and his followers

(David Bonner, Norman Horowitz) were quite successful, so that by the end of the Second World War Beadle was able to compile an overview on a series of very convincing observations, including a preliminary genetic map of over 20 markers, distributed among many linkage groups (chromosomes) (Fig.3.3). Especially informative were conclusions on metabolic pathways, reached through mutation analysis. For example, there was a series of mutants, all unable to synthesize the amino acid arginine. Genetic crosses separated these mutants into 7 distinct groups, each located at a different site on the chromosome map. Obviously, there were at least 7 different genes collaborating on the synthesis of arginine. Using diverse

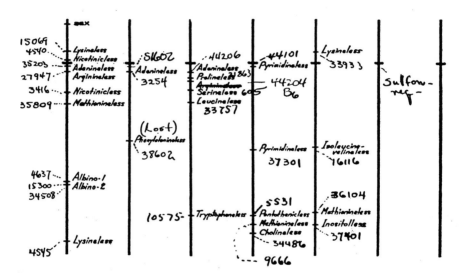

Figure 3.3: The first chromosome map of *Neurospora crassa* (Beadle, 1946).

supplements, it became clear that in 4 cases not only arginine itself could revert the mutant phenotypes to normal growth, but also two other amino acids correlated metabolically to arginine; these were ornithine and citrulline. Further, two other groups of mutants would react positively to arginine or citrulline, but not to ornithine. Still, a 7th group of mutants showed positive stimulation only when arginine was used as a supplement. The 7th gene was, apparently, responsible for the enzymatic transformation of citrulline into arginine. If genes 5 or 6 were mutated, ornithine was not changed into citrulline, but strains with defective alleles in one (or both) of these genes would attain normal growth with supplements of arginine itself or its precursor, citrulline. Defects in genes 1, 2, 3, or 4 blocked the synthesis of ornithine (Fig. 3.4). Adding ornithine to minimal medium restored the normal growth of strains mutated in these genes, since this substance could be further worked upon normally to be transformed into citrulline and then into arginine. The actual pathway of ornithine synthesis was still unknown then, but it was not difficult to realize that possibly 4 distinct reactions were involved, each one being dependent on its specific gene, respectively gene 1, 2, 3 or 4 (Beadle, 1946).

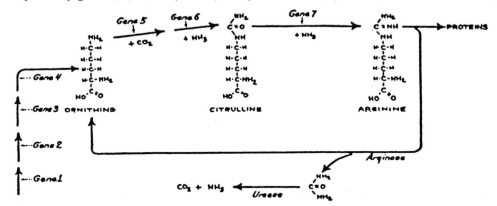

Figure 3.4: Srb & Horowitz's (1944) representation of arginine synthesis and the ornithine cycle in *Neurospora crassa* (Beadle, 1946).

In the wake of these results, Bonner and Tatum showed, among many other examples, that *Neurospora* produced tryptophan by condensing indole and serine (Fig. 3.5), indole in turn being synthesized from its precursor anthranilic acid. In the medium (supplemented with indole) utilized to grow indole-dependent mutant strains,

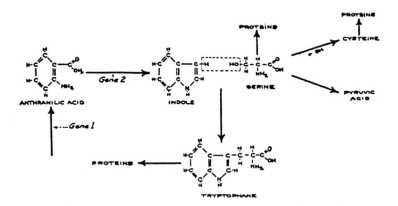

Figure 3.5: Tatum & Bonner's (1944) scheme of tryptophan synthesis, based on analysis of *Neurospora* mutants (Beadle, 1946).

large amounts of anthranilic acid were accumulated. This was due to the fact that the mutant cells unable to metabolize anthranilic acid simply excreted that substance into the medium. This specific observation was actually of general validity: many a precursor substance which, due to a genetic block in a biochemical pathway, could not be further metabolized, accumulated in increasing amounts, as if a dam had been erected. Such observations turned out to be very useful for clarifying normal metabolic pathways (Bonner, 1951).

However, the nice picture created by the "one gene – one enzyme" hypothesis was soon to be shaken by an unexpected intellectual earthquake. The physicist and bacteriophage researcher (see Chapter 4), no-nonsense sharp logical thinker, "enfant terrible", Max Delbrück, noted ironically (see discussion in Bonner, 1946) how the isolation method used would bring forth, exclusively, mutants that corroborate the hypothesis, and no others. This statement made worthless all data used hitherto as proof for the hypothesis. This state of affairs would remain until other methods were worked out, methods that would, in principle, permit researchers to refute the "one gene – one enzyme" hypothesis. Everyone had to agree that Delbrück was right. Possibly mutants could exist that affected multiple enzymes or something totally different; however, even if these were a majority of all mutants, they simply would not be detected by the reported selection method!

The embarrassment was only relieved when Norman Horowitz, also at Caltech, conceived and, together with Urs Leupold, a Swiss postdoc from Bern, went on to employ these "other methods", in their case, consisting of the isolation and characterization of temperature-sensitive mutants (Horowitz, 1948; Horowitz & Leupold, 1951). Their temperature-sensitive mutants grew normally, like the wild type, at relatively low temperatures (for example, 25 °C), but their growth was inhibited at higher ones (35 °C in the case of *Neurospora*, 42 °C for *E. coli*). This was so – as we know today – because temperature-sensitive mutant proteins had less stable tertiary structures, unable to withstand increased thermic motion. Temperature-sensitive mutants, growing well in minimal medium at low temperatures but needing complete medium when the temperature was raised, could be further tested by addition of diverse supplements to the minimal medium at conditions of higher temperature. It turned out that the majority of temperature-sensitive mutants were conditional auxotrophic mutants (prototrophs at low, while auxotrophs at elevated temperatures), their normal growth at higher temperatures being restored by the sole addition of one and only one growth factor (be it an amino acid or a vitamin). The fact that one gene determined the synthesis of one – and only one – enzyme was thus the rule rather than the rare exception.

In some other cases of temperature-sensitive mutants, though, the impaired growth in minimal medium at elevated temperatures could not be reversed merely by addition of supplements; in these cases, the affected gene product was, presumably, not directly involved in the synthesis of an organic building-block but, rather, in physiological processes involving other types of macromolecules ( Chapter 17).

Meanwhile, Tatum had established his own research group  on the east coast at Yale and was busy isolating auxotrophic strains of *E. coli*. One day in 1946, a certain medical student named Joshua Lederberg came to visit. Lederberg had decided to search for sexual processes in bacteria with the help of auxotrophic mutants. This would, possibly, help to understand the hitherto unkown metabolic pathways of these – as one thought – extremely primitive cells and, hopefully, even be of medical interest. Lederberg, under the scientific umbrella of his new mentor, plunged into work. Soon he was isolating huge numbers of mutants from the K12 strain of coli. Choosing K12, a strain which for over 3 decades had helped to fashion practical courses for medical students, turned out to be – as later shown – a real stroke of good luck. The decisive advantage brought about by working with bacteria was the availability of huge numbers of individuals, which allowed the observation of extremely rare events – as long as one found a way of doing so without much effort. [Lederberg imagined an ingenious method for discovering rare mutants, which was as surprisingly simple as it was efficient (let us remember that Beadle and Tatum had decided to test at least 5000 individual spores!); this novelty was the method of replica plating (Excursus 3-1).] The next step involved the genetic analysis of these newly isolated mutants. In the case of *Neurospora,* the facts were clear (as early as 1935): after haploid cells had fused together to produce zygotes, meiosis followed, reestablishing haploidy – the main cycle of all sexually reproducing eukaryotes. However, in the case of bacteria, sexual exchanges had never been described; it was even suspected that bacteria were void of genes, as they were then understood, conducting their primitive metabolism without a defined genomic structure. Lederberg and his auxotrophic mutants were to corroborate or disprove this supposition. Lederberg imagined that mixing two different cultures of auxotrophic mutants would be all it would take to clarify the point; bacterial cells would pair, allowing genetic recombinants to emerge. These would be detected – even if only present in extremely small numbers – by spreading the mixed culture (that is, hundreds of millions of parental cells) on minimal medium plates. Of course, only prototrophic colonies would come into view. However, the theoretically perfect experiment turned out otherwise than expected. Lederberg found indeed colonies [at the proportion of 1 prototroph per about 1 million parental cells ($10^{-6}$)]. However, there was a catch: even the control plates (that is, minimal medium plates seeded with individual parental cultures) yielded the same result. The sobering explanation was: back-mutation to wild type had occurred – no genetic recombination. If the latter were to happen at all, then at a rate a lot lower than that of back-mutations; in fact, it would be an utmost rare event. But even such extremely rare events should be able to be detected, deducted Lederberg. His reasoning was that it should be possible to isolate double or even multiple auxotrophic mutants of one bacterial strain. For that purpose, cultures of, say, leucine-dependent mutants (grown on minimal medium supplemented with leucine) would serve as a source for obtaining cultures with a further dependence, for example, on threonine. Back-mutation rates for double mutant strains correspond to the product of the single back-mutation rates. If each back-mutation appeared in a proportion of, say, $10^{-6}$, then the back-mutation rate for the double mutant would be $10^{-12}$. If the desired recombination appeared at a rate of only 1/100 of the single back-mutation, even then, it would still produce 10,000 more colonies as compared to the back-mutants in the controls with the unmixed cultures of the double mutants – this promised a comfortable safety margin. And exactly that expectation was fulfilled (Lederberg & Tatum, 1946): bacterial genetic exchanges were real. Nevertheless, if Lederberg – unknowingly at the time – instead of using *E. coli* K12 had put his hands on any other bacterial strain, he, most probably, would not have arrived at the same conclusion. We know today that sexual processes – if one can call these events by that name (see Chapter 11) – are rather the exception, not the general rule among bacteria. Soon afterwards, it was also demonstrated that cell to cell contact was the precondition which allowed sexual exchange to occur (Excursus 3-2), and Lederberg was convinced that zygotes (that is, fusioned cells) arose from the process. Lederberg went on to describe chromosomal maps in coli, defining linkage groups (Lederberg, 1947; Lederberg et al., 1951); nevertheless his results seemed extremely difficult to interpret and profoundly complex. The explanation for what appeared, at first sight, to be such an odd outcome came only years later, and turned out to be unexpectedly straightforward (Chapters 11 and 12).

Figure 3.6: At left, Joshua Lederberg (born in 1925) at the time of the discovery of bacterial sexuality, and his wife Esther, who discovered the phage λ. In the 1970s, Lederberg tried to establish a new scientific discipline, probably the only one whose research subjects were inexistent: "exobiology", the science of extraterrestrial life. Later, Lederberg directed his efforts to "emerging diseases", i. e. diseases in the coming.

It was now clear – a fundamental step forward – that genes, be they human, bacterial, or fungal, were directly responsible for the synthesis of specific enzymes (primary gene products), whose action entailed the appearance of peculiar characteristics (secondary gene action like the synthesis of pigments or of organic building-blocks). Notwithstanding, these insights did not bring one closer to grasping the real nature of the gene. Its special attributes of auto-replication and its ability to direct protein synthesis could not yet be defined in material terms. No known substance or biochemical reaction existed which was capable of explaining such almost incredible properties. The structure of the gene was still an insurmountable enigma. And, investigating as many biochemical mutants as one could get, be they bacterial or fungal, would merely emphasize this unavoidable truth. Despite the revolutionary new techniques (which, as seen, allowed extremely rare mutants and recombinants to be selected) and new experimental horizons, such as the studies on the heredity of *Neurospora*, as well as the first experiments on the sexuality of bacteria – some regarded the use of microorganisms as subjects of genetic research to be sheer absurdity –, it still was not possible to move beyond the frame of classical genetics. Nevertheless, these new approaches enriched the field with intellectually stimulating novel ideas, preparing the way for a new generation of discoveries.

George Beadle, Edward Tatum and Joshua Lederberg shared the Nobel Prize for physiology or medicine in 1958.

**Excursus 3-1**
**THE REPLICA PLATE TECHNIQUE (LEDERBERG & LEDERBERG, 1952)**

A bacterial culture previously treated with a mutagen (ultraviolet radiation or a chemical substance like hydroxylamine) is diluted and plated on a complete medium so that isolated colonies can be visualized after over-night incubation. A cylinder (with a diameter slightly smaller than that of the plate) covered with sterile velvet, to be used as stamp, is gently pressed on the plate with the colonies, and then similarly pressed on fresh plates (replica plates). These replica plates are likewise incubated over-night. On the replica plates with minimal medium, only prototrophs form colonies – no auxotrophic colonies appear. Minimal medium replica plates supplemented with substance A allow A-dependent auxotrophic mutants to develop colonies. Similarly, minimal medium with addition of substance B allows colonies of B-dependent mutants to grow. Corresponding positions on the original plate bear the original mutant colonies, which, as a next step, are picked (Fig. 3.7). Penicillin can be added to the original mutagenized culture of bacteria in minimal medium in order to raise the proportion of auxotrophs in relation to prototrophs. [Penicillin has a lethal effect exclusively on dividing cells due to its blocking of cell wall synthesis (Davis, 1948, 1950a).]

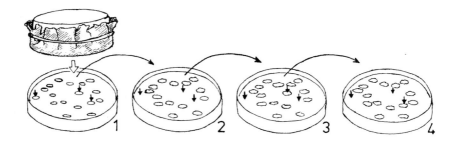

Figure 3.7: The technique of replica plating for the isolation and characterization of auxotrophic mutants. Plate 1 shows colonies originated from a suspension of mixed cultures (one prototrophic and 3 different auxotrophic strains of bacteria), which, after appropriate dilution and spreading on a plate with complete medium, were incubated over-night. A sterile velvet stamp touches the colonies and then inoculates them on 3 further plates: plate 2 with minimal medium, plates 3 and 4 with minimal medium plus different supplements (for example, the amino acid histidine and vitamin B1, respectively). Accordingly, all auxotrophic mutants fail to grow on plate 2, but a histidine-dependent auxotroph develops into a colony on plate 3, while a vitamin B1-dependent mutant grows on plate 4. Another auxotrophic mutant (the colony on the left in plate 1) was not stimulated to grow by any of the substances tested; in this case further characterization will be needed.

**Excursus 3-2**
## PROOF OF CONTACT BETWEEN PARENTAL CELLS DURING BACTERIAL CONJUGATION – AND THE DISCOVERY OF TRANSDUCTION

Bernhard Davis, a physician working on tuberculosis at the N. Y. Public Health Service (later at Harvard, he worked on social issues – Davis, 1992) reasoned that communicating U-form tubes separated by a filter whose pores would allow the culture medium, but not any bacteria in it, to freely flow from one tube to the other, could help solving the mysteries of recombination. One tube contained culture A, the other culture B. With pressure exerted alternatively on one tube or the other, the medium in the tubes could be mixed; however, no recombinants resulted when the cultures used were those of auxotrophic mutants of *E. coli* K12. The conclusion was that direct contact between parental cells was a precondition for genetic exchange (Davis, 1950b).

Shortly before he moved on to a professorship at the University of Wisconsin, Lederberg succeeded in demonstrating that genetic recombination also occurred in *Salmonella typhimurium*. This result was attained in collaboration with his 19-years-old graduate student, Norton Zinder. [Zinder, later, as professor at the Rockefeller University, N.Y., discovered the first bacteriophage whose genetic material was ribonucleic acid (RNA) (Loeb & Zinder, 1961; Zinder, 1975).] Davis' method, when applied to *Salmonella* cultures, revealed an unexpected event: even without direct contact between parental cells, genetic recombinants did appear (Zinder & Lederberg, 1952). A new phenomenon was unearthed: transduction. Soon afterwards it was demonstrated that this occurrence was due to the spontaneous presence of bacteriophages (bacterial viruses) in the bacterial cultures. These viruses, apparently, propagated in these cultures, occasionally encapsulating in their tiny coats not only their own genetic material, but also bits of genome from their bacterial host – certainly incorporated by accident. If such surrogate virus particles (bacterial genetic material disguised in a virus coat) happened to infect a bacterium, "transduction" – as the phenomenon was dubbed – could take place: that is, integration of marker-genes into the genome of a recipient host bacterium (Chapter 11).

# CHAPTER 4

## PHAGES

Let us once more go back to Hermann Muller; he observed in 1926 that X-rays caused mutations in *Drosophila*. The abstract gene – whatever it was – could be altered by ionising radiation. One could thus affirm that it was composed of physical matter; one could even try to guess its size. Muller surmised 2,700 atoms per gene, but due to the many arbitrary parameters he had to use, he decided not to publish his conclusions (Carlson, 1966). However, the idea was born, and one of his students followed up his reasoning.

Nikolai Timoféeff-Ressovsky was a Russian visiting scientist at the Genetics Department of the Kaiser-Wilhelm-Institut in Berlin, where, also as a guest, Muller worked for a short while in 1933. In collaboration with the physicist Karl Zimmer, Timoféeff-Ressovsky studied the mutagenic effect of X-rays on *Drosophila*. He was especially interested in the quantitative aspects of the events: the relationship between mutation rate and radiation dose. He observed that the proportion of mutants among the survivors increased linearly as a function of the dose (Fig. 4.1); apparently, there was no minimal dose under which radiation was ineffective (as was the case of poisons). As they were almost ready to publish their results, a young assistant of Otto Hahn from a neighboring institute joined them: Max Delbrück. The three enthused colleagues talked endlessly about the putative material nature of the still totally abstract Mendelian gene, the invisible target of X-ray quanta. In 1935, the so-called green pamphlet (reprints had a catchy green cover) sprang from these discussions, propagating a most interesting notion: mutations are brought about by modifications of a molecule! This molecule could itself be the gene or at least one of its essential components. Genes were thus defined as molecular structures, and mutations resulted from changes in these structures. Such mutations were caused either by spontaneous chemical events or by experimentally applied radiation energy.

From a chemical point of view, however, it remained quite mystifying how to propose a molecular structure befitting the incredible features of a gene – its ability to direct the appearance of one of thousands of different characteristics, its ability to auto-replicate, and to maintain these fundamental properties even after a mutation had occurred.

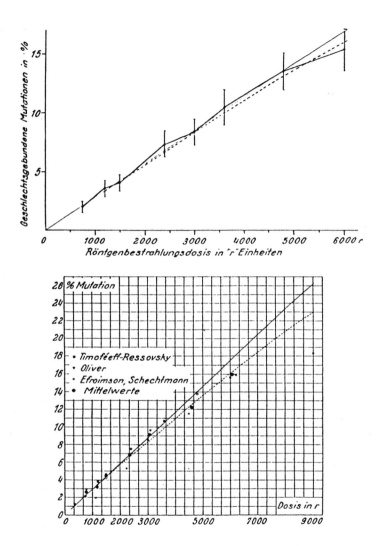

Figure 4.1: From the "green pamphlet" of Timoféeff-Ressovsky, Zimmer & Delbrück (1935). A strict proportionality between the dose of radiation and the incidence of mutations among the surviving *Drosophila* flies is to been seen.

Even so, Delbrück boldly „calculated", that the activation energy necessary for a mutation to occur was about 2 eV (Fig. 4.2). Thus it would be possible for a given gene to remain stable for thousands of years, at body temperature, but it would nevertheless be able to eventually change from one energy level to another one.

And, in the same year as the green pamphlet, Wendell Stanley, working at the Rockefeller Institute in Princeton, featured the crystallization of the tobacco mosaic virus (Stanley, 1935) – an era-marking event (for which Stanley was laureated with the Nobel Prize in 1946). What no one had reckoned to be feasible, the crystallization of a virus – a self-replicating unit, then considered to be an independent, "feral" gene –, was tantamount to identifying it as a chemically defined structure. This new direction of thought pointed away from the image of the gene as an abstract entity, or a forever inscrutable living structure.

.... Die wichtigste Eigentümlichkeit dieser Zusammenhänge, die durch die Tabelle zum Ausdruck gebracht wird, ist die, daß sehr geringe Änderungen der Aktivierungsenergie ganz gewaltige Änderungen der Reaktionsgeschwindigkeit im Gefolge haben. Z. B. ist eine Änderung der Halbwertszeit von 1 Sekunde auf über

Tab. 14. Zusammenhang zwischen der Reaktionsgeschwindigkeit und: dem Verhältnis von Aktivierungsenergie zu mittlerer Energie der Temperaturbewegung pro Freiheitsgrad $\frac{U}{kT}$, den Absolutwerten von U bei Zimmertemperatur (U in eV), und dem Temperaturquotienten für 10° C.

| $\frac{U}{kT}$ | W in sec$^{-1}$ | $\frac{1}{W}$ | U in eV | $\frac{W_{T+10}}{W_T}$ |
|---|---|---|---|---|
| 10 | $4,5 \cdot 19^9$ | $2 \cdot 10^{-10}$ sec. | 0,3 | 1,4 |
| 20 | $2,1 \cdot 10^5$ | $5 \cdot 10^{-5}$ sec. | 0,6 | 1,9 |
| 30 | 9,3 | 0,1 sec. | 0,9 | 2,7 |
| 40 | $4,2 \cdot 10^{-4}$ | 33 min. | 1,2 | 3,8 |
| 50 | $1,9 \cdot 10^{-8}$ | 16 Monate | 1,5 | 5,3 |
| 60 | $8,7 \cdot 10^{-13}$ | 30000 Jahre | 1,8 | 7,4 |

1 Jahr nur mit einer Erhöhung der Aktivierungsenergie von 0,9 auf 1,5 eV (um 70 %) verbunden. Da die bei Molekülen bekannten Aktivierungsenergieen zwischen noch weiteren Grenzen gelegen sind, kann man also von vornherein Reaktionsgeschwindigkeiten jeder Größenordnung erwarten. ....

Figure 4.2: Facsimile from the "green pamphlet" (1935) with Delbrück's conjectures on the activation energy necessary for one mutational event to occur within a given time interval.

In 1937, Delbrück obtained a Rockefeller fellowship to work at Caltech. He wanted to get acquainted with the mysteries of *Drosophila* genetics, which were being tackled by Morgan's group, established at Caltech since 1928 (see Chapter 2). However, he soon judged the matter as too complex and inadequate to bring forth crucial new insights into the physical nature of the gene. Delbrück roamed and browsed about; Caltech teemed with scientific élan. Besides Morgan, there was Linus Pauling, his hands on the nature of chemical bonds and the structure of proteins (Excursus 1-3). Some physicists were busy trying to assess the influence of X-rays on oncogeny. The theme of viruses triggering tumor growth was also intensively pursued. And experimenting with easy-to-handle bacterial viruses – bacteriophages (Excursus 4-1) – promised to be a feasible option as a model system in virus research. Emory Ellis was there studying their growth-cycle. Actually, Felix d'Hérelle, the discoverer of bacteriophages in the 1920s, and Mcfarlan Burnet, based on his work in Australia, had already described the replication cycle of these particles in the beginning of the 1930s. They had shown this cycle to encompass three steps: adsorption of the virus particle to a susceptible bacterium, replication inside this host-cell, and, finally, lysis of the cell wall, liberating progeny particles. The freed virus particles could reinitiate a new growth cycle, as soon as fresh bacterial victims were available. Nevertheless, these processes had yet to be scrutinized and described quantitatively. There was no one better than Delbrück to fulfill this task. As he heard of Ellis' work at Caltech, he immediately realized the potential of this system for resolving the principles of biological self-replication – forget about the infinitely complex and cumbersome fruit flies.

Ellis and Delbrück conceived and brought forth an experiment so simple as to be easily repeated in many a school classroom: the so-called one-step-growth-experiment (Ellis & Delbrück, 1939). Phages were added to a growing bacterial culture; after allowing adsorption to occur (some minutes), this culture was diluted in order to avoid any further phage-bacterium contact; samples were then taken at different time intervals and further appropriately diluted before being plated on petri dishes with a lawn of indicator bacteria; after over-night incubation, small craters – "plaques" – formed by phages on the lawn of indicator bacteria could be counted. The number of plaques remained constant for the samples plated during roughly the first half-hour, but then, in a matter of a few minutes, there was an approximately hundred-fold surge. Apparently, each phage replicated inside its bacterial host by this factor. During the first phase of infection, no progeny particles were liberated from the infected cells and thus it was dubbed the latent period; then all infected cells lysed more or less synchronously (Fig. 4.3), freeing newly synthesized virus particles. This genius-inspired simple experiment was a turning-point in phage research. For decades it came to be the frame for almost all newly planned phage experiments. It showed how experiments must be conceived in order to point out the elementary steps of a phenomenon – in this case, adsorption, multiplication, and lysis. (Before that, undefined numbers of phages were added to bacterial cultures of unknown titers, thus hindering a clear evaluation of the processes involved.)

As Delbrück's Rockefeller fellowship expired, he had no choice but to accept a position as an assistant at the university in Nashville, Tennessee. Returning to Germany was impossible; World War II had erupted. At about the same time (1940), he made the acquaintance of another European phage researcher, who at the last moment had managed to flee the excruciating war's events and persecutions ravaging Europe: Salvador Luria, a physician from Turin, Italy. Immediately after his arrival

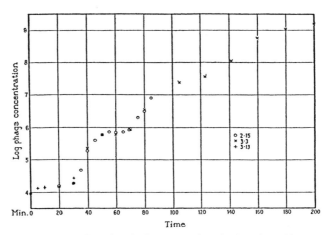

FIG. 2. Growth of phage in the presence of growing bacteria at 37°C.

A diluted phage preparation was mixed with a suspension of bacteria containing $2 \times 10^8$ organisms per cc., and diluted after 3 minutes 1 to 50 in broth. At this time about 70 per cent of the phage had become attached to bacteria. The total number of infective centers was determined at intervals on samples of this growth mixture. Three such experiments, done on different days, are plotted in this figure. The same curve was easily reproducible with all phage preparations stored under proper conditions.

Figure 4.3: Facsimile from Ellis & Delbrück's (1939) publication, describing the one-step growth of phage. The plotting reveals three successive growth cycles (synchronization markedly decreases after the first cycle).

Figure 4.4: Max Delbrück (1906-1981).

in the U.S.A., he found refuge, first at Columbia University, N.Y., and then (from 1942 on) at the University of Indiana. Luria was busy adding sensitive bacteria to an excess of phages and spreading the mixture on plates with culture medium. Among the many millions of bacteria, only those mutants which were phage-resistant were able to grow and form colonies – it seemed to be a nice selection system, pointing out rare mutational events, allowing, for example, to measure mutation rates. However, it was not a perfect system, since the percentage of mutants from similar experiments showed absurd and incontrollable variations. Luria & Delbrück pondered the issue: the phenomenon deserved closer inspection; perhaps it could help throwing light on the elusive origin of mutations. The approach to study this phenomenon – rather a detour from the set target of understanding self-replication – was the fluctuation test, now named after its creators (Luria & Delbrück, 1943) (Excursus 4-2). This test may perhaps today be considered as the most influential one in Delbrück's career because it made bacteriology accessible to genetics. It made clear that bacteria are, like all other living creatures, genetically defined beings, with generally constant characteristics; notwithstanding, due to spontaneous, although rare, mutational events, these characteristics could be altered, but the novel forms were further stably propagated to the following generations. This work dealt with a genetics of vegetative reproduction, since bacterial populations were simply huge vegetative clones encompassing sub-clones of mutated cells. At the time, there were no conjectures of sexual processes and gene transfer between bacteria.

By the way, phages themselves underwent mutations (Fig. 4.6), following the same spontaneous mechanism as bacteria – or other living creatures; this was demonstrated by Luria (1945).

And finally, a small irony of history: the statistical variance analysis, crucial to Luria & Delbrück's fluctuation test, was mathematically incorrect (although this did

Figure 4.5: Salvador Luria (1912-1991).

not affect the validity of their conclusions). A mathematical reevaluation of the fluctuation test by Lea & Coulson (1949) had shown that. Alas, these authors made mistakes of their own, finally corrected by Armitage (1953).

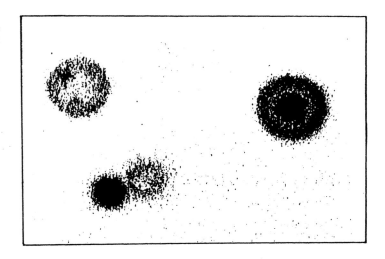

Figure 4.6: Phage mutants and recombinants. Plaques of phage T2 (magnification: about 20-fold) displaying distinctive genetic traits. Small turbid plaque: the wild type, T2 $h^+$ $r^+$ (the indicator lawn is a mixture of two coli strains, only one of them is sensitive to wild type T2); small clear plaque: a mutant that can infect both coli strains ($h$, for host range); large turbid plaque: a mutant that lyses quickly ($r$, for rapid lysis); large clear plaque: the double mutant $h$ $r$ (Hershey & Chase, 1951). If a host bacterium is doubly infected with $h$ and $r$ phage mutants (cross $h$ $r^+$ x $h^+$ $r$), recombinants $h^+$ $r^+$ (wild type) and $h$ $r$ (double mutant) arise. The rate of recombinants among all descendants reveals the distance between the markers $h$ and $r$ on the phage genetic map (Fig. 4.9).

Meanwhile – soon after World War II – Delbrück's concept of genes as molecular structures was taken up by the physicist Erwin Schrödinger, one of the pioneers of quantum mechanics. Many a physicist, befallen by a sort of professional malaise during and after the war, redirected his knowledge of physics to the understanding of biological phenomena. New perspectives opened up. For example, Schrödinger marvelled over the mystery of how a gene, responsible for the form of his nose – supposedly inherited from his grandfather – remained stable over decades at the temperature of 310 °K (37 °C). This, for a physicist, seemed astonishing indeed. Yes, such genes remained immutable even for centuries – a historical fact well documented by the inheritance of the Habsburger's facial traits, especially lips and nose. The example of the monarchs of the Austro-Hungarian Empire was obvious to Schrödinger, who was an Austrian refugee from the Nazis, living in Ireland. Schrödinger suggested that genes were so-called "aperiodic crystals", an expression coined by him, never again to be used. Aperiodic crystals were to be composed of a series of different isomeric elements – more modernly one would say building blocks – whose sequence enclosed specific information leading to the respective gene action. It rather resembled the Morse code, its "isomeres" being dots, dashes and spaces. The atoms of aperiodic crystals would lay in energy wells; and a certain level of activation energy would cause a change in their position, leading to mutations. Were these not Delbrück's ideas, propagated with Schrödinger's authority?

Figure 4.7: Erwin Schrödinger (1887-1961).

Still, these theories said nothing concrete about the material nature and chemical features of these aperiodic crystals: there was not the faintest mention of nucleic acids, for example. A substance possessing the seemingly fantastic properties that would allow it to coordinate intricate metabolic processes and to replicate itself for posterity was still totally unconceivable. Schrödinger adopted and amplified Delbrück's idea of a gene, professing the notion that gene action, besides being controlled by the established laws of physics and chemistry, was also governed by other, yet to be discovered, physical phenomena peculiar to living matter. These concepts were exposed in his booklet (Schrödinger, 1944): "What is life?" Classic biologists would find nothing revealing in it, only such well known banal facts as, for instance, a description of the cell division cycle. If they were willing to read it at all, they were mostly unimpressed; it also did not help that Schrödinger decided to handle the issues of determinism and free will at the end of his dilettantish booklet. However, it threw a hypnotic spell over his colleagues, the physicists (Perutz, 1987). Searching for the novel laws of nature, the fundamentals for the understanding of genes, became a challenge of utmost importance. It would even be worth jumping fields, learning the essentials of biology, just not to miss the search for the mechanism of self-replication.

In the summer of 1945, the first phage course took place in Cold Spring Harbor (CSH) involving a small group of phage scientists. They had become accustomed to spending their summers at CSH (Fig. 4.8), an idyllic spot on Long Island's northern shore – geographically near but in its character so distant from New York. The course, which at first attracted many of the physicists who had read "What is life?", continued to be held every summer at CSH for more than 20 years. It became the foundation of the phage school, which, in its turn, wielded a crucial influence at the onset of molecular biology – less for key experimental discoveries than for its refreshing mentality, pointing out innovative directions of thinking. The mechanism for the propagation of genetic features throughout the generations was considered top priority. It looked as though Schrödinger's "What is life?" teaching was being put into practice.

Delbrück and Bailey (1946) observed that bacteria infected simultaneously with different mutants of similar phage types yielded some genetic recombinants among progeny phages. Delbrück's own interpretation of their data was however totally incorrect. He suggested that the presence of one phage would trigger mutational events in the other. Hershey (1946) also had discovered genetic recombinants among diverse mutants of phage T4, and he was a victim of the same misinterpretation. Hershey, like Delbrück, had never attended an introductory genetics course (Delbrück was a physicist, Hershey a microbiologist). Soon after, though, the mistake was recanted, and it was easy to establish linkage groups and genetic maps (Fig. 4.9) through phage crosses (Hershey & Chase, 1951; Doerman 1952, 1953), similar to those in *Drosophila*. Viruses were not "feral genes" but self-replicating structures with their own genomes, encompassing a multitude of different genes.

The multiplication of a bacteriophage: could it really be assumed to be self-replication in its most unsophisticated form? A phage infects a host bacterium; in 30 minutes there is a progeny of over 100 new identical particles. This was detected by Ellis and Delbrück (1939). However, the unfathomed metabolism of the host cell was an essential aspect of the replication mechanism of phages. Besides, the occurrence of genetic recombination suggested that operations more complex than mere gene multiplication were involved. In Rochester, August Doermann, one of the first members of the phage school, decided to investigate these problems by examining what was happening inside the host during the latent period. To this end, he pried open infected cells by means of cyanide and massive overinfection with phages previously ultraviolet-inactivated (UV impairs their ability to replicate; nevertheless, UV-treated phages are still able to adsorb to bacterial cell walls and to drill holes into them). Doermann observed that half way through the latent period no infectious virus particles were to be found inside the broken up cells; when the first ones of these arose, there were already genetic recombinants among them. Obviously, essential steps were already underway before infectious particles matured (Anderson & Doermann. 1952; Doermann, 1952, 1953). To further unravel the processes going on in the infected cells, a more precise definition of the corresponding physiological and biochemical events had to be achieved. Seymour Cohen, who had attended the phage courses in CSH, but did not really belong to the core of the phage school, stepped in: he employed radioactive isotopes, enzymatic tests, and biochemical analysis (we shall come back to this point). Delbrück's commentary on the issue: it was not clear to him, how such experiments should solve the central question of the mechanism of self-replication. He himself was flirting – if only for a short while – with chloroplasts or other organelles (as protozoa's ciliae) as possible subjects to be studied in order to help crack the secret (Delbrück, 1949).

After years of activity, one could so describe the fruits reaped by the phage school: more incertitudes than ever! After all, what basic physical principle – according to Delbrück's wishes – could be employed to answer the central question? Immerging into the sludge of cell-free extracts in order to capture its biochemistry would result in nothing – this opinion was aired with such assiduity that it acquired the aura of a credo typically adopted by the phage school.

The notion nurtured by Delbrück, conceived by Niels Bohr, his mentor, and adopted and propagated by Schrödinger's "What is life?" views could still be typified by the following question: Was it possible that certain aspects of a cell's life, like, for example, self-replication, could only be explained by novel, still undiscerned laws of nature? A parallel was surely to be found when mechanical physics was displaced by quantum theory, opening up new perspectives in comprehending the world.

Novel laws? Maybe unfathomable laws? Vis vitalis? Certainly, Delbrück would have rejected vehemently the suspicion of vitalism. But, if his ideas came perilously close to it, they were good for one thing – so Delbrück: they had instigated at least one physicist to seriously consider biological issues. Gunther Stent (1968) suggested that the magnet attracting new adepts to molecular biology was this romantic belief in novel laws of nature which would underlie biological phenomena. Once asked whom he actually meant, Stent answered disconcerted that, besides Delbrück, there was at least still someone else. (The author remembers vividly, though, how he, as a medical student, without knowing of the events here described, was tormented by the incompetence of science to describe – even in the most general terms – the phenomenon of cell replication.)

Seymor Cohen, a relative outsider from the University of Pennsylvania, was not disturbed by such thoughts. He was a classical biochemist and as such he was convinced that the physiology of phage infection had a concrete biochemical basis susceptible to analysis. His initial results were meager but nevertheless interesting; he

Figure 4.8: The Biological Laboratory in Cold Spring Harbor, situated on the northern Long Island shore at about one hour's drive from New York, was founded with private funds more than 100 years ago. Since the beginning of the 20th century, it was financed by the Carnegie Institution. The idyllic landscape offered a modest but serene place of work for a few researchers wishing to dedicate themselves exclusively to their work. The laboratories, with the appearance of ordinary homes, are partially restructured old buildings. Genetics was well represented there from the start. But after World War II, it went through a real boom with Max Delbrück's summer phage courses and the highly appreciated symposia. Nowadays, every summer Cold Spring Harbor witnesses an invasion of hundreds of pilgrim scientists. In winter time, tranquillity is restored, with only a few permanent resident scientists remaining. Thanks to the enthusiasm of his current director, James Watson, the research facilities have expanded enormously and a new scientific publishing house has been created – maybe to the chagrin of nostalgics, who would rather remember the pioneer era, when Cold Spring Harbor was a Mecca for the gentlemen club of research individualists.

had observed that DNA synthesis was abruptly halted upon infection with phage T2, to be restarted 6 to 8 minutes later, with a many-fold increased rate. The rate of protein synthesis remained nonetheless the same during this period, but Cohen was then already conjecturing (Cohen, 1947) that marked qualitative changes would certainly occur. He suspected that crucial metabolic changes took place; the synthesis of specific coli enzymes was to be halted, but other proteins were to be produced instead (Cohen & Anderson, 1946). This information was obtained by an experiment in which cells were infected with phage and simultaneously treated with 5-methyl-tryptophan (a tryptophan analogue which blocks protein synthesis), which resulted in

the failure of infection-induced DNA synthesis (Cohen, 1948). These events could possibly be further unravelled immunologically, suggested Cohen. [This was later accomplished: specific anti-sera revealed that after phage infection some proteins were synthesized which had been absent before (Maaløe & Symonds, 1953).] The concept that phages direct protein synthesis preceding their replication was born. This knowledge was essential for the later development of virology (Excursus 4-3).

These observations indicated that new proteins, detected as the first changes after phage infection, were crucial for self-replication. Genes were themselves proteins, a belief that was – even if not openly aired – the creed to be supported by these studies on the biochemistry of the phage-infected cell.

Anyway, these experiments and the unexpected novelties unearthed by them helped to elaborate a new consensus within the phage school; learning somewhat more about the biochemistry of phage replication would not necessarily hurt!

Meanwhile a new, very promising young student had joined Luria and his group: James (Jim) Watson (Luria later said that he was the first really serious one). Jim Watson was supposed to dedicate himself to biochemistry after his doctoral work (he was just 20 years old). Luria provided a fellowship for him to work in Europe; the more relaxed European way was more propitious to creative thinking – so Watson. Sending Watson to Europe was probably the most decisive contribution made by the phage school, for the move would prove to be the perfect formula for solving the questions it formulated.

Delbrück decided, once more, to tackle replication of biological structures through theoretical-mathematical conjectures, instead of the despised biochemistry.

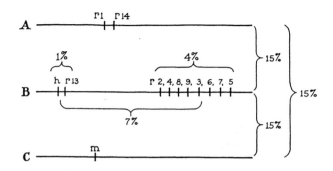

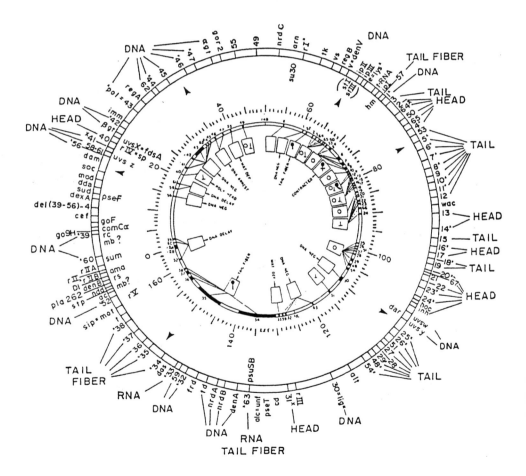

Figure 4.9 (left): At the top, a genetic map of phage T2 (Hershey & Rotman, 1949); at bottom, the genetic map of T4, a phage closely related to T2, as described later in a monograph on this phage (Mosig, 1983). If known, a reference to the gene function is given in the scheme; the scale corresponds to the sum of the recombination distances of neighboring genes and refers to the accurate gene map at the center; point "zero" at the *rII* region is arbitrarily chosen, since the map is circular.

Delbrück, then at Caltech (in 1947, Beadle, Morgan's successor, had brought him back from Nashville, as a professor), collaborated on this theme with Niccolo (Nick) Visconti, an aristocratic sunnyboy, who viewed science as a stimulating hobby. The duo developed a concept of phage multiplication based on Doermann's observations about the early production of recombinants: parental particles, or, more precisely, the corresponding undetectable vegetative phages, would divide themselves inside the host cell in successive waves of replication; after each replication, the structures would exchange genetic material with each other in a crossover-like fashion. Following the first half of the latent period, dubbed the eclipse, some vegetative phages would mature to particles, while others would remain reproducing vegetatively and crossing. The infected cell was to be compared to a pen full of rabbits, crossing freely through many generations, the descendants displaying genetic characteristics from more than two progenitors. [One can devise a three-parent-cross, for example, and analyse its descendants (see Hausmann & Bresch, 1960).] Visconti and Delbrück were bold enough to tally the different "rounds of mating": in the case of T4, they reckoned that the very first mature particles emerging during the latent period involved 2 rounds, and that at the end of the latent period each phage had gone, on average, through 5 rounds of mating (Visconti & Delbrück. 1953).

The mating theory did not make any mention of a possible material substrate of heredity; it dealt exclusively with abstract vegetative phages. It was the last of Delbrück's considerations about self-replication. Three months later Watson & Crick published their work on the double helix structure of DNA. A new universe began to unfold...

[In 1969, the Nobel Prize for Physiology or Medicine was conferred to Max Delbrück, Alfred Hershey (Chapter 7) and Salvador Luria.]

**Excursus 4-1**
**BACTERIOPHAGES**

Wherever bacteria are to be found – almost everywhere in soil, water, animals' and men's bowels – there are also bacteriophages, phages for short. Phages are viruses which attack these bacteria and propagate at their expense. For example, some water droplets from sewage, spread over a solid culture medium seeded with indicator bacteria, will cause the appearance of many holes in the bacterial lawn after a few hours of incubation (Fig. 4.10). These holes or "plaques" indicate the presence of

phages which cause the lysis of bacteria at the corresponding sites. If a plaque is
touched with a tooth pick or glass rod, and then inoculated in liquid culture and
diluted a million-fold, new plaques will be revealed, if the diluted suspension is again
spread on a bacterial lawn. These "plaque-forming-units", analysed under the electron
microscope, prove to be particles whose forms and sizes vary according to the phage,
usually within the range of 20 to 80 nm in diameter (Fig. 4.11). These phages, self-
replicating structures, are composed of roughly half DNA and half protein, a
composition that had been revealed in the 1930s and 1940s by biochemical analysis
(Schlesinger, 1934). (In the case of some very small phages, one finds RNA instead
of DNA.) Although there are surprisingly many types of phages, most of them display
a "head" with a "tail" attached to it; these two components make up the protein
fraction of the particle. The tail should rather be dubbed "trunk", since this structure
is the one involved in the process of adsorption to the cell walls of susceptible bacteria
(each phage type invades only one specific bacterial strain) upon a chance encounter.
Phage researchers were mystified when, at the beginning of the 1950s, it was
demonstrated that viral particles as such never penetrate into the bacterial cell.
Apparently, only its content made the way into the host (we know today: it's DNA),
carrying along the information for producing new generations of phages (Fig. 4.11 &
4.12).

Figure 4.10: Isolation of bacteriophages from nature. A culture of *E. coli* was spread on a plate
containing solid culture medium. On top of it, sewage was added drop by drop; after over-night
incubation at 37 °C, a whitish lawn of bacteria having developed, holes or "plaques" (a French
term originally used by d'Hérelle, which became internationally accepted) can be visualized,
corresponding to small circular areas of bacteria destroyed by the progeny of phages present in
the sewage. Each plaque represents a clone of phages, encompassing many millions of
descendants of a single phage particle.

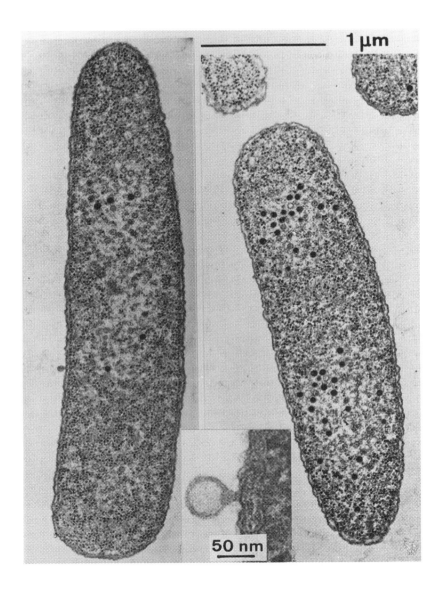

Figure 4.11: Thin slices of a coli cell infected with T7, at 10 min (left) and 12 min after infection (right). Empty, newly synthesized coats are visualized (whitish round structures). These will be packaged with phage DNA and will accumulate gradually inside the cell, as mature infectious particles appear (dark round structures). At left, and, magnified, in the middle, the emptied protein coat of a parental particle, adsorbed outside the bacterial cell wall (photo: A. Kuhn, Univ. Freiburg).

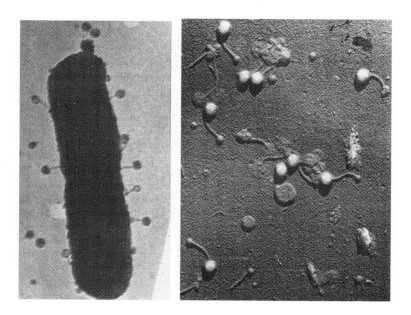

Figure 4.12: Left, one of the first electron microscopic pictures of bacteriophages (here: T5) attacking a coli cell (Anderson, 1953). Right, phages T5, treated with a sample of pure T5 receptors, isolated from cell walls of sensitive bacteria. These receptors, spherical lipoproteins, adsorb T5 tail tips (unpublished photo from 1954, received from E. Kellenberger; see also Weidel & Kellenberger, 1955).

Figure 4.13 (right): A selection of different phage types as seen by electron microscopy: a) One of the first pictures of phage T4, obtained from a lysate of a bacterial culture (Levinthal & Fisher, 1953). This picture is especially interesting since it shows also the first image of ribosomes – the beads with 20 nm diameter (their identification, though, was only accomplished years later). b) A more recent picture of the same phage, sporting its complex tail structure with 6 fibers, responsible for attachment to the wall of host bacteria (F. A. Eiserling, Univ. Geneva). c) Particles of one of the innumerable undescribed relatives of phage T1, isolated from sewage (A. Kuhn, Univ. Freiburg; Hug et al., 1986). d) A relative of phage T7, Phi1.2, which attacks special capsulated strains of E. coli (E. Freund-Mölbert, Max-Planck-Institut für Immunologie, Freiburg). e) Phage λ (F. A. Eiserling, Univ. Geneva). f) SP50, a phage specific for Bacillus subtilis (F. A. Eiserling, Univ. Geneva). g) M23, one of the smallest phages, its genome consisting of a single strand of RNA, adsorbs exclusively to sex pili of E. coli F+ (Chapter 11). Some of the phages were disrupted, revealing isolated coat-protein molecules (compare with Fig. 1.13). On top, a partial view of a bacterial flagellum (D. Lang, Univ. Texas, Dallas). h) Coli phage P2 (courtesy of Robley C. Williams, Virus Lab., Univ. California, Berkeley). The bars correspond to 50 nm.

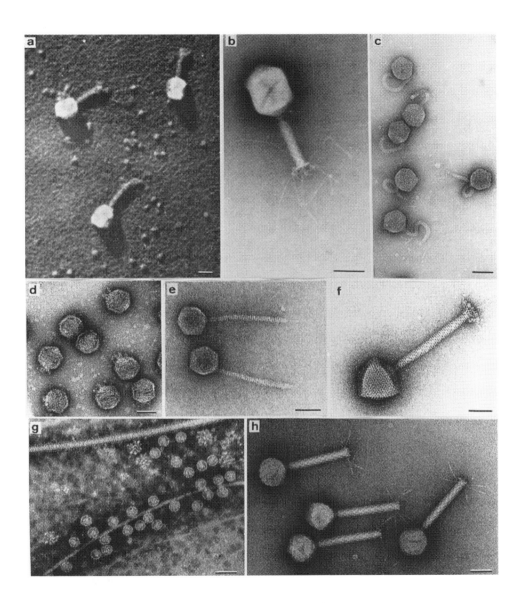

**Excursus 4-2**
**THE QUEST FOR THE ORIGIN OF MUTATIONS:**
**LURIA AND DELBRÜCK'S FLUCTUATION TEST**

In principle, there were two possible explanations concerning the processes which trigger the induction of mutations:

Hypothesis 1: the organism, in this case a bacterium, has the potential for adapting itself to a changing environment with a certain probability of success. The so acquired novel trait is then inherited by subsequent generations. This hypothesis, suggested by the French botanist and zoologist Jean Baptiste Lamarck (1744-1829), encompassed the view that, for example, the necks of the ancestors of the giraffe became ever longer because the animals always had to stretch them to reach savanna canopies whose leaves made up their main food supply. At the time when Luria and Delbrück set their experiment, this assumption prevailed among bacteriologists and medical researchers. The direct contact of originally sensitive bacterial cultures with a toxic agent, like an antibiotic, triggered the development of resistance to that toxic substance, or so they believed.

Hypothesis 2: an organism (or one of its gametes) has an *a priory* small probability to undergo any mutation – the type of mutation being independent from the environment. The addition of a toxic agent (in Luria & Delbrück's case: phages) will result in the elimination of the great majority of the cells in the culture; only a few, already pre-existing, resistant mutants will survive. These will multiply, unhindered by the rest of the cells, originating a pure culture of resistant cells. Since the growth of resistant mutants proceeds fast, the impression is given that the whole original culture adapted itself. In a bacterial culture with hundreds of millions of cells, thousands of individual diverse mutations would always exist, each mutant type comprising only a small fraction of the whole. Accordingly, every culture would encompass some mutants resistant to, say, sulfonamide, or penicillin, or streptomycin, etc, or even to toxic substances yet to be synthesized – these substances having the exclusive role of selecting agents.

The fluctuation test – often named after its creators – was designed to reveal which hypothesis was correct.

To perform the test, small numbers of sensitive bacteria are initially inoculated into each tube of a series of tubes with nutrient broth, with the aim of obtaining a series of parallel cultures. After over-night incubation, samples from each culture are spread on separate plates with solid medium containing a toxic substance (for instance, the antibiotic streptomycin). Further samples from the parallel cultures are mixed before spreading on similar toxic medium plates (Fig. 4.14).

Figure 4. 14 (right): Scheme of the fluctuation test.

- Each of many individual test tubes containing liquid
culture medium receives a small inoculum of sensitive bacteria.

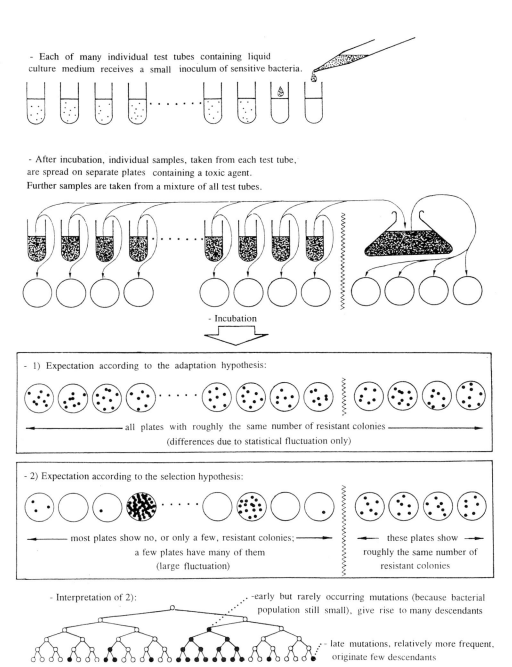

- After incubation, individual samples, taken from each test tube,
are spread on separate plates containing a toxic agent.
Further samples are taken from a mixture of all test tubes.

- Incubation

- 1) Expectation according to the adaptation hypothesis:

———————————— all plates with roughly the same number of resistant colonies ————————————
(differences due to statistical fluctuation only)

- 2) Expectation according to the selection hypothesis:

———— most plates show no, or only a few, resistant colonies; ————     ———— these plates show ————
a few plates have many of them                                         roughly the same number of
(large fluctuation)                                                    resistant colonies

- Interpretation of 2):          -early but rarely occurring mutations (because bacterial
                                  population still small), give rise to many descendants

                                  - late mutations, relatively more frequent,
                                  originate few descendants

Depending on the hypotheses, two different predictions regarding the distribution of resistant colonies are to be made.

• According to the adaptation theory, each plate should have approximately the same number of resistant colonies. Let us consider the case of one resistant cell appearing, on average, for each 10 million normal cells (mutation rate $10^{-7}$): all samples from the parallel cultures (individual or mixed ones), containing around $10^8$ cells, would yield roughly 10 resistant colonies on each plate because all bacteria on the plates, once confronted with the lethal agent, would have the same small chance of adapting themselves and passing to their progeny the new resistance trait, acquired under stress.

• If the hypothesis of pre-existent mutations and ensuing selection is accepted, the expected results are different. Let us consider that every cell in each parallel culture has the same small chance of undergoing a mutation; the number of mutated cells at the beginning of incubation is relatively small; the amount of mutants will increase with incubation time, since the total number of bacteria increases. This means that the longer the incubation time is, the more mutants will be present. Depending on the moment when a mutation occurred (sooner or later during the incubation period), the proportion of mutant cells in the original sensitive culture will be larger or smaller, respectively (Fig. 4.14). In general, the probability of a large clone of resistant cells is small (early mutations), while smaller clones are more common (late mutations). Cultures without any mutants at all should also occur.

Luria and Delbrück obtained results that corroborated the second hypothesis: in many of the parallel cultures there were none or only few mutants, whereas a few cultures had a very large tally of mutants – these being the descendants of many generations, originating from an early mutant.

**Excursus 4-3**
**PHAGE-DIRECTED PROTEIN SYNTHESIS**

The issue of phage-directed protein synthesis came only to be intensively and effectively tackled when Wyatt & Cohen (1953) discovered that the DNA of some phages, like T2 and T4, displayed the base hydroxymethyl-cytosine instead of cytosine. This base, unknown until then, did not occur in uninfected bacteria; it was newly synthesized upon phage infection. Consequently, first of all, a new corresponding enzyme (or several enzymes) had to be identified. Flaks & Cohen (1958) identified the enzyme dCMP-hydroxymethylase, appearing in *E. coli* barely 3 minutes after being infected with T2. Thereupon, Cohen and his collaborators – and others, such as Kornberg et al. (1959) – looked for further phage-directed (today one would rather say phage-coded) enzymes, bringing to light more than a dozen of them (Cohen, 1968). These enzymes were neither detectable in the mature phage particle nor inside the non-infected bacterium. Producing them was the first necessary step to bring about the synthesis of phage-specific DNA. These were the so-called early phage-directed, or phage-induced, proteins. Proteins which appeared during the second half of the latent period and contributed to the production of structural components were correspondingly dubbed late proteins. However, these fascinating novel facts regarding the time schedule of virus reproduction came too late to contribute to the development of concepts and ideas concerning the material basis of heredity. For medical virology, the notions of early and late virus-directed proteins, elaborated with the help of phage experiments, were crucial to progress in the following years.

# CHAPTER 5

# THE DOUBLE HELIX

Crucial to the discovery of the double helix by James Watson and Francis Crick was, in the first place, their involvement with the spatial structure of DNA; what was the reason for their interest? At that time, the scientific world was still permeated with the notion that only proteins possessed the structural complexity necessary to develop countless alternative genetic characteristics (Fig. 5.1). Nevertheless, DNA was surely in some way associated with this genetic material, playing a still unknown important role in heredity, as, for example, the shaping of chromosome structure. That chromosomes were the seat of the genes was known since the beginning of the 20th century and that almost all cellular DNA was located in the chromosomes was shown, if not before, then definitively by Feulgen in 1923, with his eye-catching DNA-

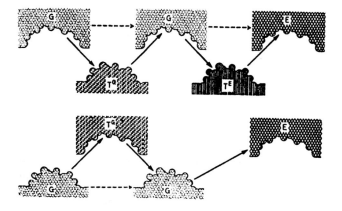

Figure 5.1: The gene (G) as a protein-like template able to transmit specific surfaces to further gene copies and to enzymes (E). The representation, created by R. A. Emerson, a pioneer of maize genetics, is taken from a textbook of the 1950s (Hovanitz, 1953).

specific staining method (Feulgen & Rossenbeck, 1924). Yet, there was no compelling evidence for DNA alone being the carrier of hereditary information. Besides, no one had ever suggested that possibility, as such an assertion would obviously be wrong: the tobacco mosaic virus (TMV) was composed solely of 94% protein and 6% RNA, no DNA at all (Schramm & Dannenberg, 1944). However, even if the role of DNA in chromosomal heredity were an unspecific one, as, for example, simply holding the mysterious genes together, any assessment of this role still would represent an important new insight, and every contribution to its understanding would be welcome – even sensational!

Watson first tried his luck in Copenhagen, with the biochemist Herman Kalckar, one of the first participants of the Cold Spring Harbor phage course.

His research on the subject of nucleotide metabolism was not, however, shaped to warm Watson's heart: one could not imagine how traditional biochemistry could explain the chemical basis of heredity.

While attending a meeting on macromolecular structures in Naples in the spring of 1951, Watson got a first exciting glimpse of a totally different approach. For the first time he saw an X-ray diffraction image of a DNA preparation. A clear, concrete diffraction pattern was visible, a fact proving that DNA had regular, analyzable structural elements. Genes could, after all, at least in some aspects, be defined structurally. The genetic material was not completely irregular, amorphous and incomprehensible! Watson had never heard of the seminar speaker before. How could he – a certain Wilkins, from London.

Maurice Wilkins, a physicist who had worked on the Manhattan project, turned his postwar interests to biological questions. By his own account, reading Schrödinger's "What is life?" had inspired him to do so. More by chance than by design, Wilkins came across a DNA preparation of the best quality at that time; he observed, still by chance, that upon touching the viscous DNA suspension with a glass rod, one could pull out hair-fine threads which immediately dried out, being thin enough to originate relatively sharp diffraction patterns. These were, by far, the best images yet – much better than those made by Astbury many years before (Fig. 5.2). But they were not good enough for drawing definite conclusions regarding the structure of DNA.

Watson tried hard but was unsuccessful in engaging Wilkins in a discussion. However, Watson had firmly decided to stick to the X-ray diffraction analysis of DNA, a technique which he, nevertheless, still had to learn. Through tricks on tracks already laid down, he was able to change his scholarship in Copenhagen for one in Cambridge, U.K. His mentor, Luria, had actually just met John Kendrew, who, as Max Perutz' junior partner, was working at the Cavendish Laboratory on the X-ray diffraction analysis of the structure of myoglobin (Excursus 1-4).

There, Watson met Francis Crick, who shortly before had restarted, under Perutz' guidance, his doctoral work, which had been interrupted by the Second World War. Crick was also a physicist who had worked on a military project during the war, designing magnetic sea mines; he also had read "What is life?"; he also was fascinated by the idea that important life processes, especially heredity, could be approached on the basis of precise physical concepts. He always abhorred the flight into mysticism and theism as a means of comprehending nature – and biology, with its complex and inexplicable phenomena, among all natural sciences, provided a most fertile ground where such thoughts could flourish. He decided to undertake a counter offensive. First of all, one should become better informed on the central category of biological molecules: the proteins!

His thesis adviser, Max Perutz, was the right man for that. For many years, he had dedicated himself to hemoglobin, the blood oxygen carrier. This protein could be easily crystallized in a reproducible process, but the X-ray diffraction patterns originating from such crystals remained – for many years to come – unanalyzable (Excursus 1-4). It wouldn't take long before Crick, the bright young fellow, tried to explain to his laboratory colleagues that they did not quite have a grasp on the subject..., that, somehow, they were thinking along hopeless lines.

Crick and the newcomer Watson quickly recognized that both of them were impelled by the same notions concerning the fundaments of biology: important were only the basic phenomena, and not the infinite details in which most biologists and biochemists used to delve. And what was more fundamental than the common characteristic of all living beings, namely the capacity to reproduce themselves and to replicate their structures, the most varied and complex ones? An intense cooperation developed between the 35-years old, extremely intelligent, flamboyant doctoral candidate Crick and the 23-years old, equally highly intelligent but a little diffident, American postdoc from Chicago, a pair seemingly taken out of a novel (Watson, 1968). They quickly agreed to try to clarify the DNA structure. Only, this was not part of their official project, not their task. Also, they neither possessed any sample of this substance to experiment with, nor the know-how to prepare one (nowadays a banality in any medical or biology beginners course). So they tried what was possible with the available data: these were, to start with, the X-ray DNA diagrams obtained by William Astbury before the Second World War. He had subjected DNA to X-ray diffraction analysis, out of mere curiosity, without any special concept in mind, as he did with other materials like wool fibers, horn, etc. These first, and for a long time only pictures, did not disclose much – nevertheless, a periodicity of 3,4 Å (0,34 nm) could be derived from them, corresponding to the nucleotide bases which apparently were stacked one upon another. Much better images would soon be obtained, by a specialist in X-ray crystallography, Rosalind Franklin, working in Wilkins' Laboratory at King's College in London. (Shortly before, in Paris, she had been doing successful work on the analysis of the X-ray diffraction patterns of graphite.) The idea of hiring her came from John Randall, a controversial personality within the Medical Research Council, who tried to pursue research of life phenomena through physical methods –  an attempt which yielded at best modest results. Rosalind understood, probably rightly, that Randall had asked her to study, as her own project, the spatial organization of the DNA molecule as determined by X-ray diffraction analysis. Wilkins, on the other hand, believed, probably also with good reason, that Rosalind was supposed to assist him on the analysis of X-ray diffraction patterns (with which at first he was not very familiar). This was the beginning of a protracted conflict. They could not stand each other; no communication was possible on a scientific level, not to mention a personal one.

During a seminar, given by Franklin by the end of 1951, Watson saw the best DNA X-ray diagrams yet, and also got hold of some important data concerning the dimensions (height and diameter) of the so-called unit cell, the smallest repeating grouping of atoms within a crystal. From those figures, one could derive clues about the molecular structure, possibly a thread with double or triple strands, since the known DNA density did exclude the possibility of a single-stranded molecule. More than that bit of information was not available. But it was enough to prompt Watson and Crick to try a model, which they presented to their colleagues at King's College: it was a complete flop!

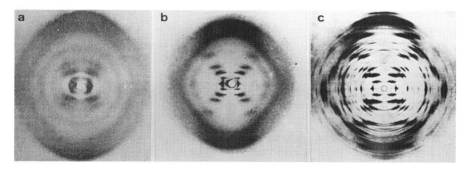

Figure 5.2: X-ray diffraction images of DNA. *a*) an old picture from Astbury, used by Wilkins et al. (1953) to demonstrate that Watson & Crick's model was compatible with the X-ray diffraction data derived from it. *b*) the pivotal picture made by Rosalind Franklin (Franklin & Gosling, 1953), which, offically, should not have been available to Watson and Crick; the data originating from it were nevertheless passed on to them by Wilkins, and also by Perutz – without Franklin's knowledge. *c*) a later, technically improved picture, taken in Wilkins' laboratory, which, nevertheless, did not contribute any new insight (Langridge et al., 1960a).

This humiliation in December 1951 had practical consequences: Bragg and Randall simply prohibited Crick and Watson to occupy themselves with DNA. Crick should finally get earnestly involved with his doctoral thesis regarding the structure of hemoglobin, while Watson ought to dedicate himself to the project of the X-ray diffraction analysis of TMV. Watson soon was thoroughly successful with his task – although this, for him, was only of secondary importance. He demonstrated that the

– in molecular terms – huge TMV particle displayed the form of a screw; he had learnt from Crick how to differentiate such spiral structures by means of X-ray diffraction diagrams. Crick developed a somehow unsophisticated mathematical formula to describe such structures – he was no expert on this subject –, but he was absolutely correct, as the mathematician and crystallographer W. Cochran professionally and elegantly corroborated. This resulted in a publication (Cochran, Crick & Vand, 1952), documenting Crick's solid accomplishment.

Linus Pauling, at Caltech, was also tempted by the challenge of DNA structure. DNA, as one more apparently important biological molecule, was enticing, after the huge success story of the α-helix. But, as he had no suspicion regarding the decisive role of DNA, his efforts were not full-blown. Besides, Pauling also depended on data from other sources, since, like Crick and Watson, he had neither good DNA preparations nor the proper analysis technology. His insights were based on the original X-ray diffraction diagrams of Astbury & Bell (1938) and of Astbury (1947). Carried away by the triumph of the α-helix, he embraced the problem, together with his collaborator of many years, R.B. Corey, producing a model within a short time.

His son, Peter Pauling, was at that time in Cambridge as a young visiting scientist; through him, Crick and Watson received the news that father Pauling was interested in DNA, with a manuscript already in the making. As the two impeded DNA researchers became aware of this (Peter, amicably, maybe naively, had forwarded a copy to them), they were deeply relieved: the model (Pauling & Corey, 1953a,b) was absolutely nonsensical: three polynucleotide strands were intertwined, the sugar-phosphate backbones placed inside, tightly packed, just at the limit of acceptability; the bases protruded loosely to the outside, similar to the amino acid side chains of the α-helix. The idea was, obviously, to get rid of the four different, inconvenient bases (as Pauling and Corey formulated, to allow them to interact with proteins). The model did not have any merit other than being spatially viable and in accordance with those ancient, not very informative, X-ray pictures of Astbury. It did not clarify anything – not even that DNA was an acid. The bases pointing outwards suggested rather basic characteristics for Pauling's construct. Besides, Pauling had apparently forgotten what he, together with Delbrück, had postulated in an absolutely correct theoretical proposition formulated in 1940: the duplication of the genetic information must be based on the principle of mutual (spatial and electrostatic) matching of two complementary structures. This work had been conceived in order to contest a concept by Pascual Jordan (also a renowned quantum physicist, see Excursus 1-5). Jordan (1938) had postulated that a quantum mechanical attraction between similar structures was the foundation of the mystery of gene replication; however, Pauling and Delbrück had shown that such forces as those proposed by Jordan – if they existed at all – would be too weak to be effective (Pauling & Delbrück, 1940; Pauling, 1974). What else remained unexplainable by Pauling's model? Erwin Chargaff, an Austrian immigrant and classical biochemist at Columbia University, New York, had determined the relative proportions of the nucleotide building blocks of DNA by means of the newly developed method of chromatography (Fig. 5.3). He noted that in higher organisms, like vertebrates, there was roughly a 1:1:1:1 proportion for the four heterocyclic DNA bases, adenine (A), cytosine (C), guanine (G) and thymine (T) – an observation that led to Levene's fatal tetranucleotid hypothesis. This was also valid for many bacterial strains, like *Escherichia coli*. Still, many other organisms had proportions of bases which markedly deviated from this finding. What was striking, though, and Chargaff pointed it out, was the fact that in each DNA preparation the molecular ratios of purines corresponded to that of pyrimidines. And more, that the amount of the purine base adenine corresponded to

In order to show examples far removed from mammalian organs, the composition of two desoxyribonucleic acids of microbial origin, namely from yeast[3] and from avian tubercle bacilli[4], is summarized in Table IV.

*Table III* [2]

Composition of desoxypentose nucleic acid of man (in moles of nitrogenous constituent per mole of P).

| Constituent | Sperm | | Thymus | Liver | |
|---|---|---|---|---|---|
| | Prep. 1 | Prep. 2 | | Normal | Carcinoma |
| Adenine   . . . | 0·29 | 0·27 | 0·28 | 0·27 | 0·27 |
| Guanine   . . . | 0·18 | 0·17 | 0·19 | 0·19 | 0·18 |
| Cytosine   . . . | 0·18 | 0·18 | 0·16 | | 0·15 |
| Thymine   . . . | 0·31 | 0·30 | 0·28 | | 0·27 |
| Recovery  . . . | 0·96 | 0·92 | 0·91 | | 0·87 |

*Table IV* [5]

Composition of two microbial desoxyribonucleic acids.

| Constituent | Yeast | | Avian tubercle bacilli |
|---|---|---|---|
| | Prep. 1 | Prep. 2 | |
| Adenine   . . . . . . . | 0·24 | 0·30 | 0·12 |
| Guanine   . . . . . . . | 0·14 | 0·18 | 0·28 |
| Cytosine   . . . . . . . | 0·13 | 0·15 | 0·26 |
| Thymine   . . . . . . . | 0·25 | 0·29 | 0·11 |
| Recovery  . . . . . . . | 0·76 | 0·92 | 0·77 |

*Table V*

Molar proportions of purines and pyrimidines in desoxypentose nucleic acids from different species.

| Species | Number of different organs | Number of different preparations | Adenine/Guanine | | | Thymine/Cytosine | | |
|---|---|---|---|---|---|---|---|---|
| | | | Number of hydrolyses[3] | Mean ratio | Standard error | Number of hydrolyses[3] | Mean ratio | Standard error |
| Ox[1]  . . . . . . . . | 3 | 7 | 20 | 1·29 | 0·013 | 6 | 1·43 | 0·03 |
| Man[2]  . . . . . . . | 2 | 3 | 6 | 1·56 | 0·008 | 5 | 1·75 | 0·03 |
| Yeast. . . . . . . . | 1 | 2 | 3 | 1·72 | 0·02 | 2 | 1·9 | |
| Avian tubercles bacillus | 1 | 1 | 2 | 0·4 | | 1 | 0·4 | |

[1] Preparations from thymus, spleen, and liver served for the purine determinations, the first two organs for the estimation of pyrimidines.

[2] Preparations from spermatozoa and thymus were analysed.
[3] In each hydrolysis between 12 and 24 determinations of individual purines and pyrimidines were performed.

The results serve to disprove the tetranucleotide hypothesis. It is, however, noteworthy—whether this is more than accidental, cannot yet be said—that in all desoxypentose nucleic acids examined thus far the molar ratios of total purines to total pyrimidines, and also of adenine to thymine and of guanine to cytosine, were not far from 1.

Figure 5.3: Chargaff's data on the variation of base ratios in the DNA of different species (Chargaff, 1950). One should point out that these data were not as clear cut with respect to the postulate "A = T and G = C" as frequently surmised.

that of pyrimidine base thymine (A = T); the same applied to guanine and cytosine (G = C). The relation of adenine to guanine in each DNA preparation was the same as that of thymine to cytosine, and necessarily the same as the relation of adenine plus thymine to guanine plus cytosine. The proportion A + T/G + C, nevertheless, varied within relatively wide limits, namely between 0,5 and 2. It was still to be clarified wether a deeper meaning could be assigned to such base proportions – that was the most that Chargaff dared to speculate about his data (Chargaff, 1950). Such deeper meaning was not evident in Pauling's model; it happened that Pauling had never read Chargaff's papers. Anyway, Pauling's manuscript had its impact. Now, that the Californian rival had addressed the subject, the ban imposed on Crick and Watson was to be reconsidered. Why on earth could Caltech, but not Cambridge? Bragg (Excursus 1-4) and Randall decided to loosen the reins on Watson and Crick, depending on Wilkins' acceptance; Wilkins acquiesced, although reluctantly. Rosalind Franklin was not consulted; she did not possess any "rights of her own" on DNA.

This was the starting point for Watson and Crick. Crucial were three categories of data and considerations:

First, the chemical data. Important were Chargaff's observations on the base ratios of DNA preparations from various organisms; these were the results that Pauling had ignored. Chargaff himself was the source of Crick and Watson's knowledge of his work: during a visit to Cambridge, he met the two and discussed it with them. Later, Chargaff recalled with horror the – in his opinion – absurd fact that anyone should try to comprehend the structure of DNA without even knowing the chemical formulas of its bases (Chargaff, 1974, 1979). To know this, though, was no easy task, since most textbooks represented them in their enol forms instead of the keto forms (Fig.5.4). Watson and Crick stumbled on the right formulas by chance during a visit by David Donohue. Donohue should have known the correct formulas; he was an expert who for several years had investigated this type of chemical bond in Pasadena in collaboration with none other than Linus Pauling.

Still other data of a purely chemical nature were an essential precondition for elaborating the double helix structure of DNA: it involved the confirmation that its nucleotides were linked to each other covalently through 5'-3'-phospho-diester-bonds, thus forming long strands (Fig. 2.16 and 5.5). This fact had been shown, a short time before, by Lord Alexander Robertus Todd (Nobel Prize for chemistry, 1957), also in Cambridge (Brown and Todd, 1952). [Todd later asserted (one cannot say, how serious he was): Watson and Crick did not discover the structure of DNA – I did; Watson and Crick only discovered the spatial arrangement of that structure.]

The second crucial fact contributing to Watson and Crick's insight were the results of the X-ray diffraction analysis obtained by Franklin. Patterns, as those seen by X-ray diffraction, reveal absolutely nothing neither to the layman nor to most scientists, excluding the trained eyes of the expert, who can glean clues from them as to the arrangement of atoms composing the matter traversed by X-radiation. Franklin

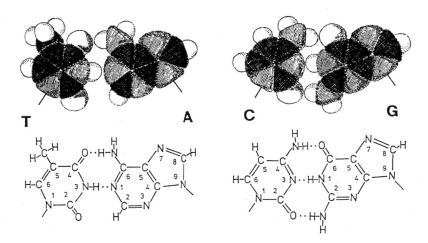

Figure 5.4: The four heterocyclic DNA bases, and how the sterically similar ("isomorph") base pairs, A-T and G-C, fit together through the postulated hydrogen bonds; for that, T, C and G must take – as one knows today – the more common keto form ($> C = O$) and not the rare enol form ($\geqslant$C-OH) represented in most older textbooks.

by then had obtained pictures which were highly informative. She had found, for instance, that, depending on the degree of humidity, the DNA threads (in which the strands were aggregated in crystal-like fashion) would adopt one of two different structures, A or B. If the two forms were mixed, unclear X-ray patterns would result. To that, and many other details, one ought to pay attention.

Nevertheless, the secret of the genes was not foremost in Franklin's mind. As a non-biologist, the DNA structure represented to her, indeed, an interesting challenge, but this was no reason why she should thrust herself into constructing unrealistic hypotheses. Thus, Crick and Watson's worries that someone else could precede them, converged more on Linus Pauling in California than on Wilkins or Franklin in their very neighborhood. And, in their eyes, Pauling himself was now out of the race. Through obscure paths, Watson and Crick acquired important details from Franklin's work (See Fig. 5.2 and the cryptic last sentence on the fourth last paragraph in Watson and Crick's publication, Fig. 6.1). Critical was a report by Franklin to the Medical Research Council, which Perutz (acting as a referee) gallantly passed to Crick and Watson without the knowledge of Franklin: the diameter of the apparently spirally arranged DNA thread was 20 Å; the density of the material allowed the assertion that more than one polynucleotide strand were constituting the spiral thread; the thickness of the bases, arranged vertically to the axis of the spiral, was 3,4 Å; one turn of the helix was 34 Å (a stack of 10 bases); this 34 x 20 Å unit cell did not display any polarity (that means, turning the structure upside down did not affect the diffraction pattern obtained). And that was all.

What finally proved to be conclusive was the third point: model building. This ultimately decisive idea came directly from Linus Pauling. With a simple paper model, he had recognized the possibility of the α-helix structure for polypeptides (Excursus 1-3). This basically simple method consisted in first constructing a molecular model out of paper or wire, in which the relative dimensions of the atoms and the binding angles between them were proportional to the real values; then one would try to assemble the models, spatially, into larger structures. (Nowadays spherical models, or, better, computer programs, are available for these purposes.) As clear cut and intuitively promising this method was, it was mostly rejected by the scientists of the time, probably because it suggested a capitulation of the mind in favor of handiwork: shouldn't one be able to calculate mathematically all the coordinates of a model? In practice, though, a few manipulations with a model could disprove month-long calculations. Crick and Watson ordered wire models of each nucleotide from the institute's workshop. As it took too long for them to be ready, Watson cut the models of the four bases out of cardboard, and began to play with them. Suddenly, on his table he saw the contours of G fitting those of C, and A matching T; in addition, the pairs had almost the same profile (Fig. 5.4). When the wire models finally arrived, they worked feverishly on the construction of their model of a single turn of the double helix, with a height of 2 meters (Fig 5.5). The triumph was complete – now, only now, by inference from its structure, one could assert convincingly that the mediator of genetic information from one generation to the next was DNA. It only could be DNA...

Figure 5.5: The DNA double helix. *a*) A simplified spherical model (Feughelman et al., 1955). *b*) Base pairs with hydrogen bonds (solid lines), without the sugar-phosphate backbone. In this representation, the distances between the planes of the base pairs are streched out; the 36° twist between the neighboring base pairs is maintained. *c*) Representation of both DNA strands as chemical formulas, to demonstrate their inversed, antiparallel polarities. *d*) drawing of the original wire model of the DNA double helix, about 2 m high, built by Watson and Crick in 1953.

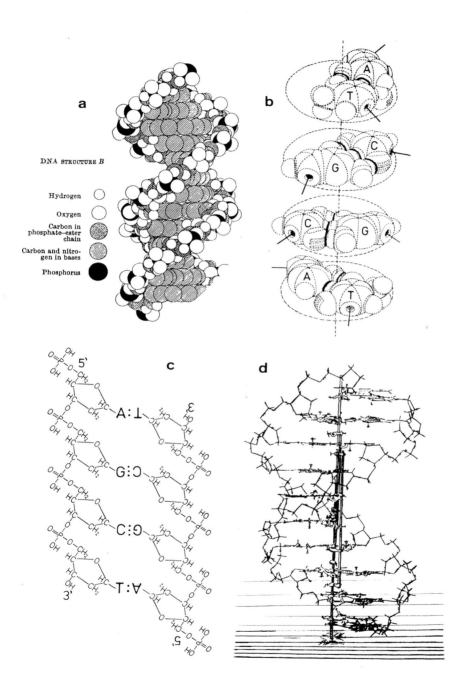

a

DNA STRUCTURE *B*

Hydrogen

Oxygen

Carbon in
phosphate–ester
chain

Carbon and nitro-
gen in bases

Phosphorus

b

c

d

Figure 5.6: Maurice Wilkins (born in 1916).

Figure 5.7: Rosalind Franklin (1921-1958).

Figure 5.8: James Watson (born 1928) and Francis Crick (born 1916), in 1953, and in 1993.

# CHAPTER 6

# DOUBLE HELIX: THE ANTICLIMAX

Unravelling the structure of DNA ought to be considered the most important discovery in biology – so it can be argued. What was unfathomable before, namely, a plausible mechanism for the material basis of heredity, emerged now spontaneously, displaying itself to anyone caring to look at it. The double helix, as a self-revealing molecule, made understandable, even to a child, how the genetic information was dealt with: it was kept as an encoded text represented by the sequence of its 4 bases, and it was perpetuated throughout the generations by the pulling apart of the two strands and subsequent synthesis of complementary ones. Watson & Crick guaranteed for themselves the priority for the enlightening discovery of "the secret of life", without actually formulating it: their extensively commented upon sentence "It has not escaped our notice... " did the trick. This sentence, scolded by some as snobbish, seen by others as too coy, was actually an all-saying – nothing-saying compromise between the coauthors: Crick supposedly did not wish to make an open declaration in the first paper (Fig. 6.1), while Watson feared that the obvious could be usurped by someone else, so scooping the fruits of their efforts.

All directly involved (Crick, Watson, Wilkins, Franklin and her graduate student Gosling) – as well as those indirectly involved (Bragg and Randall) – finally agreed to the publication of simultaneous but separate papers: one concerning the structural model as conceived by Watson and Crick (1953a), the other containing a rather odd discussion by Wilkins, Stokes & Wilson (1953) and asserting that the model did not conflict with the available X-ray diffraction data which Franklin and Gosling (1953) published in their paper. Nothing was more fair, claimed Randall in a patronizing judgement. (Randall was the one who had organized Wilkins' lab, supported and promoted him personally, invited Franklin to join in and "gave" her DNA as a research theme – only to take it away from her afterwards: a very unusual, but not isolated attitude in the history of science.)

Figure 6.1: Reproduction of the first and decisive publication by Watson and Crick (1953a).

# MOLECULAR STRUCTURE OF NUCLEIC ACIDS

## A Structure for Deoxyribose Nucleic Acid

WE wish to suggest a structure for the salt of deoxyribose nucleic acid (D.N.A.). This structure has novel features which are of considerable biological interest.

A structure for nucleic acid has already been proposed by Pauling and Corey[1]. They kindly made their manuscript available to us in advance of publication. Their model consists of three intertwined chains, with the phosphates near the fibre axis, and the bases on the outside. In our opinion, this structure is unsatisfactory for two reasons: (1) We believe that the material which gives the X-ray diagrams is the salt, not the free acid. Without the acidic hydrogen atoms it is not clear what forces would hold the structure together, especially as the negatively charged phosphates near the axis will repel each other. (2) Some of the van der Waals distances appear to be too small.

Another three-chain structure has also been suggested by Fraser (in the press). In his model the phosphates are on the outside and the bases on the inside, linked together by hydrogen bonds. This structure as described is rather ill-defined, and for this reason we shall not comment on it.

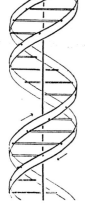

We wish to put forward a radically different structure for the salt of deoxyribose nucleic acid. This structure has two helical chains each coiled round the same axis (see diagram). We have made the usual chemical assumptions, namely, that each chain consists of phosphate diester groups joining β-D-deoxyribofuranose residues with 3',5' linkages. The two chains (but not their bases) are related by a dyad perpendicular to the fibre axis. Both chains follow right-handed helices, but owing to the dyad the sequences of the atoms in the two chains run in opposite directions. Each chain loosely resembles Furberg's[2] model No. 1; that is, the bases are on the inside of the helix and the phosphates on the outside. The configuration of the sugar and the atoms near it is close to Furberg's 'standard configuration', the sugar being roughly perpendicular to the attached base. There is a residue on each chain every 3·4 A. in the z-direction. We have assumed an angle of 36° between adjacent residues in the same chain, so that the structure repeats after 10 residues on each chain, that is, after 34 A. The distance of a phosphorus atom from the fibre axis is 10 A. As the phosphates are on the outside, cations have easy access to them.

The structure is an open one, and its water content is rather high. At lower water contents we would expect the bases to tilt so that the structure could become more compact.

This figure is purely diagrammatic. The two ribbons symbolize the two phosphate—sugar chains, and the horizontal rods the pairs of bases holding the chains together. The vertical line marks the fibre axis

The novel feature of the structure is the manner in which the two chains are held together by the purine and pyrimidine bases. The planes of the bases are perpendicular to the fibre axis. They are joined together in pairs, a single base from one chain being hydrogen-bonded to a single base from the other chain, so that the two lie side by side with identical z-co-ordinates. One of the pair must be a purine and the other a pyrimidine for bonding to occur. The hydrogen bonds are made as follows: purine position 1 to pyrimidine position 1; purine position 6 to pyrimidine position 6.

If it is assumed that the bases only occur in the structure in the most plausible tautomeric forms (that is, with the keto rather than the enol configurations) it is found that only specific pairs of bases can bond together. These pairs are: adenine (purine) with thymine (pyrimidine), and guanine (purine) with cytosine (pyrimidine).

In other words, if an adenine forms one member of a pair, on either chain, then on these assumptions the other member must be thymine; similarly for guanine and cytosine. The sequence of bases on a single chain does not appear to be restricted in any way. However, if only specific pairs of bases can be formed, it follows that if the sequence of bases on one chain is given, then the sequence on the other chain is automatically determined.

It has been found experimentally[3,4] that the ratio of the amounts of adenine to thymine, and the ratio of guanine to cytosine, are always very close to unity for deoxyribose nucleic acid.

It is probably impossible to build this structure with a ribose sugar in place of the deoxyribose, as the extra oxygen atom would make too close a van der Waals contact.

The previously published X-ray data[5,6] on deoxyribose nucleic acid are insufficient for a rigorous test of our structure. So far as we can tell, it is roughly compatible with the experimental data, but it must be regarded as unproved until it has been checked against more exact results. Some of these are given in the following communications. We were not aware of the details of the results presented there when we devised our structure, which rests mainly though not entirely on published experimental data and stereochemical arguments.

It has not escaped our notice that the specific pairing we have postulated immediately suggests a possible copying mechanism for the genetic material.

Full details of the structure, including the conditions assumed in building it, together with a set of co-ordinates for the atoms, will be published elsewhere.

We are much indebted to Dr. Jerry Donohue for constant advice and criticism, especially on interatomic distances. We have also been stimulated by a knowledge of the general nature of the unpublished experimental results and ideas of Dr. M. H. F. Wilkins, Dr. R. E. Franklin and their co-workers at King's College, London. One of us (J. D. W.) has been aided by a fellowship from the National Foundation for Infantile Paralysis.

J. D. WATSON
F. H. C. CRICK

Medical Research Council Unit for the
Study of the Molecular Structure of
Biological Systems,
Cavendish Laboratory, Cambridge.
April 2.

[1] Pauling, L., and Corey, R. B., Nature, 171, 346 (1953); Proc. U.S. Nat. Acad. Sci., 39, 84 (1953).
[2] Furberg, S., Acta Chem. Scand., 6, 634 (1952).
[3] Chargaff, E., for references see Zamenhof, S., Brawerman, G., and Chargaff, E., Biochim. et Biophys. Acta, 9, 402 (1952).
[4] Wyatt, G. R., J. Gen. Physiol., 36, 201 (1952).
[5] Astbury, W. T., Symp. Soc. Exp. Biol. 1, Nucleic Acid, 66 (Camb. Univ. Press, 1947).
[6] Wilkins, M. H. F., and Randall, J. T., Biochim. et Biophys. Acta, 10, 192 (1953).

A short time later, Watson and Crick (1953b) brought forth a second publication, exposing in some depth their considerations on the consequences of the DNA structure for genetics. For example, they assumed that errors in base-pairings would lead to mutations (see also Excursus 6-1). In what concerned replication, though, they had no opinion as to whether it was spontaneous or enzymatically driven. Crick commented later on this matter: "This showed a gap in our overall grasp of molecular biology, our tentative suggestion that DNA synthesis might not need an enzyme..." (Crick, 1988).

The discovery of the double helix prompted Stent (1978) to try to demonstrate that scientific discoveries are much more similar to an act of artistic creation than generally accepted; it is not so that concrete facts were rationally collected by objectively thinking scientists in such a strict fashion as not to allow for a large subjective component; much to the contrary, the personal contribution, and the way a discovery took place are determinants of its influence on the scientific establishment. Who would have disentangled the DNA structure had Watson and Crick not had the chance to do so in the spring of 1953? When and how would that have occurred? Crick, as well as Pauling, his scientific rival, suggested that other minds would have tackled it some 2 or 3 years later; but certainly, the course of events would have been totally different – surely not so dramatic as to be a suiting theme for an autobiographical literary work (Watson: The Double Helix, 1968). But, truly, the real events will never be fully retrieved; the descriptions made by those directly involved and by historiographers may help to grasp some factors, but may confound and obscure some others. This is especially so, because many details were only considered after the fact. If one asks, for instance, what were the direct consequences of Watson and Crick's publication for science, the opinions will diverge widely. It has often been spoken of as a radical revolution in the fields of biology, genetics and biochemistry. In trying to avoid subjective impressions, let us assess the impact of Watson and Crick's breakthrough by analyzing the immediately successive publications at the time. The result is meager. For example, in the review articles on nucleic acids, the spatial structure of DNA as proposed by Watson and Crick is indeed generally referred to, but mostly just for the record, at the end of an article (see, for example, Brown & Todd, 1955). Years went by till the significance of the discovery of the DNA double helix was generally recognized (see, for example, Gaster, 1990).

Paradoxically, it was Delbrück, the romantic mind in quest of new laws of nature, who right away embraced the Watson-Crick DNA model. ("One cannot imagine that Nature would not have made use of such a phenomenal discovery" he is supposed to have stated.) Delbrück himself organized the distribution of copies of the Watson-Crick publication among the participants of the 1953 Cold Spring Harbor Symposium (not so easy a task in the pre-quick-copy era), and as coordinator he made the last-minute arrangements to include Watson in the symposium program (theme: virus), so that a report on the work could be presented.

So, this was the big revelation? The long sought for "novel laws of nature" were nothing but hydrogen bonds! In its own way, this was, after all, a fantastic insight! After the initial shock, Delbrück and his intellectual follower, Gunther Stent, got intensely involved with this fresh theme. For example, they theorized on how the thousands of helical turns of a double strand, apparently impossible to disentangle, could nevertheless be replicated (Excursus 6-2).

At the same time, Delbrück's interest in heredity essentially vanished with the advent of the double helix. Unravelling the details, albeit certainly important and interesting enough, would be well cared for – geneticists and biochemists would surely accomplish this task.

It did not take long for Delbrück to settle on a completely different research field, one yet fully to be clarified: stimulus processing. Comprehending how a cell reacts coherently to specific outside stimulation became his new quest. This fundamental problem represented the clue for understanding the functioning of the brain and the nervous system, probably the last of life's riddles yet to be solved. Delbrück's approach to the new challenge paralleled his choice in the field of gene replication: phages as the minimal units of reproduction. He picked, as the minimal unit of sensory perception, the fungus *Phycomyces*, a one-celled organism excitable by light and gravitational stimuli. 30 years of dedication – twice the time spent with phages – guaranteed Delbrück a small group of followers (almost none of them phage workers), though it all ended with only a handful of aficionados (see Cerdá-Olmedo & Lipson, 1987). It is noteworthy that Delbrück, as one of the few to immediately point out the significance of Watson and Crick's achievement, was also the one who distanced himself of the subsequent progress in molecular biology.

Not so Watson and Crick. Watson dedicated himself for a short while, without really succeeding, to the resolution of RNA structure. Afterwards, he made a name as the author of a bestselling textbook and as an administrator of the Cold Spring Harbor Laboratory. Also, in 1988, he became the head of a world-wide project for sequencing the human genome, only to resign four years later for not entirely clear reasons (Excursus 23-1). Crick – we will follow his path in this book – became, as no other, a crucial and decisive participant of the subsequent developments in molecular biology well into the late 1960s. Then, he veered his interests in a new direction – comprehending the brain.

Wilkins – would not he have been the discoverer of the double helix within a few years had these two adventures not stormed the field? – persisted in analyzing details of all atomic coordinates of the Watson-Crick model (Wilkins, 1956; Langridge et al., 1960a,b). It was actually a humiliating endeavor, lingering on for seven or more years, culminating in relative insignificance, although he, Wilkins, also was a Nobel Prize laureate in 1962 (Pauling saying, undeservedly so!), together with Watson and Crick. Rosalind Franklin, though, had been left out; the honor of the prize is bestowed exclusively on living persons, and she, 35-years-old, had succumbed to cancer in 1958 (see Sayre, 1975; Piper, 1998).

Chargaff was horrified by the lack of style and manners. How was it possible
that two such arrogant clowns got to make such a craze in science? ...That such
dwarfs threw so huge a shadow showed how advanced the day was... the old scientific
era with its humble, hard-working apostles, who started their careers with almost
religious vows of poverty, was coming to an end... he conjured till his death in 2002,
at the age of 96, the advancing decadence of science... That was the topic of his
decades-long criticism of science (for instance: Chargaff, 1974, 1978, 1998): Repent,
the end is nigh!

**Excursus 6-1**
**A COROLLARY OF WATSON AND CRICK'S MODEL: THE INSIGHT THAT
ERRORS IN BASE-PAIRING CAN LEAD TO MUTATIONS**

The hypothesis that inaccurate base pairing would trigger mutations, already raised by
Watson and Crick (1953b), was later extensively pursued, especially by Ernst Freese,
at Harvard. DNA base analogues were characterized as mutagenic and their possible
mechanism of action analyzed. To this end, Freese took advantage of so-called *rII*-
mutants of phage T4. These mutants – in contrast to the wild type phages – are not
able to reproduce in a special coli strain, *E. coli* K12(λ) (see Excursus 8-2), so that
the few wild type back-mutants within a lysate of T4*rII* can be easily and selectively
detected.

The base analogue 5-bromo-uracil is accepted by the cell as thymine, and may
replace it almost totally. The mutagenic action of 5-bromo-uracil is according to
Freese (1959a,b,c) based on the higher probability of this base occurring in the
tautomeric enol form (endo-alcohol), rather than the usual keto-form (Fig. 5.4 & 6.3).
In this enolic configuration, 5-bromo-uracil would pair with guanine, not with
adenine, as is usually the case, so that in the subsequent round of replication a pair of
A-T would be replaced by a G-C pair (Fig. 6.2).

The purine analogue 2-amino-purine would direct a similar change. This
analogue is only rarely incorporated into DNA, but doing so, it would undergo
pairings with thymine as well as with cytosine.

Nowadays, other mechanisms, as wobble base pairing, are accorded a higher
significance regarding the triggering of mutations, as compared to the tautomeric
replacement of bases [see for example, Morgan (1993) and Fig. 6.3]. Nevertheless,
that does not undermine the principle derived by Watson and Crick (1953b) by just
looking at the double helix: occasional erroneous base pairings are a molecular
foundation of gene mutations.

Freese pointed out the fact that mutations emerging from the action of 5-bromo-
uracil or 2-amino-purine were different from most spontaneously occurring mutations.
This hints at the existence of mutational mechanisms others than base substitution.

Benzer (1957) had already demonstrated that mutations could result from the loss of blocks of genetic material (Excursus 8-2). And mutations may also occur by the addition of one or more bases (Chapter 8).

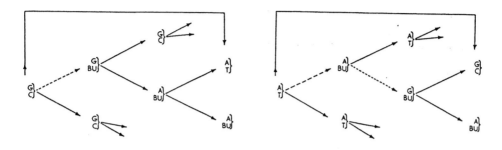

Figure 6.2: Two types of mutational changes brought about by pairing errors (Freese, 1959c). Left: incorporation error; during replication of DNA, C normally pairs with a newly incoming G to form a G-C pair. In the presence of BU (5-bromo-uracil), though, the G from the parental strand will pair with it. BU is actually an analogue of T (which normally pairs with A), but because it adopts more often than T the enol form, in which it pairs with G instead of A, a mismatch will be more frequent. In the next round of replication, A will replace G as a partner for BU. From this moment on, A will pair normally with T, and so a G-C pair will be replaced by an A-T pair, probably causing a mutation. Right: replication error; BU is incorporated as a T analogue, pairing with A. In the next round of replication, because of the higher chance to adopt the enol form, BU may often erroneously pair with G. Incorporation of G leads to the replacement of the original A-T pair by a G-C pair.

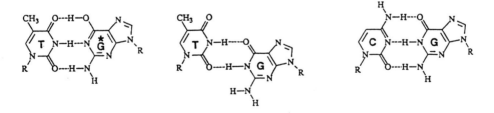

Figure 6.3: Two possible erroneous pairings, leading to mutations; examples based on T-G pairing, able to occur because G may take the rare enol form (on the left) or because G slips into a position favoring the formation of two hydrogen bonds, a so-called wobble base pairing (middle). For comparison (on the right), the correct G-C pairing (Morgan, 1993).

**Excursus 6-2**
**THE REPLICATION MODE OF DNA**

Watson and Crick's model did not leave room for doubts: both complementary DNA strands of the double helix were intertwined, intertwisted – a full turn encompassing 10 base pairs (the strands represented a so-called plectonemic double helix as opposed to a paranemic one, in which the two screw-like structures simply lay closely side by side). Delbrück and Stent (1957) envisaged the difficulties posed by this molecular twisting when it came to separate both strands in the course of their replication (one of the strands is occasionally colloquially called the Watson, and the other the Crick). They considered two possibilities: either the strands were untwisted in their total length during replication (Fig. 6.4) – and that would amount to a huge rotational speed for such long molecules –, or else the strands were cut at intervals, the segments so derived would then be partially untwisted and replicated, the segments being subsequently joined again (Fig 6.5). Considering the second possibility, one could assume that, during replication, the newly synthesized DNA strands and the old ones could get mixedly joined, so that, after the process, stretches of DNA strands were built of alternating old and newly synthesized DNA material. This scenario would correspond to a "dispersive" mode of replication, where building blocks from a "parental" duplex were evenly distributed among all progeny structures. If during replication, though, the original double helix was untwisted in its total length, without any breaks in the sugar-phosphate backbone, then each single strand would keep its integrity and individuality; this last scenario would correspond, according to Delbrück and Stent, to a "semi-conservative" replication mode. One could still propose a third possible replication mode, with one full double strand being synthesized side by side to the parental one, so that no blending of old and new strands occurred. The latter would be a "conservative" mode (Fig 6.6). These were the hypothetical, theoretical possibilities. How to distinguish among them experimentally? Delbrück and Stent were at a loss.

However, two of Delbrück's followers at Caltech, Matthew Meselson and Franklin Stahl, conceived an experiment, which – in addition to the fluctuation-test (Excursus 4-2) – stands today, simply, as another classical example of an experimental test corroborating a theoretical hypothesis. The three different hypothetical DNA replication mechanisms led to three distinct predictions for the experiment devised by Meselson & Stahl (1958). Their approach was the following: coli cells were inoculated into liquid medium prepared with the heavy isotope of nitrogen, $^{15}N$; $^{15}NH_4Cl$ being the sole nitrogen source, a culture of $^{15}N$-labeled bacteria resulted. After many generations, incorporation of $^{15}N$ by the cells was inhibited by the addition of normal ammonium chloride ($^{14}NH_4Cl$) in overwhelming excess. The culture was then sampled from time to time; treating the samples with a detergent set their DNA free. The so obtained homogenate was then added to a solution of cesium chloride whose density of 1.7 corresponded to that of DNA (CsCl,

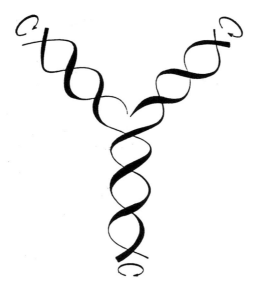

Figure 6.4: The replication of DNA, as based on Watson and Crick's scheme, implies a continuous rotation of the newly replicated arms of DNA as well as of the yet to be replicated strands (Delbrück & Stent, 1957).

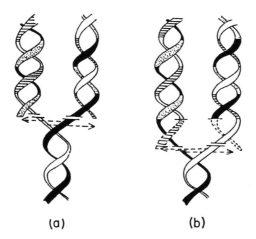

(a)          (b)

Figure 6.5: It would also be possible to assume that during the replication of DNA its sugar-phosphate backbone was continuously broken and rejoined (Delbrück & Stent, 1957).

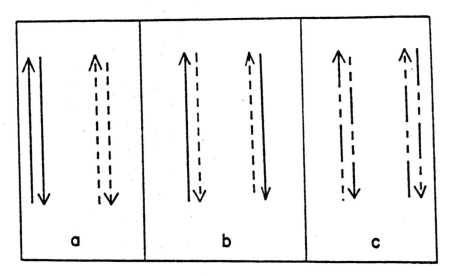

Figure 6.6: Scheme of *a*) conservative, *b*) semi-conservative and *c*) dispersive hypothetical modi of replication for the DNA double helix (Delbrück & Stent, 1957).

having a molecular weight of 168.5 is about three times as dense as the otherwise similar table salt, NaCl). The suspension was subsequently ultracentrifuged for 20 hours at more than 40,000 rotations/min, yielding a centrifugal force of about 140,000 times the earth's gravity. Due to this, the cesium ions gradually migrated towards the bottom of the centrifuge tube, producing a concentration gradient with increasing values towards the bottom. The DNA molecules were trapped in the region of CsCl density corresponding to their own density, forming there a band of molecules too heavy to move up and too light to descend. The location of these bands could be monitored through measurements of UV absorption. Clearly separated positions of DNA molecules from $^{15}$N cultures and from $^{14}$N normal ones – an obviously essential precondition for further fine analytical differentiation – assured a reliable outcome. After one cycle of cell division, the DNA molecules banded exactly at the middle of $^{15}$N- and $^{14}$N-DNA. The hypothesis of conservative replication could thus be excluded, since, in this case, two bands corresponding to the parental heavy DNA helices and the newly synthesized light ones, respectively, would be expected. Discerning between the hypothetical dispersive and semi-conservative replication modi required further analysis of the subsequent replication cycles. In the case of a dispersive replication modus, the density of the DNA molecules would, in the course of many replication cycles, gradually approach that of the $^{14}$N-DNA. If the semi-conservative modus was the correct one, though, the amount of DNA banding at the middle after two divisions would remain constant, accompanied by a band of $^{14}$N-DNA. Meselson-Stahl's experimental results reflected exactly this last hypothetical scenario (Fig. 6.7). Their procedure also pointed to astonishing new research possibilities for the field of molecular biology, brought about by the new technique of cesium chloride density gradient centrifugation (Excursus 6-3; Chapter 19).

A very detailed historical appaisal of the Meselson-Stahl experiment ("The Most Beautiful Experiment in Biology") has appeared in book form (Holmes, 2001).

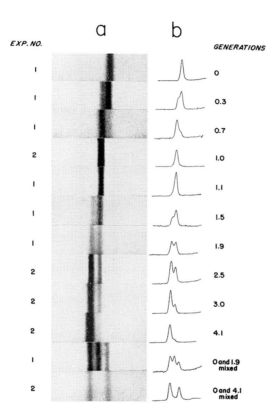

Figure 6.7: Pattern of ultraviolet absorption by the bands of DNA in the Meselson & Stahl experiment (1958). *a*) Photographic reproduction of the critical region of the centrifuge tube (the tubes take a horizontal position inside the centrifuge; the region of denser CsCl solution – closer to the bottom – is on the right. *b*) Densitometric monitoring of the dark bands on the photographic emulsion.

Coli cells were first cultured in the presence of $^{15}NH_4Cl$ as nitrogen source; at time 0, an overwhelming excess of $^{14}NH_4Cl$ was added. The culture was sampled at time intervals (here indicated as rounds of cell divisions), and the DNA profiles of the samples analyzed after 20 hours of centrifugation at 140,000 x g in a CsCl gradient.

Excursus 6-3
DENATURATION, RENATURATION AND HYBRIDISATION OF DNA

Even before Watson and Crick's (1953a) model was conceived, Stephen Zamenhof at Columbia University, N.Y., observed that when DNA in solution was heated, unusual alterations occurred; for example, the viscosity of the solution suddenly dropped at temperatures around 80 °C. Zamenhof realized (Zamenhof et al., 1953) that this temperature inflicted drastic structural alterations upon the DNA molecule. Years later, at the Albert Einstein College of Medicine in the Bronx, N.Y., Julius Marmur and his co-workers figured out how to obtain the opposite effect: to allow a previously heated DNA preparation to cool slowly. This procedure almost fully restored the original viscosity. The tendency to regain the original viscosity was not so pronounced, though, if the cooling process was a quick one, brought about by immersion in ice.

Interpreting these pieces of information became an easy task as soon as the publication of Watson and Crick's model was out: the increased molecular kinetic energy at about 80 °C was sufficient to break all hydrogen bonds between the two strands of the DNA double helix. The relatively stiff double helix fell apart, yielding two single strands which, as free chains, were unrestricted to move about in the solution, fold up or roll up, resulting in the observed drop of viscosity. The double helix was denatured. Amazing was the capability of the single strands, under conditions of slowly falling temperatures, to recompose a double helix. Clearly, the complementary single strands, by diffusing within the solution, were able to find each other again. Probably, first, some short segments of complementary single strands met by chance, allowing new hydrogen bonds between complementary bases to be reestablished, thus forcing the two strands to remain together, and finally, progressively additional hydrogen bonds were rebuilt all along the lengths of the strands. The DNA double helix regained thus its original form: it was renatured. The possibility of verifying this astonishing interpretation experimentally did not escape Marmur and his co-workers. Meselson & Stahl's new technique of CsCl density gradient centrifugation seemed ideally suited for their purpose, although it had to be slightly altered. Phages, with their relatively short and less complex DNA, seemed to be the ideal sources of material for this project. Lysates of phage T7 (see Fig. 4.13) were obtained from bacterial cultures grown in $^{15}$N-, as well as from bacteria from normal $^{14}$N-medium. DNAs from both cultures were isolated. The $^{15}$N-T7-DNA was mixed with $^{14}$N-T7-DNA. Denaturation by heat followed, originating a suspension with 4 different single strands, namely, $^{15}$N(heavy)-"Watsons", $^{14}$N(normal)-"Watsons", $^{15}$N-"Cricks" and $^{14}$N-"Cricks". A renaturation procedure would bring together the complementary single strands, whether these were heavy or normal was dependent exclusively on chance. The expectation for the renatured helices, in what concerned their density, was 1 part dense $^{15}$N-DNA to 2 parts DNA of intermediate density ($^{15}$N-"Watsons" joined to $^{14}$N-"Cricks" and $^{15}$N-"Cricks" linked to $^{14}$N-"Watsons") to 1 part

$^{14}$N-DNA of normal density. Employing the CsCl density gradient centrifugation method, Marmur's group was able to confirm precisely the validity of this expectation (Schildkraut et al., 1962). And they went even further, substituting $^{14}$N-T3-DNA for $^{14}$N-DNA of T7, a relative of phage T3, followed by denaturation and subsequent renaturation. In this way, they obtained DNA double helices of intermediate density. Nevertheless, such "hybrid" DNA structures – T3 strands paired to T7 strands – were apparently not as stable as the homogeneous double strands, since denaturing the hybrid DNA could be attained by temperatures lower than 80 °C. The implication was that T7 DNA and T3 DNA, although similar enough to engage in pairing, had, nevertheless, marked differences in base sequences so that not all the bases had corresponding partners on the heterologous strand. This resulted in a restricted stability due to incomplete hydrogen bonding. Hybrid DNA duplexes, also called

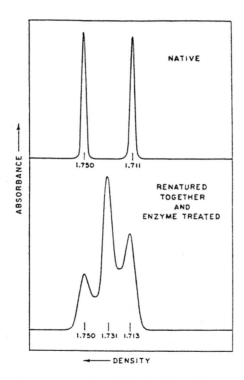

Figure 6.8: Densitometric representation of bands from $^{15}$N-T7-DNA (left), $^{14}$N-T3-DNA (right) and "hybrid"-DNA (middle). This banding pattern was the outcome of the following procedure: a mixture of $^{15}$N-T7-DNA and $^{14}$N-T3-DNA was first heated and subsequently slowly cooled, a process which leads to denaturation and subsequent renaturation of the DNA molecules. As a next step, an ultracentrifugation of this DNA solution in a CsCl gradient was carried out (24 h, 15,000 x g) (Schildkraut et al., 1962).

heteroduplexes (Fig. 6.8), were the less stable, the smaller the so-called base sequence homology was between the two strands. In other words, the tally of bases without corresponding ones on the opposite strand determines the degree of instability. [It was later established that between T3 and T7 the base sequence homology amounts to about 85 % (Davis & Hyman, 1971).]

Assessing the degree of stability of heteroduplexes turned out to be a relatively simple way of revealing the evolutionary proximity of DNA sequences. Already the pioneering work of Britten and Kohne (1968) at the Department of Terrestrial Magnetism, Carnegie Institution, Washington D.C. (the original Institution maintained its name even after the field of research diverted from earth magnetism to molecular biology), contributed to characterize many phylogenetic pathways. Nowadays, refined sequencing techniques allow much more accuracy in this branch of research. For example, Sibley & Ahlquist (1984) established, somewhat surprisingly, that between humans and chimpanzees there is a base sequence homology of about 99 %, a value found usually only within the same genus, for closely related species.

The principle of spontaneous formation of DNA double helixes, based on fitting base sequences of single-stranded DNAs (but also of DNA-RNA helixes) is nowadays crucial for gene technology, and it is utilized in innumerable variations. Specially important are techniques which employ solid surfaces (for example, sheets of nitrocellulose or glass), on which single strands of DNA can be fixed; complementary strands, as radioactive probes, or labeled with fluorescent dyes, are then deployed to identify and locate specific sequences (see also Excursus 21-3).

The search for such sequences has, in the last few years, been facilitated enormously through the use of so-called microarrays in which, on a small surface of less than a few $cm^2$, hundreds or even thousands of samples of different single strand sequences can be placed by means of robots. A fully automated monitoring technique allows industrial scale hybridization tests, which have many applications in basic research as well as in medical screening programs for the detection of hereditary disease genes.

# CHAPTER 7

# AVERY? HERSHEY?

We have already pointed out the often evident discrepancy between the weight accorded to experimental discoveries, as judged by contemporary scientific publications, and the impact granted them in later historical descriptions and didactical approaches.

Two well known scientific accomplishments linked to the discovery of DNA as the genetic substance stand out as representative examples of such contrasting judgements.

Transformation in bacteria is one of these examples. It was described in 1927 by the Londoner physician Frederick Griffith, who studied pneumococcus infections in mice, and whose findings were interpreted by Oswald Avery and his collaborators in 1944. These scientists, at the Rockefeller Institute in New York (today, Rockefeller University), working with pure solutions of DNA extracted from pneumococci able to form polysaccharide capsules, demonstrated that these solutions were able to propagate this special characteristic to other strains of pneumococcus, originally uncapsulated. These originally uncapsulated strains were thus "transformed" into capsulated ones by the action of the DNA solution (Excursus 7-1).

Today, these experiments are often regarded as the decisive evidence for DNA being the material carrier of the genetic information. However, who, among the contemporary scientists, was convinced of that? Certainly not Avery and his co-workers. They actually showed that proteins could not be the substances responsible for the transforming activity present in their solution, since the finest of detection methods failed to pinpoint them; in addition, heating the preparations did not impair the transforming activity, while treatment with DNAase suppressed it immediately and totally. But, who knows? It could very well be that a putative gene-protein was especially heat-stable! Perhaps it was protected by the DNA, which, as a supporting substance, was imbued with an important, though unspecific function! The ridicule suffered by Willstätter 15 years before (see Chapter 1) was still quite alive in their memories: Avery et al. (1944) were overcautious and extremely sceptical. Only privately, in a now famous letter to his brother, did Oswald Avery hint that his DNA was actually like a gene... something like a virus... In the 1960s, Chargaff claimed having his interest for the chemical analysis of DNA awakened by the impact caused by Avery's observations. Nevertheless, Chargaff's publications, and especially the most representative of them, the one compiling the most extensive description of his results (Chargaff, 1950), conveys exactly the opposite impression: Avery's work deserves not more than a cursory, inexpressive mention, together with many other publications totally forgotten today.

Lederberg (1986, 1987) also affirms that Avery's findings were of decisive significance to him; however, a printed corroboration of this assertion is not to be found anywhere before Watson & Crick: Lederberg mentions indeed Avery et al. (1944) but in an incidental and sceptical way (Zinder & Lederberg, 1952).

And one can assume with a good degree of certainty that even Watson and Crick themselves were not especially impressed or influenced by Avery's research — if they knew it at all before 1953. No reference to him is to be found, neither in their first published works (Watson & Crick, 1953a,b) — which would be understandable — nor in the later detailed publication from Cold Spring Harbor in June (Watson & Crick, 1953c). Wilkins et al. (1953), as well as Franklin & Gosling (1953), did not cite Avery, although one ought to remember that these authors did not have a special attachment to molecular genetics; nevertheless, Wilkins et al. (1953) did mention that their results from X-ray diffraction (in collaboration with H. Ephrussi-Taylor) suggested helical molecular forms for the transforming principle ( Excursus 7-1), sperm heads and phages.

No one could claim that Avery had published in unknown, obscure scientific journals — as Mendel did 100 years before. Even if the Journal of Experimental Medicine — where Avery's original work first appeared — did not exactly count as one of the journals most read by molecular biologists, subsequent important results from his laboratory were presented at and published by Cold Spring Harbor, the Mecca of young molecular biologists (McCarty, Taylor & Avery, 1946). Despite this, no one listened to or read these presentations, since Avery did not belong to the circle of insiders in the CSH clan.

A symposium edited as a commemorative tome for the 50th anniversary of Mendel's rediscovery, sporting the title "Genetics in the 20th Century" (Dunn, 1951), typifies this state of affairs: none among the thirty renowned coauthors found even one acknowledging word for Avery; no one even actually mentioned him, except Alfred Mirsky — Avery's neighbor at the Rockefeller Institute — who quoted him exclusively to emphasize that genes were certainly proteins which had escaped detection by Avery's methods.

Later, when Delbrück was interviewed by Olby (1974) on how he could explain that molecular biologists ignored Avery for such a long time, he defended himself by affirming that it was just not true that Avery's findings had remained unknown. And Delbrück added that he himself, in Nashville, came to know Roy C. Avery, Oswald's brother, personally, and was therefore confronted by Oswald's famous letter, containing his personal suspicion on how DNA resembled a gene... a virus... And Delbrück continued, saying that Avery's findings, at the time, did not lead to anything; one just could not fit them into any current theory. Experimental results which do not apply to a theory remain worthless [Here we have a classical example to which a half-serious maxime, attributed to the physicist Arthur Eddington, applies: One should never believe any experiment until it has been confirmed by theory (see Weinberg, 1992)]; and a possible, even most plausible theory at that time – and in this point Delbrück may not have been that wrong – was to accord the transforming DNA solely a triggering function, responsible for turning on the switch of a synthesis process, without necessarily, strictly speaking, carrying any real genetic information. And besides all that, bacteria were not as yet totally established as genetic subjects, comparable to other living beings; it could well be that transformation represented an oddity referring solely to these organisms "containing only primitive protoplasm". And finally, the last argument: DNA could not possibly be the only carrier of genetic information, since Bawden and Pirie (1937) had already shown that the tobacco mosaic virus was built of RNA and protein – no DNA!

A second finding pointing to the significance of DNA as carrier of genetic information, before Watson & Crick's work, was that of Hershey & Chase (1952). While the physician and biochemist Avery stood outside the inner circle of molecular biologists, Alfred Hershey was, together with Luria and Delbrück, one of the founding fathers of the phage school. Hershey conceived an experiment inspired by some phage electron micrographs (Excursus 4-1) and by some observations made by Thomas Anderson, which revealed that phage DNA was released from its protein coat after virus particles had been subjected to osmotic shock (Excursus 7-2).

What the experiment showed – and the authors asserted no more than that – can be read in their publication (Hershey & Chase, 1952): by violently shaking a liquid culture of phage-infected bacteria in a Warring-blender, one could – without impairing the infection process – remove roughly 80% of the phage proteins from the host bacteria, whereas about 65% of the phage DNA remained associated with them. In other words, once the infection had started, only 20% of phage proteins but 65% of phage DNA remained linked to the host bacterium. The blender experiment, which was to become very famous, demonstrated that what remained associated with the infected bacterial cells after shearing was a phage DNA fraction 2,5-fold more concentrated, as compared to undisturbed phage particles. What fresh fundamental insight could be derived from these observed facts? Absolutely none, considering what Friederich Miescher, the discoverer of DNA, had already demonstrated at the beginning of the 20th century in work on salmon from his native Basle's Rhine waters: In sperm, DNA was not just 2,5-fold, but over a thousand-fold enriched, as compared to the protein fraction in other tissues (Fig. 7.1)!

No one had ever stated that all kinds of proteins were crucial for genetic perpetuation of generations. Even if 90, or 99, or even 99,9% or more phage proteins could be dispensed with, without affecting the infection outcome, even that would not

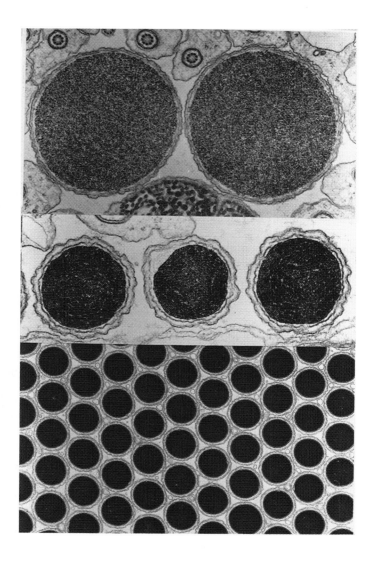

Figure 7.1: Final phases of spermatogenesis (example of scorpion). After the spermatid (that is, the direct product of meiosis in a male individual) sheds almost all of its cytoplasm (not shown), the condensation of the nucleus follows, here shown in 3 phases. The process results in a more than 1000-fold reduction in the chromosomal packing volume, as compared to the interphase nucleus. Related to the initial cell, the DNA concentration is increased many thousand-fold. Amplification: top: ca. 7,500 times; middle and bottom: ca. 15,000 times. (from Phillips, 1974).

be as informative as the fact that DNA was highly accumulated in sperm. It could always be argued that solely a few special protein molecules were essential for exerting the genetic function.

In this context, Avery's findings were a lot more convincing because his extraction methods for obtaining the transforming principle removed all possibly detectable traces of proteins. Of course, the real sceptics could still hide behind the unfounded *ad hoc* assertion: proteins with genetic functions are special and impervious to procedures affecting the usual ones.

Why was Avery not quoted by Hershey & Chase (1952)? In considering that both lines of work were directed towards the crucial and critical problem of the material basis of genetic information, this would have been just natural. The most plausible answer is that Hershey was not aware of Avery's research.

And still, one last, curious point. Why did Hershey's experiment permeate into practically all textbooks of genetics (even our own: Bresch & Hausmann, 1972)? Although, seen objectively, it did not carry persuading scientific arguments, could it have nevertheless stimulated Watson and Crick's fantasy or motivation, directing them to a swifter discovery of the double helix? This presumption, too, lacks veracity; when the news of Hershey's results reached them across the Atlantic, they had been delving into the hectic quest for solving the DNA structure for quite some time already. A strictly objective assessment of the renowned Hershey & Chase (1952) experiment reveals a complete lack of both, logically compelling novel insights, and mentionable practical consequences for further research. If, despite all that, it persists in the history of molecular biology as one of the most referred to and described experiments, being even considered as a decisive key-experiment, then certainly because of *a*) an indisputable didactical sex-appeal, especially if one veils the crude data in a well-meaning rhetoric disguised as "didactical freedom", and *b*) because of Delbrück's enthusiastically patronized publicity campaign in favor of his buddy from the phage school (actually, considering Delbrück's notorious scepticism, an atypical attitude), and mouth-to-mouth propaganda. All that lent the work an aura of significance, which remained uncriticized even after the extremely detailed and difficult-to-read original publication came out. All that is very understandably human; however, let us not boast about scientific objectivity...

Summarizing, a definitive proof that DNA was actually the carrier of genetic information did not exist before Watson & Crick's glorious accomplishment. But the fact that DNA was somehow involved in genetic transmission was already known at least since Feulgen & Rossenbeck (1924) identified DNA as being an essential component of the chromosomes (Feulgen et al., 1937).

Watson & Crick themselves were initially not fully convinced of the veracity of their model. Indeed, Watson's nightmare for a brief period of time was that all would turn out to be a total flop. This doubtful state of mind did not prevail, though; soon he was asserting, self-assuredly: "It would be surprising to us, however, if the idea of complementary chains turns out to be wrong" (Watson & Crick, 1953b).

Again, the compelling evidence for DNA being the sole carrier of genetic information was delivered by the double helix model itself, simply by its beauty! In contrast, for example, to the inconclusivness concerning the function of insulin, even after 8 years of intensive work on its sequence (Ryle et al., 1955; Sanger, 1959), the Watson & Crick model, conceived in merely a few months, seemed to scream directly into the ears of those ready to listen: I am the carrier of the genetic information, and I propagate it to daughter molecules through the base complementarity of my two strands. No one could resist the hypnotizing power wielded by this assertion (however, see Chargaff, 1963).

**Excursus 7-1**
**THE TRANSFORMING PRINCIPLE IS DNA**

Before the onset of the antibiotic era, serological typing of pneumococci, the causative agents of pneumonia, was of utmost importance. Accurately matching antisera were the clue for successful therapies. There were many different antigen specificities, determined by the diverse types of polysaccharide capsules of the varied pneumococci strains. [By the way, Heidelberger & Avery (1923) were the first to accord antigenicity to polysaccharides – pneumococci polysaccharides; till then, antigenicity was assumed to be solely an attribute of proteins.] The polysaccharide capsules protected the infecting pneumococci from phagocytosis by the host macrophages (white blood-cells), thus endowing them with resistance and corresponding pathogenicity. In the laboratory, colonies of capsule-forming pneumococci could be recognized by their glittering smooth surfaces (hence the denomination *S*, for smooth), whereas their counterparts, capsule-free mutants, had colonies with a rough appearance (hence dubbed *R*, for rough). By cultivating pneumococci in laboratories, the researchers were confronted daily with transmutations from *S* to *R* types; the underlying mechanism being nevertheless unknown. [Only after the Luria-Delbrück test (Excursus 4-2) would it be possible to comprehend that *R* type cells were mutants which, in a culture medium, were endowed with a selective advantage over *S* type cells, since outside the animal body capsule synthesis was superfluous.] Transmutations from *R* type to the original *S* type could also be achieved with a certain regularity, if large amounts of *R* cells (non-pathogenic, sensitive cells) were injected into mice; after their becoming sick and dying, *S* cells could be isolated from their bodies. This was so, because the eventual *S* back-mutants existing within the population of *R* cells displayed, within the animal host, the selective advantage of being resistant to macrophage phagocytosis. The transmutation of one *S* type into another *S* type had as yet never been observed, though – exception accorded to Fred Griffith! Griffith injected mice with a mixture of pneumococci consisting of heat-killed *S* cells of serum type *SIII* and live non-pathogenic *R* cells originated from *SII* type cells. The mice so treated died after a few days and from them Griffith isolated live *SIII* pneumococci. Although Griffith (1928) judged the phenomenon weird and interesting, he did not follow it up any further. Instead, he delved into other aspects of medical bacteriology concerning his beloved subjects, the pneumococci and staphylococci, till 1941, when a Nazi bomb reaching his London laboratory tore his body to pieces. He had considered it to be a nuisance having to interrupt his work all the time in order to take refuge in bomb shelters whenever the sound of bomb alarms filled the air.

However, at the Rockefeller Institute, N.Y., the working place of Oswald Avery all of his life, the transformation of *R* cells into *S* cells was given special attention for many years. First, Griffith's observations were reproduced; then, Dawson & Sia (1931), Avery's co-workers, found a way to achieve the same transformation in the test tube – no mice involved –, and Alloway (1932) obtained transformation of *R* cells through cell-free extracts of *S* cells, opening up the way for fractionation of these suspensions. Progressive purification of the critical fractions finally led to the isolation of the "transforming principle" (TP). The experimental procedures were exceptionally tiresome; especially despairing was the unreproducibility of the results: sometimes it worked, most of the time it did not. Several times over the years, all involved were repeatedly tempted to throw everything out of the window (see, for example, Dubos, 1976); no publications were produced – unthinkable in today's terms. However, after many doubtful moments they had reached their goal: Avery, MacLeod & McCarty (1944), extremely scrupulously, had, totally unexpectedly, characterized the TP as being DNA. Contamination by proteins was proven not to occur using three methods: first, chemical methods assured the absence of proteins; second, heat treatment was expected to destroy all proteins, unless they were exceptionally heat-resistant; third, even the finest of serological methods were not able to show any evidence of pneumococcal proteins (the only ones possibly present in the preparations). This last procedure also applied to the pneumococcal polysaccharides (capsule substance). Lipids could also be excluded as the TP, since repeated alcohol and ether extractions did not render it inactive. However, treating TP with even the smallest traces of DNAase, led to immediate impairment of its activity. In addition, after ultracentrifugation and electrophoresis of the material, the TP overlapped precisely the DNA position. In sum, the project was an utmost thorough and undisputable piece of research, absorbing Avery's full attention and dedication throughout his last active working years. [Avery, a bachelor by conviction, was already 67 years old (Fig. 7.2) by the time of the publication of his now famous work.] Avery's work deservingly represents an example of conscientious research and technical competence.

Nevertheless, the only assertion that the three authors allowed themselves was: "The evidence presented supports the belief that a nucleic acid of the deoxyribose type is the fundamental unit of the transforming principle of Pneumococcus Type III."

Rollin Hotchkiss, also at the Rockefeller Institut, was the first to point out that transforming activities were not solely confined to the example provided by Avery but

Figure 7.2: Oswald Avery (1877-1955).

could also be observed in cases of resistance to penicillin and streptomycin (in pneumococci) (Hotchkiss, 1951, 1957). Later he extended his observations to many cases of auxotrophic markers in *Bacillus subtilis* and *Haemophilus influenzae* (see also Chapter 19).

Nowadays, transformation by DNA is also routinely achieved in cells of higher organisms: transgenic animals, today so crucial for many projects in basic and applied research, are obtained through DNA transformation of egg cells (for example, in the case of pigs and siring bulls).

**Supplement to Excursus 7-1**
**AFTER ALL: WHAT IS THE HISTORIC TRUTH? – GRIFFITH (1928) VERSUS CANTACUZENE & BONCIU (1926)**

Griffith's 1928 publication was, as described here, the event that triggered further research at the Rockefeller Institute. However, was Griffith the real discoverer of transformation, as pictured in all extensive historic evaluations (see for example Olby, 1974; Portugal & Cohen, 1977) and innumerable textbooks? In 1926 Cantacuzene & Bonciu had collected the exudates from scarlet fever patient's throats, freeing them from all pathogenic streptococci by filtration; they described then, how this sterile filtrate could stably confer to other streptococci a characteristic typical of pathogenic strains, namely the ability of being specifically agglutinated by reconvalescent sera. Similar observations by many other authors (see Travassos, 1979) reiterated and supplemented those of Cantacuzene and Boncius. Overall, one gets the overwhelming impression that all these cases are examples of transformation. Nevertheless, these works fell into oblivion – perhaps because they were published in French, or worse, in Portuguese (Travassos, 1930)? Joshua Lederberg once arguing over this theme, implied that these pioneering works had actually been extensively referred to in reviews (see Fig. 7.3); however, by searching for the facts, one finds little support for Lederberg's affirmation. Correct, though, is Lederberg's remark: "scientists generally do not long sustain a historic perspective" (Fig. 7.3).

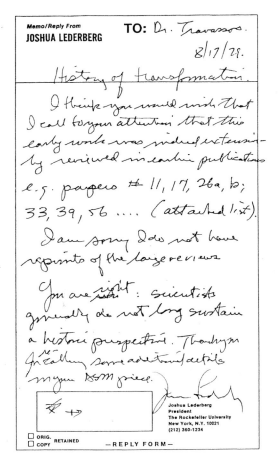

Figure 7.3: Reproduction of a letter from Lederberg to a colleague (L.R. Travassos, Escola Paulista de Medicina, São Paulo) regarding the priority in the discovery of transformation.

**Excursus 7-2**
**HERSHEY & CHASE'S (1952) BLENDER EXPERIMENT**

Alfred Hershey, a microbiologist at Cold Spring Harbor, was one of the phage school's founding fathers. He had already in 1946 mapped the first phage genes (Fig. 4.9) and investigated the physiology of phage infection – rendered specially gratifying after radioactive isotopes became available for biological research after the war. Two previous observations made by phage researchers seem to have influenced him: first, the impressive fact that some phages were endowed with long "tails", essential for adsorption to the host bacteria; and second, that upon sudden dilution of phage lysates from salt solution into water, the phage particles, suffering an osmotic shock, burst open, yielding free DNA and empty coats (Anderson, 1950). What was the role of these two phage components during infection? The fragile adsorption process, involving solely the tail tips, led to the suspicion that phage particles, once adsorbed, could be mechanically sheared away from the host bacterial cell. The fate of the two distinguishable components, DNA and protein, could be followed up by specific radio-isotope labeling. Hershey and Martha Chase, his technical assistant, produced two parallel cultures of *E. coli* and infected them with phage T2. $^{32}$P-phosphate was supplied to one culture. The other received $^{35}$S-sulfate. Because phosphate is incorporated into DNA, whereas sulfur is built into proteins (via methionine and cysteine), these two macromolecule fractions could be differentially labeled in two parallel phage lysates. From parallel bacterial cultures infected with $^{32}$P-labeled or $^{35}$S-labeled phages, samples were taken at different times after phage addition. These

Figure 7.4: Alfred Hershey (1908-1997) and Martha Chase.

samples were then briefly but vigorously shaken by means of a regular kitchen blender and then immediately centrifuged. The bacteria were collected at the bottom of the centrifuge tubes, whereas the much smaller phages remained in the supernatant. Radioactivity measurements of sediment and supernatant showed that, after the shaking procedure, roughly 80% of the $^{35}$S was found in the supernatant – with correspondingly 20% being dragged with the bacteria to the bottom – while 65% of the $^{32}$P appeared associated with the sedimented bacteria – the corresponding 35% remaining in the supernatant (Fig. 7.5); the findings indicated that a three times higher proportion of viral DNA, as compared to viral protein, was centrifuged down along with the bacteria.

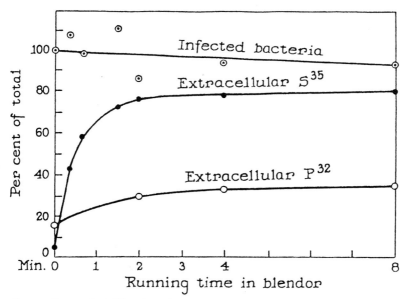

FIG. 1. Removal of S$^{35}$ and P$^{32}$ from bacteria infected with radioactive phage, and survival of the infected bacteria, during agitation in a Waring blendor.

Figure 7.5: Facsimile from Hershey & Chase (1952), as a résumé of their results: infected bacteria survive the vigorous shaking by the blender (detected as plaque-forming units); the subsequent centrifugation drags 65% of the phage DNA and roughly 20% of protein to the bottom of the centrifuge tube. The notations P$^{32}$ and S$^{35}$ (today, $^{32}$P and $^{35}$S) correspond, respectively, to the radioactively labeled phosphate in DNA and sulfur in proteins.

# CHAPTER 8

## THE CODE CRAZE

Whereas the principle of DNA replication was unambiguously revealed by the mere contemplation of its proposed structure, it was clear that the mechanisms by which DNA was supposed to direct protein synthesis would remain thoroughly mysterious for a while to come. How was it possible for DNA to determine the amino acid sequence of proteins, if between the bases of DNA and the amino acids of proteins there was no steric or chemical affinity whatsoever?

Well, George Gamow was no chemist. He was a bright astrophysicist... and a jester. His interests gravitated particularly around the sun (G. Gamow: A Star Named the Sun) and, generally, the whole universe. He was the one who, together with his graduate student Ralph Alpher, described in 1948 the origin of the cosmos as the Big Bang. For the publication of their results he persuaded his friend and colleague Hans Bethe to pose as a co-worker – just to create the troika: Alpher, Bethe, Gamow (1948). [Somehow, he also managed the publication to come out on the 1st of April; see Bernstein (1993).]

Through pure chance, Gamow came to know Watson & Crick's paper of May 30, 1953 (Watson & Crick,1953b). He read and grasped it right away: here was something novel! The foundation for solving the "mystery of heredity" was convincingly explained in a few sentences; but simultaneously a new charade emerged – which he also immediately recognized – : how to translate the 4-base-language of DNA into the 20-amino-acid-writing of proteins? Obviously, a code had to exist, a secret writing with four letters (the 4 bases of DNA), able to transfer from DNA the information necessary to establish the specific alignment of the 20 different amino acids in the diverse proteins. Each gene should be considered as a long number written in a 4-symbol-system. This was the uncontestable logical conclusion which sprang from the molecular DNA structure proposed by Watson & Crick. Each number was supposed to have a counterpart word in the language of proteins, which were permuted assemblages of the 20 available letters, the 20 amino acids. How to relate the numbers of a 4-symbol-system to words composed of letters taken out of an alphabet of 20 symbols? This was a crucial riddle. These thoughts led Gamow to boldly postulate a direct correspondence between the two notation systems (correct, as later demonstrated), based on the key and lock principle (totally wrong). According to him, each amino acid adapted specifically to a rhomboid concavity on the outside of the DNA double helix, each concavity being encompassed by 3 base pairs. The amino acid had only to align to its fitting concavity to be subsequently linked to its neighbor by a peptide bond. However, the molecular dimensions raised doubts: three base pairs alongside the double helix occupied a stretch of about 10 Å, whereas the distance between two amino acids in a peptide measured only 3,6 Å. This posed no hindrance to Gamow (1954) who right away decided to settle this deadlock by overlapping the rhombes: the base pairs 1, 2, and 3 would code for one amino acid, whereas the pairs 2, 3, and 4 determined the next, and the pairs 3, 4, and 5 decided the third (Fig. 8.1) and so on. Every biochemist grasped immediately the inconsequence of Gamow's code, dubbed diamond-code (because of the rhomboid form of the concavities), which was exposed in a wild theoretical essay (Gamow, Rich & Yčas, 1955). Besides the incompatibilities between the diamond code and the biochemical reality, there was the established fact that protein synthesis occurred in the cytoplasm, and not in the nucleus (where more than 99% of the cell's DNA was located): protein synthesis, thus, could not possibly take place directly on the DNA.

Gamow's proposition might not have been the right one but at least it pointed to a riddle awaiting to be solved: that of a hypothetical code. For mathematicians, information experts and secret service code breakers, it opened up a fresh and exciting challenge! (Biochemists involved with the theme of protein synthesis remained, for years to come, aloof to this newly arisen problem; see next Chapter.) In 1954 in Berkeley, Gamow founded a fancy elitary club: the "RNA Tie Club", whose goal it was to disclose the paths leading from nucleic acids to protein production. Gamow named himself head of the club, and limited its members to 20 researchers, each representing one amino acid (women were not admitted). It remains difficult to find out who, apart from Crick, Watson, Sidney Brenner, Alexander Rich, and Leslie Orgel, belonged to the select circle; Edward Teller, the neurotic father of the hydrogen bomb, a friend of Gamow, was surely among them, having, he too, elaborated a code, never to be published (the Russians, who knows, could get hold of it, and derive some advantages in the incipient Cold War). The club members exchanged information and ideas by means of informal circulars, where complex codes and infinite code variants were exposed; this way their elaborations dodged the strict constraints imposed by the norms of the regular scientific journals. It did not take long for Crick to disavow Gamow's overlapping code. As Gamow himself realized, overlapping the code words, or codons, would lead to unavoidable limitations in what concerned the possible sequence of the amino acids in the peptide; according to it, one had to postulate that some of the 20 amino acids never occurred in a direct sequence and only two of them could possibly appear three times in a row.

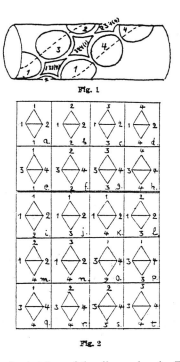

Figure 8.1: The first of Gamow's sketches of the diamond code. The pairs 1-2 and 3-4, placed horizontally, correspond to the base pairs of the double helix, the positions above and below, on the diamond, correspond to one of the neighboring bases. The 20 schemes outlined here would correspond to the 20 amino acids, according to Gamow. He did realize that for some rhomboid figures two mirror-symmetric images would be possible, but he did not accord any significance to that (Gamow, 1954).

And Crick's analysis of the known amino acid sequences – only two, at that time, those of insulin and β-corticotropin – inflicted the coup de grâce to Gamow's overlapping code. In 1957, as further amino acid sequences were disclosed, Brenner concluded that, apparently, no restraints whatsoever existed with regard to the possible combinations of amino acids in proteins, hence eliminating all overlapping codes.

Still, the possibility remained for the existence of only partially overlapping codes, (bases 1, 2, and 3 coding for the first amino acid, bases 3, 4, and 5 for the second one, bases 5, 6, and 7 for the third one, and so on). The sequence analysis of proteins seemed to open up a theoretical way for breaking the code – no one at the time fathomed this being a preposterous undertaking.

Mutations in sequences of overlapping or partially overlapping codes should have verifiable consequences. In such structures, replacements of two (or even three) neighboring amino acids on the respective peptide should always (in the case of overlapping codes) or often (in partially overlapping codes) occur. However, the analysis of a first case, exemplified by the sickle cell hemoglobin, revealed that exclusively one amino acid was changed as a consequence of one mutation (Ingram, 1956, 1957; Excursus 8-1).

Obviously, non-overlapping codes had to be given serious consideration. And then, Brenner (1957) pointed out that the disparate molecular dimensions of the codons, on the nucleic acids, on one side, and its counterpart amino acids, on the other side, did not actually pose any obstacle to the concept of the non-overlapping code, if one assumed that solely the newcomer amino acid contacted directly with the codon template; the already synthesized segment of the polypeptide could drift away from the template as soon as the new amino acid was incorporated. [This scenario had already been depicted by Dalgliesh (1953) (Fig. 8.2), but as an outsider he was only remembered much later.]

Non-overlapping codes would require certain restrictions: not any base sequence, perhaps with a false or nonsensical meaning, but solely the right codons within a sequence were to be read and recognized. For that, the reading had to proceed from a fixed starting point, and all codons had to encompass the same number of bases, so that an automatic counting could ensue (reading frame). A representation of commas [Crick's (1963) terminology] could also be visualized, interspersed among the actual codons. One of the four bases could pose as a symbol for the comma, for example, the other three remaining free to be used as actual code symbols. Assuming codons with three bases, this scenario would provide for 3 x 3 x 3 = 27 diverse code words, more than enough for the 20 amino acids, and signals, denoting the beginning and the end of a peptide chain. But, how to demonstrate the validity of such conceivable but totally hypothetical possibilities?

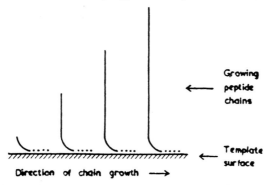

Figure 8.2: Illustration of C. E. Dalgliesh's concept, that no spatial limitations hinder the incorporation of amino acids directly on a template, if it is assumed that the product being synthesized, i.e., the growing polypeptide, leaves the template before full completion of synthesis (Dalgliesh, 1953).

Crick, Griffith & Orgel (1957) concocted an alternative to the comma code: the comma-less code! This was so inherently elegant as not to demand any experimental demonstration of its veracity; it postulated codons with three bases each (no novelty in that), so disposed that the sequences, if read in the wrong frame, automatically failed to provide a sense. In Crick, Griffith & Orgel's scheme, out of the 64 potential codons, exactly 20 remained to represent the 20 amino acids. Could that be due to mere chance? Truly, it looked too perfect to be wrong!

Then, Delbrück's mistrust of the nice scheme arose; actually the same doubts had already assaulted Crick himself. It was not enough to declare as senseless all codons on a DNA strand which were out of frame. The DNA double helix, chemically considered, did not have any polarity; based on this fact, one had to assume that the complementary strand was also embued with sense (however, try to organize a collection of sentences equally meaningful if read from one end to the other or vice versa). If the codons complementary to the sense codons were to be assumed senseless, the number of sense codons had to be reduced to 10, if one wanted to remain faithful to the 3-base-code system. However, a 4-base-code could come to the rescue of the comma-free code (as Delbrück dubbed Crick's comma-less code); such a code was embued with a novel characteristic: transposability; the idea that the genetic message was possibly carried on both strands. This proposition presupposed that the complementary codons had the same meaning; so, for example, $5'ACG3' = 5'CGT3'$ (Golomb, Welch & Delbrück, 1958). Golomb and Welch, two Danish mathematicians, were undaunted by the calculations necessary to deal with all potential theoretical combinations concerning the transposable code, which they accurately handled in no less than 16 theorems. Alas, it did not occur to anyone, that the hypothesis of the transposable code fully ignored the polarity of the polypeptides – as a result of which amino terminal and carboxy terminal were not interchangeable.

Crick kept his mind occupied with all imaginable theoretical characteristics of a genetic code. Hitherto, all proposed alternative codes assumed the validity of the so-called colinearity hypothesis, which presupposed a direct correlation between the linear information carried by the nucleic acids and the sequence of amino acids on the polypeptides. Although it seemed quite logical, it was as yet an unproved hypothesis (Excursus 8-2). All the variants already mentioned here aside – with or without commas, overlapping or not, etc –, other conceivable alternative types of codes had to be taken into account. Still to be addressed were the incognita: *a)* was the code degenerate; that means, do synonymous codons exist; i.e., is each amino acid represented by more than one codon? *b)* could the code be perhaps ambiguous; in other words, could the code possibly possess a double meaning, one codon corresponding to two or more amino acids? *c)* would the code be universal; i.e., did one and only one code exist for all living beings? Crick, Griffith & Orgel's fanciful comma-less code intrinsically eliminated synonymous codons, since it counted with no more than 20 codons standing for the 20 amino acids. And precisely this point became its Achilles' heel.

Belozerski & Spirin (1958), from the Moscou Academy of Sciences, published measurements of the DNA and RNA base compositions of various microorganisms (Fig. 8.3); Sueoka (1961), using the same microorganisms studied their amino acid composition. The analysis revealed that their RNA bases and protein amino acids displayed a quite extensive homogeneity. However, the base composition of their DNAs demonstrated great variability – similar to what Chargaff had already observed. For example, the percentage of A + T in *Mycobacterium* DNA was 30% (with 70% G + C), whereas in *Clostridium* these values were nearly reversed (Fig. 8.3). Insecurity was sown among the theoretical cryptoanalysts and code breakers. Crick (1963) recognized, hidden behind the data collected by Belozerski & Spirin, the possibility of a code with a degenerate modus, i.e., different synonymous codons (the universality of the code was never really doubted). And this was tantamount to disavowing the so praised comaless code.

Robert Sinsheimer (1959), at Caltech in Pasadena, tried a last theoretical circumvention of the impasse, suggesting a two symbol code. He proposed that A and C were equivalent (both possessed a 6-amino function), as well as T (or U) and G (these were endowed with a 6-keto function).This was the trick that should allow researchers to cope with the most disparate relations of (A + T) to ( G + C). But the proposition came with a price tag: an unhandy code of five bases pro codon.

Still, all these hypothetical codes (overlapping, non-overlapping, comma, frame, etc.) did not address the problem of the lack of stereo-specific complementarity between the bases of the nucleic acids and their cognate amino acids of the proteins. This incongruity was solved by Crick (1955, 1958) with a simple – and, as later revealed, correct – stroke of genius: he postulated a mediator molecule, or adaptor, able to act as a bridge, and so overcoming the lack of direct affinity between bases and amino acids. The adaptor's structure allowed for distinct sterically fitting regions: one adaptable to its specific codon, another one adjustable to its respective amino acid. As promising as this novel adaptor approach was, it turned out to be treacherous for the theoretical cryptoanalysts: postulating an adaptor lent wings to fantasies involving as many and as varied codes as desired. One might infer that a theoretical solution of the charade of the code was an unreachable undertaking.

And the Russians had already ruined the whole code cracking game – if not through their singular observations (these could be circumvented, see Sinsheimer), then by simply spoiling all the fun.

Crick, the quintessential theoretician, was forced to revert to an experimental trial, and a tricky one at that. What aim he originally pursued never became quite clear and it was never published. [According to him (Crick, 1988), it was the matter of a complex code with "looped-out bases".] Soon afterwards, though, his experiments provided evidence – later validated by other experiments of a totally different nature – for the assumption that the code was translated from a fixed starting point, following a reading-frame of three nucleotides (bases), each triplet codifying one amino acid; the

Table 1.  COMPOSITION OF BACTERIAL DEOXYRIBONUCLEIC ACIDS

| Species | Bases (moles per cent) | | | | $\dfrac{Pu}{Py}$ | $\dfrac{G+T}{A+C}$ | $\dfrac{G+C}{A+T}$ |
|---|---|---|---|---|---|---|---|
| | $G$ | $A$ | $C$ | $T$ | | | |
| Clostridium perfringens | 15·8 | 34·1 | 15·1 | 35·0 | 1·00 | 1·03 | 0·45 |
| Staphylococcus pyo-genes aureus | 17·3 | 32·3 | 17·4 | 33·0 | 0·98 | 1·01 | 0·53 |
| Pasteurella tularensis | 17·6 | 32·4 | 17·1 | 32·9 | 1·00 | 1·02 | 0·53 |
| Proteus vulgaris | 19·8 | 30·1 | 20·7 | 29·4 | 1·00 | 0·97 | 0·68 |
| Escherichia coli | 26·0 | 23·9 | 26·2 | 23·9 | 1·00 | 1·00 | 1·09 |
| Proteus morganii | 26·3 | 23·7 | 26·7 | 23·3 | 1·00 | 0·98 | 1·13 |
| Shigella dysenteriae | 26·7 | 23·5 | 26·7 | 23·1 | 1·01 | 0·99 | 1·15 |
| Salmonella typhosa | 26·7 | 23·5 | 26·4 | 23·4 | 1·01 | 1·00 | 1·13 |
| Salmonella typhi-murium | 27·1 | 22·9 | 27·0 | 23·0 | 1·00 | 1·00 | 1·18 |
| Erwinia carotovora | 27·1 | 23·3 | 26·9 | 22·7 | 1·02 | 0·99 | 1·17 |
| Corynebacterium diphtheriae | 27·2 | 22·5 | 27·3 | 23·0 | 0·99 | 1·01 | 1·20 |
| Aerobacter aerogenes | 28·8 | 21·3 | 28·0 | 21·9 | 1·00 | 1·03 | 1·31 |
| Mycobacterium vado-sum Kras. | 29·2 | 20·7 | 28·5 | 21·6 | 1·00 | 1·03 | 1·37 |
| Brucella abortus | 29·0 | 21·0 | 28·9 | 21·1 | 1·00 | 1·00 | 1·37 |
| Alcaligenes faecalis | 33·9 | 16·5 | 32·8 | 16·8 | 0·98 | 1·03 | 2·00 |
| Pseudomonas aeruginosa | 33·0 | 16·8 | 34·0 | 16·2 | 0·99 | 0·97 | 2·03 |
| Mycobacterium tuber-culosis BCG | 34·2 | 16·5 | 33·3 | 16·0 | 1·03 | 1·01 | 2·08 |
| Sarcina lutea | 36·4 | 13·6 | 35·6 | 14·4 | 1·00 | 1·03 | 2·57 |
| Streptomyces griseus | 36·1 | 13·4 | 37·1 | 13·4 | 0·98 | 0·98 | 2·73 |

Abbreviations: $G$, guanine; $A$, adenine; $C$, cytosine; $T$, thymine; $Pu$, purine bases; $Py$, pyrimidine bases.

Table 2.  COMPOSITION OF BACTERIAL RIBONUCLEIC ACIDS

| Species | Bases (moles per cent) | | | | $\dfrac{Pu}{Py}$ | $\dfrac{G+U}{A+C}$ | $\dfrac{G+C}{A+U}$ |
|---|---|---|---|---|---|---|---|
| | $G$ | $A$ | $C$ | $U$ | | | |
| Clostridium perfringens | 29·5 | 28·1 | 22·0 | 20·4 | 1·36 | 1·00 | 1·06 |
| Staphylococcus pyo-genes aureus | 28·7 | 26·9 | 22·4 | 22·0 | 1·25 | 1·03 | 1·05 |
| Pasteurella tularensis | 29·3 | 27·3 | 21·0 | 21·9 | 1·33 | 1·07 | 1·03 |
| Proteus vulgaris | 31·0 | 26·3 | 24·0 | 18·7 | 1·34 | 0·99 | 1·22 |
| Escherichia coli | 30·7 | 26·0 | 24·1 | 19·2 | 1·31 | 1·00 | 1·21 |
| Proteus morganii | 31·1 | 26·0 | 23·7 | 19·2 | 1·31 | 1·01 | 1·21 |
| Shigella dysenteriae | 30·4 | 25·9 | 24·4 | 19·3 | 1·29 | 0·99 | 1·21 |
| Salmonella typhosa | 30·8 | 26·1 | 24·0 | 19·1 | 1·32 | 1·00 | 1·21 |
| Salmonella typhi-murium | 31·0 | 26·1 | 23·8 | 19·1 | 1·33 | 1·00 | 1·21 |
| Erwinia carotovora | 29·5 | 26·5 | 23·7 | 20·3 | 1·27 | 0·99 | 1·14 |
| Corynebacterium diphtheriae | 31·6 | 23·1 | 23·8 | 21·5 | 1·21 | 1·13 | 1·24 |
| Aerobacter aerogenes | 30·3 | 26·0 | 24·1 | 19·6 | 1·29 | 1·00 | 1·19 |
| Mycobacterium vado-sum Kras. | 31·7 | 23·8 | 23·5 | 21·0 | 1·25 | 1·12 | 1·23 |
| Brucella abortus | 30·2 | 25·4 | 24·9 | 19·5 | 1·26 | 0·99 | 1·23 |
| Alcaligenes faecalis | 30·9 | 25·7 | 24·1 | 19·3 | 1·31 | 1·01 | 1·22 |
| Pseudomonas aeru-ginosa | 31·6 | 25·1 | 23·8 | 19·5 | 1·31 | 1·05 | 1·24 |
| Mycobacterium tuber-culosis BCG | 33·0 | 22·6 | 26·1 | 18·3 | 1·25 | 1·05 | 1·45 |
| Sarcina lutea | 32·7 | 23·2 | 24·2 | 19·9 | 1·27 | 1·11 | 1·32 |
| Streptomyces griseus | 31·1 | 23·3 | 25·2 | 19·9 | 1·22 | 1·04 | 1·29 |

Abbreviations: $G$, guanylic acid; $A$, adenylic acid; $C$, cytidylic acid; $U$, uridylic acid; $Pu$, purine nucleotides; $Py$, pyrimidine nucleotides.

code was degenerate; i.e., different triplets may be codons for the same amino acid (Chapter 10). What had Crick discovered that allowed such insight?

To understand Crick's project, we have to go back to the *rII* mutants of T4 (Excursus 8-2). These mutants, obtained by employing some chemical substances, for example the base analogues 5-bromo-uracil or 2-amino-purine, could be triggered to mutate back to the wild type by the use of the same mutagenic substances (Excursus 6-1). Another type of mutation could be brought about by proflavin. This type could undergo spontaneous or proflavin-induced – but not base analogue-induced – back-mutations. A plausible way to explain this observations seems to be that proflavin was no substitute for bases in the DNA and also did not alter already incorporated bases (as hydroxylamine or nitrite did); instead, it possibly edged itself between two DNA bases without being incorporated, leading to the loss of a base or to the addition of one extra base in the course of DNA synthesis (Lerman, 1961; Fig. 8.4).

Crick, himself, single-handedly, isolated over 100 proflavin-induced *rII* mutants. Starting with these, he obtained what seemed to be back-mutants (possessing the wild type phenotype), either spontaneously or proflavin-induced. From crosses between such apparent "wild type back-mutants" with the original wild type, he often got recombinants of the *rII* type: the "wild type back-mutants" were indeed not veritable back-mutants, but so-called phenocopies of the wild type [able to propagate on K12 (λ)]. The wild phenotype re-emerged because the second *rII* mutation was obviously able to counteract the effect of the first mutational event, i.e., a so-called "suppressor mutation". Crosses with different proflavin-induced *rII* mutants (only mutations occurring within a small section of the *rII* B gene were used in this project – see Fig. 8.5) led to the distinction of two categories: mutants which suppressed each other (that means, recombinants from crosses between them re-acquired the wild phenotype) belonged to two different categories, whereas mutants not suppressing each other (recombinants between them retained the *rII* phenotype) were of the same category. These two categories of mutants were arbitrarily denoted *plus* and *minus* types. Double mutants of type (+ −) evoked mostly the wild phenotype, whereas the (+ +) or (− −) double mutants still behaved like an *rII* mutant.

How to account for these occurrences? It was assumed that the genetic information for the synthesis of the *rII* B protein (hypothetical at the time) was read from a fixed starting point at the beginning of the gene, and that a constant number of bases within this gene represented a codon for each corresponding amino acid; the binding of proflavin to one DNA strand forced the removal or the insertion of one base, and this, in its turn, led to misreading due to the alteration of the original reading frame. It could be postulated that mutants from the (+) type derived from insertions, whereas those from (−) type resulted from the deletion of one base. [Or

Figure 8.3 (left): Facsimile of Belozerski & Spirin's (1958) tables displaying large variations in the base composition of the DNAs from diverse origins, whereas the RNAs showed only small differences (compare to Chargaff's rule for the base composition of DNA, Fig. 5.3).

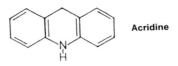

Acridine

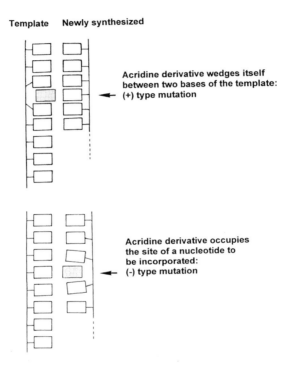

Template    Newly synthesized

Acridine derivative wedges itself
between two bases of the template:
(+) type mutation

Acridine derivative occupies
the site of a nucleotide to
be incorporated:
(-) type mutation

Figure 8.4: Scheme of the formation of frameshift mutations caused by acridine derivatives, like proflavin. These polycyclic compounds, shaped as flat as the bases, can occupy for a short period of time a nucleotide's place on the template or on the newly synthesized strand, without being permanently incorporated. Also many other polycyclic hydrocarbons, such as those of tobacco smoke, act similarly (see, for example, Phillips, 1983), causing globally hundreds of thousands of lung cancers yearly (see for instance, Cook, 1993).

Figure 8.5 (right): *a*) Representation of the changes caused by (–) type and (+) type mutations (respectively, deletion or insertion of one base), on the reading frame of a triplet code with fixed starting point; from Crick et al. (1961), *b*) The genetic map of the left end of the B-cistron [Crick et al. (1961); compare to Fig. 8.7] with the frameshift mutants isolated by Francis Crick (FC-series), and the additional frameshift mutants of types (+) or (–), isolated as

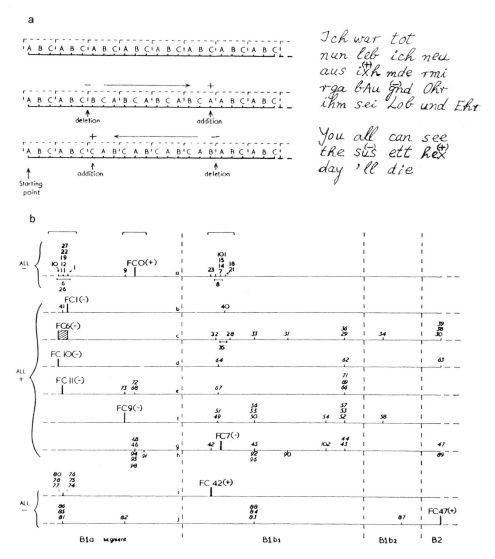

suppressors of the FC-mutants located on the same line. [Designating the mutants FC earned Crick the criticism of being conceited; that apparently disturbed him so much as to impel him to explain his doing so in his memoirs (Crick, 1988): "Unfortunately I could not remember for certain which letters had already been used, so I decided to rename my mutant... the real explanation was that I have a rather fallible memory."] Top right: common text illustrations (one in German) of frameshift suppression.

vice-versa; to what mutational mechanism corresponded what type of mutation was irrelevant to the reasoning. But, truly, the correspondence between the ( +) type mutations and base additions, and (–) type mutations and base losses was later validated.] It was tempting to hypothesize that the reading frame in the segment between two mutually suppressing proflavin-induced mutations was displaced from the point of the first mutation (backwards by the insertion of one base or forwards by the deletion of one base), but corrected again at the site of the second mutation, acting as a suppressor. If the region underlying the erroneous reading frame was so small as not to impair the function of the *rII* B protein (quite an acceptable postulate), then, the fact that two proflavin mutants of opposite types [one ( +) the other (–)] suppressed each other could very well be understood in accordance with the assumption of a frame code and a fixed starting point for reading. This hypothesis, in conjunction with the interpretation of the mode of action of ( +) and (–) mutant types, allowed the elaboration of a further hypothesis, which then could be checked experimentally for its validity – an elegant example of the nature and usefulness of scientific hypotheses.

If codons really encompassed a constant number of bases, then suppressing a frameshift mutation could be achieved, not only by another frameshift mutation of opposite sign, but possibly also by multiple frameshift mutations of the same sign, which, summed up, equalled the loss or insertion of a complete codon. Three was the hypothetical number of bases making up a codon (since the resulting 64 possible combinations would be more than enough for the codification of the 20 amino acids). If this supposition was correct, then triple mutations, either ( + + +) or (– – –), should evoke the original wild phenotype (presupposing that the addition or the loss of the corresponding amino acid took place at a non-essential site, so as not to affect the protein function).

Obtaining such mutants from crosses was indeed somewhat laborious but not an insurmountable task. And it turned out to be a worthwhile undertaking: triple mutants of the same type behaved as a wild type on the indicator bacteria! The assumption was corroborated: the genetic code proved to be a frame code, composed of 3 bases per codon, and endowed with a fixed reading start for each gene. After supplementary experiments on the same system confirmed and extended the first findings, one of the most elegant and elucidative works of the pioneering era of molecular biology was published : Crick, Barnett, Brenner & Watts-Tobin (1961).

Crick (1988), looking back, thought, in a slightly depressive mood, that this work, as a tour de force to demonstrate genetically the triplet nature of the code, had had practically no influence on the further course of events, since within a short time the same insights were to be arrived at by totally different, purely biochemical experiments (Chapter 10). He avoided mentioning the other, meaningful and brilliant side of the work which showed, maybe for the last time, in a beautiful way, the contrasting spirits of biochemistry and molecular genetics. These contrasts would inexorably fade away from then on.

**Excursus 8-1**
**ONE MUTATION ALTERS HEMOGLOBIN IN ONLY ONE AMINO ACID**

In 1955, Francis Crick was keen to demonstrate that one mutational event led to the alteration of a sole amino acid in a polypeptide. He first tried his luck with lysozyme, a protein derived from chicken egg white or from tear drops (Crick, 1988), but he then shifted his subject of choice to the sickle cell hemoglobin (Hb-S). Pauling et al. (1949) had already described the sickle cell anemia as an example of a molecular disease, after having shown the contrasting behavior of Hb-S and normal hemoglobin (Hb-A) when subjected to an electric field. Vernon Ingram, who actually worked with Max Perutz, was persuaded to work on the project. For that, he created a new technique designated "fingerprinting". The two forms of hemoglobin (Hb-A and Hb-S) were initially digested enzymatically by trypsin. Trypsin splits polypeptides behind the amino acid residues arginine or lysine, so degrading the original polypeptide to a collection of smaller peptides. The resulting digest was then processed by paper electrophoresis, a treatment that separates peptides of different electrical charges. As a next step, the paper was turned by 90° before being subjected to a chromatographic process, which separated the still partially overlapping peptides, in this instance according to their solubility in the chromatographic solution (a mixture of butyl-alcohol, acetic acid and water in a 3:1:1 proportion). The final step consisted in staining the isolated peptides, creating a characteristic blot pattern for each protein, so individual as to be compared to a fingerprint (Fig. 8.6). The fingerprints from the normal and the sickle cell hemoglobins were distinct in one small peptide (Ingram, 1956). This differential peptide was cut out of the paper and further analyzed: it turned out that solely one amino acid, a glutamic acid from Hb-A, had been replaced in the Hb-S by one valine (Ingram, 1957, 1958). It did not take long before dozens of hemoglobin variants had been similarly analyzed, revealing that in each case the culprit was only one amino acid (Ingram, 1962).

**Excursus 8-2**
**LINEARITY ... COLINEARITY**

Like beads on a necklace – that was the classical concept for the arrangement of genes on the chromosomes (see, for example, Carlson, 1966). The alternative alleles were to be compared to the isotopes of an element, and mutations were substitutions of one of the beads by a slightly altered one. However, already before the onset of the molecular biology era, the plain image evoked by the bead comparison became ground for suspicions. Compact and indivisible genetic units could not account for such observations as certain position effects, occurring in the fruit fly *Drosophila*. In this example, the resulting phenotype of two mutations closely located on the

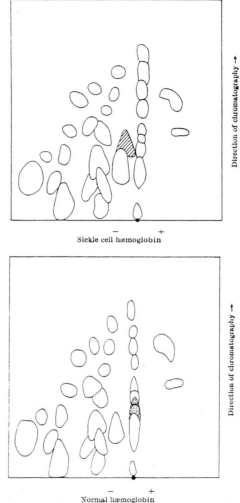

Fig. 1. 'Finger prints' of human normal and sickle-cell hæmo-
globins. Electrophoresis at *p*H 6·4, chromatography with *n*-butyl
alcohol/acetic acid,water (3 : 1 : 1). The shaded and the stippled
spots are those belonging to the peptide showing the difference

Figure 8.6: The fingerprint pattern from normal and sickle cell hemoglobin. The only
difference is found at the position of the indicated (shadowed) peptide. [It was later shown that
this alteration is due to the replacement in sickle cell hemoglobin of glutamic acid in the sixth
position of the normal β-chain by a valine (Ingram, 1962.)

chromosomal map depended on their being physically positioned on the same chromosome (both mutations inherited from the same parent) or else on separate homologous chromosomes (each mutation inherited from a different parent). The first scenario corresponded to a *cis* situation (one chromosome fully wild type and the other carrying the two mutations: $m_1 m_2 / + +$). The other scenario would be the *trans* situation (each homologous chromosome carrying one of the mutations: $m_1 + / + m_2$). The resulting phenotype for the *cis* position was that of the wild type, whereas the outcome of the *trans* situation was that of the mutant phenotype. Such odd observations could be interpreted by assuming that the gene products of the wild type alleles had to act tandemwise in order to be effective – similarly to the enzymes in Beadle's synthetic chains (Chapter 2); the two gene products, being practically non-diffusible, would necessarily have to be located very closely on the same chromosome (that is, in a *cis* position). Alternatively, one could argue that the gene, taken as the physical entity responsible for the appearance of a phenotypic characteristic, was actually subdivided in individual mutable sites and that between these sites genetic recombination could take place. Hence, genetic recombination could occur not only between the beads of the necklace but also within each individual bead. This implicitly meant attributing linearity to these very beads.

Milislav Demerec (1895-1966), an East European immigrant, the somewhat stingy (at the time of scant research grants, he counted every penny that he spent), long-time-director of the Cold Spring Harbor Laboratory, thought that approaching these themes through *Drosophila* was a real waste of resources. In order to observe rare recombinants between such pseudo-alleles, tens of thousands of flies had to be bred and analyzed. This made him consider the newly fashionable bacteria as his subjects of choice: first, *E. coli*, and then, as Zinder & Lederberg (1952) described the phenomenon of transduction (Excursus 3-2), *Salmonella*. He isolated – among others – a collection of Cys⁻ *Salmonella* mutants (cystein-dependent); cultures of each of these were infected with the phage P22. The lysates obtained were further employed for transduction experiments. The reasoning was: if all Cys⁻ mutant bacteria underwent exactly the same genetic change, no prototrophic recombinants would ensue. However, from the 25 Cys⁻ *Salmonella* mutants, almost all displayed recombinants when crossed with each other, in any combination (Demerec et al., 1955). There seemed to be a few too many variants for all to result from the metabolism of cystein, so, instead of interpreting this outcome through the "one gene – one enzyme" hypothesis, it seemed more reasonable to postulate that the inner structure of a gene responsible for a specific function, as the synthesis of an enzyme, was divided in sub-units, independently mutable and interchangeable; in other words, the genes were composed of smaller units able to undergo intragenic recombination.

Demerec's work paved the way for Seymour Benzer. Benzer, one more physicist deeply impressed by Schrödinger's "What Is Life?", had attended the phage course at Cold Spring Harbor with the intent of getting familiar with biology. The idea of constructing chromosomal maps of phages excited his fancy. The so-called *rII* locus of phage T4 (Fig. 4-9) seemed especially thrilling. Within this locus, many T4 mutants were mapped, as they were detectable by their larger than normal plaques on the normal host *E. coli* K12; these mutants, however, failed to form plaques on the strain of *E. coli* K12(λ), a susceptible host for the wild type T4. The system seemed ideally suited for selective screening. It paralleled Demerec's *Cys⁻* mutants, which could recombine among each other, yielding *Cys⁺* cells selectively detectable on minimal medium – even if they appeared at a rate of 1:100 million or smaller. *rII* mutants could be similarly crossed with each other, producing wild type phages, *rII⁺*, identifiable as the only ones forming plaques on a lawn of indicator strain K12(λ). Soon, Benzer had his refrigerator, located in the laboratory for Biophysics at Purdue University in Lafayette, Indiana, stuffed with thousands of test tubes containing different *rII* lysates. Most *rII* mutants, when crossed with each other, yielded wild type recombinants, whose proportion among the total progeny – even if extremely small – could be scrutinized through the number of "plaque-forming-units" (PFU) on K12(λ) as compared to the total number of PFU on K12 [obviously, the crossing lysate had to be diluted many hundred times more when plated on K12 as compared to platings on K12(λ)]. To each individual *rII* mutant, a specific site on the *rII* locus could be accorded; such sites were mostly represented by a single point on the genetic map (see for instance, Benzer, 1961). However, some mutants never recombined with a series of other mutants, although these other mutants still produced wild type recombinants when crossed with each other. The former mutants were interpreted as resulting from block mutations, deletions, represented on the genetic map as bars overlapping point mutations and smaller block mutations with which they failed to yield recombinants (Fig. 8.7).

Now, what was the *rII* locus really? Did it represent only one gene, or was it perhaps an assemblage of genes? Experiments, analogous to the *cis-trans* tests in *Drosophila*, helped cast some light on the matter: a bacterial cell from the *E. coli* K12(λ) strain, if infected simultaneously with two different mutants of a phage, could be regarded as a transient equivalent of a diploid heterozygous cell of a higher organism. Mutant genes in *trans* or *cis* positions could be thus introduced experimentally into a bacterial cell. In the *trans* scenario, two different mutations were set apart in two phages as if it were on two separate chromosomes. The other possibility, the *cis* position, implied both mutations being carried by the same phage, a double-mutant (constructing such double-mutants through recombination was not the easiest of undertakings, but, nevertheless, a feasible one), whereas the additional phage was the wild type. The infection of K12(λ) with mutants in *cis* position was not expected to deliver fresh insights, being designed primarily as an experimental control. This was the case because co-infection with wild type would supposedly complement any other phage, be it a single or a double mutant (wild type *rII⁺* behave as dominant over *rII*). The assumption was that the wild type *rII⁺* phages expressed a

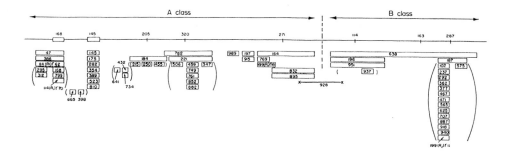

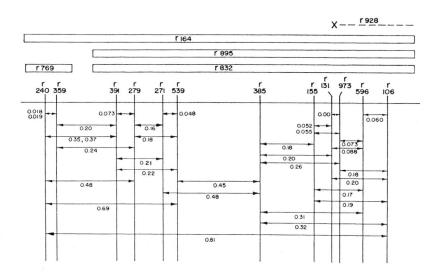

Figure 8.7: The genetic map of *rII* mutants of phage T4. Above: scheme of cistrons A and B, with the block mutants (deletions) recognized as such because no recombination was observed in crosses between these and a series of point mutations. The point mutations, when crossed with each other, always yielded recombinants, as shown below for the region of block mutant *r*164. The numbers represent the percentage of recombinants obtained from the respective crosses; compare with the position of the *rII* locus on the genetic map of Fig. 4.9 (Benzer, 1957).

gene function which, although superfluous in K12, was essential for growth in K12(λ). As for the *trans* situation, according to the mutant pair performing the double-infection, two distinct outcomes were possible: *a*) the two mutants complemented each other; i.e., together they were able to overcome the growth impairment in K12(λ), characteristic for individually infecting mutants, or, *b*) no complementation was elicited, that is, even under the condition of double infection, the incapacity of producing progeny on K12(λ) was sustained. The *rII* mutant phages could thus be considered as belonging to one of two different categories: *rIIA* or *rIIB*; phages belonging to the same category did not complement each other, whereas complementation occurred when the phages were representatives of two distinct groups. Not just one, but two different gene functions (perhaps enzymes) were apparently required for the successful growth on K12(λ): one coded by the gene *rIIA*, the other by gene *rIIB*. Alas, the terminology "gene" did not please Benzer's fancy, because genes could also be defined as mutable sites, and according to this definition there were scores, even hundreds of *rIIA* and *rIIB* genes. Thus, distinctions had to be made (Benzer, 1957). What underwent mutation was not the gene as a whole, but – as a physicist he was surely inspired by the atom's constituents – the muton. And what was the smallest interval between two mutons, detectable through recombination? The recon. Finally, the functional unity, evidenced by the *cis-trans*-test? The cistron ("transtron" would be perhaps an equally fitting name because only the *trans*-test was relevant for the subdivision in two groups).

What emerged so clearly from Benzer's experiments was that the gene, the functional unit, was a linear structure, composed of many individual mutable sites – thus, constructing a mutation map of a gene should be feasible. The combination of the fact that polypeptides, in their primary structure, also displayed a linear character, with the "one gene – one enzyme" hypothesis (which assumed a direct correlation between gene and protein) led to the assumption that the linearity of the gene had a straight correspondence to the linearity of the polypeptides. This was the colinearity hypothesis.

To prove the validity of this hypothesis in an elegant manner was the dream of quite a few molecular biologists at the time. Alas, the *rII* system of phage T4 turned out to be unsuitable, since the products of *rIIA* as well as *rIIB* remained uncharacterized, and this state of affairs would prevail for over a decade. [And when the deadlock was finally lifted – *rIIA* and *rIIB* were shown to code for membrane proteins (Weintraub & Frankel, 1972) – the interest in them had already vanished; the issue had been solved long ago.]

Benzer's experiments – which actually were expanded to encompass a detailed analysis of the action of mutagens (see for example, Benzer, 1961) –, although so exciting at the time, could be summarized as a genetic confirmation of what the mere contemplation of the DNA double helix made manifest: the gene must be linear, and it can mutate at different locations. Further, it provided us with the legacy of the terms muton, recon, and cistron. Soon it became obvious that the muton was nothing else than a base pair of the double helix, whereas the recon represented the interval between two neighboring base pairs. These neologisms, considered to be fashionable jargon soon after their creation, did not endure the pressure of time. Although the cistron survived the muton and the recon for many years, it also succumbed to old age, being replaced by what it always had been: the gene.

Confirming colinearity of amino acid sequence and genetic information experimentally appeared to be, at least theoretically, an easy-to-solve task: it would be enough to map different mutations of a gene and show that the alterations caused by them in the primary structure of the corresponding polypeptide, that is the resulting amino acid substitutions, were aligned in the same order as the mutants on the map. However, tackling this task experimentally proved to be quite elusive. As already mentioned, the *rII* system had to be dismissed. Also, other systems, contrary to expectation, did not help to advance the cause. Grotesque, was Crick's assessment of the situation... Actually, it became progressively obvious that no plausible options existed. Such experiments would only be meaningful, if the colinearity hypothesis turned out to be wrong. And it just could not be anything but right, as emphasized by Charles Yanofsky's group at Stanford. They reached that conclusion by comparing the genetic map of the tryptophan synthetase A locus from *Salmonella typhimurium* with its counterpart polypeptide (Yanofsky et al., 1964).

The intuitive notion that a direct relationship between the sequence of bases in the DNA and the order of the amino acids in the proteins should exist, lingered on for many years. Alas, a novel phenomenon, detected in higher organisms, came to light: alternative splicing of the same genetic information, giving rise to variable message combinations (see Fig. 21.12). Reality was thus quite more complex than initially realized. In spite of that, the central principle remained recognizable even for special instances: the genetic information is based on a linear array of symbols, the nucleotide bases, which, for their part, determine the sequence in which the protein building blocks, the amino acids, are going to be aligned in the polypeptide.

As Crick (1988) observed in a pensive mood, all these long efforts to demonstrate colinearity were a consequence of the incapacity to sequence the bases of DNA, now a daily routine.

# CHAPTER 9

## PROTEIN SYNTHESIS *IN VITRO*. HOW DID IT ALL START?

The interest that Paul Zamecnik developed for protein synthesis derived – according to his own statement – from his involvement, in 1938, at the Massachusetts General Hospital in Boston, with a grotesquely fat female patient, admitted to the clinic in order to shed pounds through controlled dieting. The woman died after only a few days of treatment, leaving open to conjectures the real reason of her death: perhaps excess fat, ... maybe a dearth of muscles? Muscles! Proteins! But how in this world were proteins synthesized? Fritz Lipmann, working one floor below, was the one who was dedicated to the synthesis of fats, his work would make him a Nobel laureate in 1953.

Proteases, those digestion enzymes able to degrade proteins to smaller peptides or even down to amino acids, were among the first enzymes to be analyzed. The stomach's pepsin and pancreas' chymotrypsin had already been described by Emil Fisher. Since enzyme reactions were two-way processes, it was to be expected that protein synthesis also would be boosted by proteases or similar enzymes (see, for instance, Loftfield et al., 1953). Zamecnik (1984) reminisced on what a shock it was for him when Lipmann, in 1942, challengingly inquired whether it was feasible to consider the mechanism of protein synthesis to be totally independent from that of proteolysis, thoroughly different?!

The isotope-labeling technology, introduced immediately after the Second World War, was soon to be employed to help to disentangle biochemical aspects of protein synthesis, thus enormously boosting the development of this field of research.

[14]C-labeled amino acids, when added to thin slices of liver tissue in buffer or nutrient medium, not only revealed the occurrence of protein synthesis, but also its dependence on cell respiration and oxidative phosphorylation (Frantz et al., 1948). The next step to be considered was – what else but – doing experiments with cell-free extracts: that would open the way to discerning the intermediary products on the path from free amino acids to complete peptide chains.

After some years of excruciatingly minute work, it became apparent, as described by Zamecnik and Keller (1954), that the incorporation of [14]C-amino acids into acid-precipitable material, i.e., proteins, depended: first, on the presence of ATP or an ATP-generating system (for example, phosphopyruvate and pyruvate-kinase); second, on the so-called microsomes – the old denomination for eukaryotic ribosomes, held together in large numbers by membrane debris of the endoplasmatic reticulum (Fig. 21.2); and third, on the supernatant of a one hour, 100,000 x g ultracentrifugation procedure that prompted the sedimentation of microsomes. (Ultracentrifugation was another method belonging to the list of post-war novelties.)

Whereas $^{14}$C-amino acids did not react at all with the microsomes, even in the presence of ATP, the supernatant of the 100,000 x g ultracentrifugation step provided the first clue to the unravelling of the biochemical pathways leading to proteins: this supernatant apparently contained an enzymatic activity (thermo-labile, not dialyzable and protease-sensitive), able to catalyze the synthesis of amino acyl-AMP, parting from free amino acids and ATP, liberating pyrophosphate in the process.

The amino acyl-adenylate was thus recognized as a potential go-between product: it consisted of an amino acid endowed with the right energetic level needed for completion of peptide bonds; these amino acids could then easily polymerize into polypeptides, since they were – as one would later say – activated. [The fact that energy was a requirement for the process of peptide bonding was definitely ascertained by Borsook & Huffman (1938) in demonstrating that the equilibrium of a dipeptide hydrolysis reaction was greatly biased towards the free amino acids.]

These first experiments settled some points of contention, because instead of amino acid activation, one could alternatively imagine the activation of the carboxy-terminal of a long peptide (as a parallel to the activation of γ-glutamyl-cystein, ocurring in the glutathion synthesis pathway), or else, that the activation was a coenzyme-A-dependent reaction, as the synthesis of fatty acids was, or...

Adding microsomes to the supernatant with activated amino acids resulted in those being covalently bound into proteins at the ribosomes.

The type of bonding was surely peptidic because it could be reversed by protease action. Obviously, the site where the proteins were being synthesized was the ribosome!

Another point to be clarified referred to the effect of the addition of all other amino acids to the cell-free homogenate (aside the $^{14}$C-labeled one; for instance, leucine). It was to be assumed that this would contribute to enhance the process of incorporation into protein, because proteins were normally constituted of 20 different amino acids, and a single amino acid would not keep the process running. Zamecnik's group, however, failed to detect such stimulation – in contrast to results of other laboratories (Peterson & Greenberg, 1952). A most probable interpretation for these divergent results was that Zamecnik's crude homogenates contained already enough available free amino acids.

One further important issue was to determine how the RNA from the ribosome fraction was related to the ribosomal protein. No clear distinction between the regular protein constituents of the ribosomes and the newly synthesized other cell proteins had yet been made. "In our experiments and in those of Gale & Folkes (1953) ribonuclease destroys the incorporation ability (Today we know that RNAase mainly hydrolyzes mRNA, a process that at the time had not yet been identified.); this is a suggestive, but by no means conclusive, evidence of a relationship." With this sentence Zamecnik & Keller (1954) ended their article. Contrary to later claims, no questions like "Does RNA originate from DNA?", or "Does RNA carry genetic information?" were then raised – at least not as seriously as to deserve a written formulation. Quite to the contrary, it seems astonishing today that no reference to DNA was made as its being the ultimate source of information for determining the specificity of amino acid sequences in polypeptides. Ten years after the "one gene – one enzyme" hypothesis, and one year after Watson and Crick, the idea of DNA as a carrier of genetic information had not settled yet in the biochemists' minds. At the time, what was regarded as crucial was mainly the issue: where does the energy for assembling peptides come from?

That question had been at least partially answered by then. Still, a score of old and new doubts kept afflicting the biochemists' thoughts, such as for instance: did small peptide subunits form first, to be then assembled into larger chains? Loftfield et al. (1953) had shown that free amino acids were directly incorporated into proteins; the inference was that, apparently, no go-between products existed. Alas, this had not yet been unambiguously proven. And still unsettled was the point of whether amino acids already incorporated into a peptide were exchangeable against free ones newly added to the system. No way, was the answer. However, the opinion that the amino acids of a protein were in a dynamic equilibrium with free amino acids was still widespread. To accept the concept that a protein did not continuously exchange its components with its surrounding environment seemed to stand as a preposterous paradox, regarding the principles of metabolism directing all of life's events. Besides that, all those experiments performed on living animals at Columbia University for many years by Rudolf Schoenheimer, a highly respected physiologist, emigrant from Nazi-Germany, and a pioneer of heavy isotope labeling ($^2$H and $^{15}$N), did not disentangle that issue any further (see, for instance, Schoenheimer, 1942).

And one other point came up, deriving from the solution to the last one: does the synthesis of a peptide start at the amino- or at the carboxy-terminal? A definitive answer to this question would not come soon, but the time sequence involved in the synthesis of a protein molecule, based on the example of hemoglobin, indicated clearly that the assembly of proteins was initiated at the amino-terminal, proceeding towards the carboxy-terminal (Dintzis, 1961; see Excursus 9-1).

Meanwhile, after years of trial and error with cell-free extracts – unfractionated or else separated in 100,000 x g supernatant and microsome fraction –, a significant breakthrough had been achieved. Hoagland et al. (1958), in Zamecnik's group, detected in the 100,000 x g supernatant of the cell-free homogenate from liver tissue a novel and important component of the protein synthesis apparatus: a fraction of small RNA molecules, covalently bound to the activated amino acids.

Prior to that, some disappointing observations had been made, as follows: first, it did not suffice, for protein synthesis to occur, to bring together $^{14}$C-amino acid, its cognate purified activating enzyme, ATP, and washed ribosomes; second, a labeled amino acid in the cell-free extract, if already activated, could not be displaced on its way to the protein by adding an excess of unlabeled amino acids. Obviously, something crucial was missing.

At this point, it came to Hoagland and Zamecnik's attention that Marianne Grunberg-Manago, working with Ochoa's group in New York, had detected, in a cell-free homogenate, the formation of radioactive polyadenylate (an unusual type of RNA with adenine as its only base) parting from $^{14}$C-ATP. (Actually, ATP was not the decisive substance, but adenosine diphosphate, ADP, which was present in the ATP preparation as an impurity; see Chapter 10.) Hoagland and Zamecnik reasoned that their cell-free homogenate, so nicely suitable for protein synthesis, should actually also be able to synthesize RNA, perhaps even of the regular type, instead of this odd, probably meaningless, poly-A. ...And so they added $^{14}$C-ATP to the cell-free liver extract, and isolated the RNA after a short incubation time. This RNA was radioactive; ATP was indeed being incorporated into RNA. The result was interesting enough for the experiment to deserve being repeated, this time with a control. It had to be checked whether the washing procedure for removing the free, not polymerized ATP was reliable. Accordingly, a parallel experiment was carried out to settle this point, using $^{14}$C-leucine instead of $^{14}$C-ATP. Since $^{14}$C-leucine is not incorporated into RNA, a corresponding sample should contain no radioactivity if the washing procedure was efficient. However, it did – at an incredibly high level. This radioactivity remained linked to the RNA even after extensive washing; leucine, consequently, had to be covalently linked to RNA! But to which RNA? The covalent binding of leucine to RNA occurred in the supernatant, i.e., in the absence of ribosomes. The implication of these results was that RNA was present not solely in ribosomes. It was known that roughly 10-15% of all RNA was to be found in the fraction not sedimenting with 100,000 x g, but this material had been considered as some sort of unsedimented left-over of the main RNA fraction. It was now clear that it was a totally novel RNA category, a relatively small molecule – as shown soon afterwards, it encompassed merely 75-90 nucleotides – whose possible function remained involved in mystery. This "soluble" RNA (sRNA), could very well be endowed with an active function in the process of protein synthesis, since, as previously seen, that synthesis did not proceed in a system containing pure activating enzyme, plus amino acids, plus ATP, plus ribosomes; indeed, something was still missing! On the other hand, it could also very well be that the RNA-bound amino acid was just provisionally attached to this singular sRNA, as a sort of storage step, in the wait of its future definitive use...

Anyway, trying to clarify this issue would be certainly worthwhile. Soon afterwards, kinetic analyses with radioactive amino acyl-sRNAs showed that they were directly involved in the process of protein synthesis: the radioactivity of these amino acyl-sRNAs, derived from the bound $^{14}$C-amino acids, decreased at the same rate as the radioactivity of freshly synthesized proteins increased (Fig. 9.1). The path from free amino acid to protein went through its activation by ATP and – further – through its covalent binding to sRNA! The latter transferred the amino acid to the ribosome, and thus it received the new denomination of transfer RNA, tRNA.

The next logical step to be pursued was to establish the exact role played by tRNA. Soon, the existence of at least 20 different tRNAs, one for each amino acid, was inferred, because if the supernatant's binding capacity was exhausted for a certain amino acid, another amino acid, subsequently added, would keep binding to existing tRNAs at an unchanged rate. Actually, this observation had already been made by Hoagland et al. (1958). Alas, it was neither discussed any further nor accorded any significant meaning. The real weight of this observation was only appreciated much later (Zamecnik et al., 1960), after two other groups (Preiss et al., 1960; Lipmann et al., 1959) had discovered that for each of these couples of unequal partners, the amino acid and "its" tRNA, a specific enzyme existed in the supernatant, responsible for coupling them together into amino acyl-tRNA.

In the meantime, Crick had heard of these experiments and realized that the tRNAs were no more and no less than the long dreamed of adaptors – indeed somewhat larger than the tri- or oligo-nucleotides hypothesized by him, but anyway... He first believed (Crick, 1958) that a molecule as large as tRNA would not be able to diffuse quickly enough towards the ribosomes – but, after a while he conceded that perhaps it could... Hoagland and Zamecnik finally were convinced that the concept of the adaptor function for their tRNA (Hoagland et al., 1959) was the most adequate. In any case, the corresponding terminology was then adopted – but to which extent this fact related to their awareness that the transfer of genetic information from nucleic acids to proteins was of central importance, is difficult to assess (see, for instance, Hoagland, 1959).

What Hoagland and Zamecnik quickly realized, however, is the fact that the new biochemistry, and the refreshing genetic perspectives opened up by it – at first so unfamiliar to them –, made an intense use of microorganisms rather than utilizing liver tissue from a freshly killed ("sacrificed", as scientists would rather say) rat.

Thus, a postgraduate student in their group, M. R. Lamborg, dedicated himself for three years to the task of grinding coli cells with alumina (aluminum oxide powder) in a mortar and testing the so obtained coli extract for its ability to incorporate radioactively labeled amino acids. The main challenge was to keep the proportion of intact cells (whose rate of synthesis was many times higher than after the destruction of their cell structure) so low that it could be considered as negligible. When this critical point was finally attained, Lamborg's laboratory neighbor, Jim Watson, who had made acquaintance with Lamborg's preparatory work before his results were published, recognized it as extremely interesting, and quickly improved the system with the help of A. Tissières, and even more quickly published the method (Tissières & Watson, 1958). This hasty action granted them the priority, before Lamborg & Zamecnick (1960) had time to react.

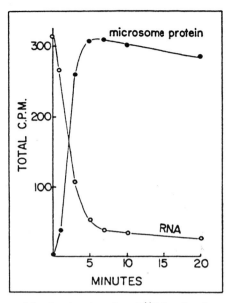

Figure 9.1: Time sequence of the *in vitro* transfer of [14]C-leucine from tRNA to the polypeptide on the ribosome (Hoagland et al., 1958). [14]C-leucine was bound to RNA through the action of ATP and the ribosome-free material present in the supernatant from the ultracentrifugation of a homogenate of liver cells. A ribosome suspension was then added to these activated amino acids (together with GTP and ATP); this mixture was incubated at 37 °C. Samples were withdrawn at different times and their ribosomes centrifuged down. The radioactivity (C.P.M.) present in the RNA fraction of the supernatant and in the ribosomes was assessed. Obviously, originating with the charged tRNA, a direct incorporation of amino acids into the growing peptide occurred on the ribosome. [C.P.M., counts per minute, is a commonly used measurement of radioactivity, corresponding directly to the marking of the counter (in early works the Geiger counter; at present, scintillation counters) within a minute; depending on the sensitivity of the counter and the type of sample, 1 CPM may signify 1,5 to 5 DPM, atomic disintegrations per minute.]

Nevertheless, neither Tissières & Watson had worked out the perfect method for performing the ideal experiments for protein synthesis. For one, the system was not stable enough – the crucial ability of incorporating radioactive amino acids into acid-precipitable material decreased quickly, and aside from that, it could not be kept stably frozen. This latter problem was an especially annoying disadvantage, because not only was it extremely arduous to obtain cell-free extracts anew for every experiment, but there was also an impairment of the reproducibility and comparability of the results if somewhat different extracts had to be used for repeated experiments.

This challenge was confronted and overcome by another pair of young scientists working at the National Institutes of Health in Bethesda, Maryland. By playing around with other details, they established a method leading to a cell-free homogenate of *E. coli* which was stable, which could be durably stored after freezing and in which the incorporation of radioactive amino acids into acid-precipitable material was possible. This amounted to the theme of their publication: Matthaei & Nirenberg (1961). If they only could have suspected the avalanche they would trigger with it...

**Excursus 9-1**
**POLYPEPTIDE SYNTHESIS PROCEEDS FROM THE AMINO- TO THE CARBOXY-TERMINAL**

Reticulocytes, immature red blood cells, were for decades – and still are – one of the preferred starting materials for investigating the details and the different aspects of protein synthesis. Reticulocytes are obtained from mammalian bone marrow (most commonly from rabbits), being highly praised because of the straightforward advantage in synthesizing almost exclusively one single protein, hemoglobin, the red blood pigment. The first main question directed to this system concerned the mode of synthesis of proteins: does a nascent polypeptide grow linearly from one end? And if the answer was affirmative, from which end, the amino- or the carboxy-terminal? Or, perhaps, did it grow from diverse internal points of the future polypeptide in all directions? Did small peptides form first, to be then, in a second step, joined together?

Howard Dintzis and his co-worker at the Massachusetts Institute of Technology, M. A. Naughton, added the radioactive amino acid leucine to a reticulocyte suspension and halted the protein synthesis after a few seconds. After prying the cells open, all ribosomes, some bound to still unfinished proteins (see Chapter 10), were harvested by centrifugation; the supernatant contained the proteins which were synthesized to completion during the radioactive pulse. These latter were digested with trypsin and analyzed by the method of fingerprinting (see Excursus 8-1) (Naughton & Dintzis, 1962). In the case of a very short pulse, only the amino acids of the carboxy-terminal were radioactive; the longer the pulse duration, the more peptides with radioactive amino acids in the direction of the amino-terminal were found. [The peptide sequence of the hemoglobin molecule was inferred from previous work (see, for instance, Braunitzer et al., 1961).] These results implied that the amino-terminal was synthesized first; the carboxy-terminal, consequently, was the last to be added (Fig. 9.2). Canfield & Anfinsen (1963) chose lysozyme, an enzyme synthesized in chicken's ovarian tubes, to perform similar experiments, which led to a more precise documentation of the studied features. For instance, the appearance of radioactivity in different sections of the polypeptide was then correlated to its amino acid sequence.

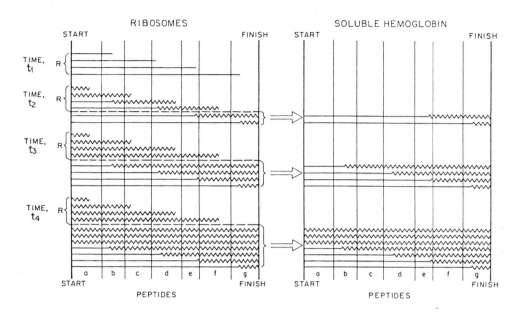

Figure 9.2: Scheme of the sequential synthesis of polypeptides (based on the example of hemoglobin; Dintzis, 1961; see also Naughton & Dintzis, 1962). If $^{14}$C-amino acids are given to reticulocytes, the radioactivity accumulates first in the sequences nearest to the carboxy-terminal, permitting the inference that the synthesis had started on the other end, the amino-terminal (smooth lines: non-radioactive region; zigzag lines: radioactive region; R: incomplete peptide, still bound to the ribosome).

# CHAPTER 10

## THE CODE *IN VITRO*

The cell-free *E. coli* system worked out by Matthaei & Nirenberg (1961) improved markedly the incorporation of radioactive amino acids, as compared to the method with liver cells. Cell extracts could be obtained in large, workable amounts and freezed in smaller portions, so that comparable conditions could be granted for many experiments. The system was so convenient that it incited even further optimization, as it was still far from ideal; a real, biochemically measurable net synthesis of protein mass, for instance, was not attainable. The rate of protein synthesis was minimal. There were only traces of newly synthesized proteins and these could only be detected through their radioactivity. A significant improvement consisted in pre-incubating the cell homogenate with DNAase, a measure which brought the inherent background incorporation of the system to an almost complete halt after roughly 20 min. Thus, the effect of specific substances added to the system could be more clearly detected.

Meanwhile, it had been established (see Hoagland et al., 1958) that two types of RNAs existed in the cell: the one present in the ribosomes, and the soluble one. The amount of soluble RNA was still to be optimized. If a pre-incubated coli homogenate was supplemented with additional sRNA, an approximately three-fold increase in the incorporation rate resulted. This initial boost was followed by saturation, not to be overcome by adding more sRNA. Consequently, extra sRNA was, from then on, routinely supplied to the system.

Still further improvements were attempted. In that quest, RNA from some other sources was used to substitute for soluble RNA. And the RNA easiest to get was obviously that from ribosomes, amounting to 80% of total cellular RNA. Ribosomal RNA, separated from its proteins – biologically nonsensical – had simply the advantage of ready availability. This RNA had a further, clear, although relatively modest, positive effect on the incorporation of $^{14}$C-valine. And, if instead of coli rRNA one added rRNA extracted from yeasts, an even more marked improvement was observed. The efficiency of the system could be still further improved by using tobacco mosaic virus RNA (TMV-RNA).

Apparently, any free RNA, quite independently from its origin, was able to stimulate *in vitro* protein synthesis. Hoagland & Zamecnik were not blessed with this insight, despite all those many years of toiling with their system. The merit for conceiving the idea of adding some extra – source irrelevant – RNA to their promising system belongs solely to Nirenberg & Matthaei. And they came to it through uncertainty and curiosity rather than by the inspiration of a concrete concept and the consequent necessity of its experimental validation – at least, in hindsight, such is the prevailing impression. Notwithstanding, a further type of polynucleotide was on the waiting-list to be tested, a polyuridylic acid, a synthetic RNA composed solely of uracil, which repeated itself indefinitely from nucleotide to nucleotide. Nirenberg and Matthaei's colleagues from the same institute, Leon Heppel and Maxine Singer, were the helpful suppliers of the strange RNA.

This poly-U, which had been produced by Heppel's group in the course of their studies on RNA synthesis, was actually supposed to be used by Nirenberg & Matthaei as a mere control for their experiments – at least that is the emphatic testimony of some who closely followed the developments at the time. Certainly, such polyanion, though similar to RNA but deprived of any conceivable physiological function, could not wield any influence on the complex machinery of protein biosynthesis. Matthaei was the one who recognized its unexpected, sensational effect when poly-U was added to the cell homogenate: $^{14}$C-phenylalanine became incorporated into acid-precipitable material with a hitherto unseen high efficiency; 17 other amino acids tested were not affected at all! Surely, poly-U was the causative agent of the highly specific polymerization of phenylalanine, yielding a repetitious, biologically meaningless polypeptide, poly-phenylalanine – as meaningless and repetitious as the polynucleotide poly-U itself. Nirenberg & Matthaei (1961) described this effect as "remarkable". Template RNA, already perceived as essential by Hoagland & Zamecnik, was hitherto vaguely defined and still equated with ribosomal RNA. It was supposed to direct in a still unidentified way the specific assemblage of proteins on the ribosomes; Nirenberg & Matthaei's unexpected, "remarkable" results initiated the clarification of that role. The fact that poly-U was no constituent of the ribosomes led to the suspicion that perhaps the template arrived on the ribosomes, coming from elsewhere outside... But, those thoughts were indeed mere suppositions, utter suspicions... The term "messenger" had already reached Matthaei & Nirenberg (1961) by word of mouth. It was reasonable to assume that their artificial template had something to do with it: anyway, it was an RNA that somehow was promoting the assemblage of proteins... Whose work could be linked to it, so that it could be accordingly cited? Certainly those people doing research in the field of RNA synthesis: Hurwitz et al. (1960), Stevens (1960), Weiss & Nakamoto (1961a); and for the sentence "the template for protein synthesis is called messenger RNA", Matthaei & Nirenberg invoked Volkin, Astrachan & Countryman (1958), Nomura et al. (1960), Hall & Spiegelman (1961). None of all these scientists were, however, familiar with the term "messenger" ...

Those who were better informed – the people from Paris, Cambridge, U.K., Caltech, MIT or Harvard (see Chapter 14) – were unknown to Nirenberg and Matthaei.

And how about a certain code, supposedly worked upon by theoreticians, and about which one had already heard something? One could dare to theorize a code word for phenylalanine, inferred from the experimental results with poly-U; it should contain exclusively uracil! At least that much deserved written documentation by Nirenberg and Matthaei (1961), and an oral presentation at the International Biochemical Congress in Moscow in 1961. [Zamecnik (1979), in his reminiscences, misplaced absentmindedly this congress to Stockholm. ...perhaps, after all that time, his interest in that matter had slackened.] Nirenberg's presentation was meant for a small public, being accordingly scheduled as a secondary event; it would have remained utterly unnoticed if not for Matthew Meselson's attention. Meselson's curiosity was honed by the experiments designed to prove experimentally the existence of messenger RNA (see Chapter 14), carried out in his own laboratory at Caltech by Brenner & Jacob. He thus recognized immediately that these "remarkable" experiments with poly-U as template were actually absolutely sensational. Meselson passed the word to Crick, who was present as congress co-organiser. Crick went out of himself with excitement and arranged immediately for Nirenberg to repeat his presentation for a larger public. In his reminiscences, Crick defined the public as "electrified"; most probably he had unconsciously projected his own emotions onto the other congress participants; Benzer, documenting this situation later, produced a photo from this very same audience, most participants seeming to be asleep (Crick, 1988).

Figure 10.1: Marshall Nirenberg (born in 1927).

The unexpectedly enthusiastic reaction of Crick certainly wielded an important influence on Nirenberg & Matthaei's resolve to pursue the novel perspective: perhaps synthetic templates would open the way to cracking the genetic code. Experiments after experiments followed... and not solely in Nirenberg's laboratory...

Marianne Grunberg-Manago, a postdoc from Paris studying in New York, had found – as so often in science, by pure chance, since she was looking for something else – an enzyme able to synthesize RNA, starting from nucleoside-diphosphates, liberating inorganic phosphate in the process. This enzyme was certainly the one responsible for RNA synthesis in the cell, or so she affirmed (Grunberg-Manago et al., 1955); and even before new evidence established that this was actually not the case, her boss, the Spaniard Severo Ochoa, had already been bestowed with the Nobel Prize in 1959. Actually, from the beginning one should have been suspicious of this enzyme; this polynucleotide phosphorylase polymerized nucleotides in an utterly haphazard fashion, originating random sequences, without the prerequisite of a template (see Chapter 16). (By the way, this enzyme's physiological function is still not fully understood.) This was the reason why Ochoa's laboratory was filled with all imaginable synthetic RNAs, all synthesized by the mysterious polynucleotide phosphorylase. And then, like a bolt from the blue, it was realized that by means of these shelved synthetic molecules of RNA one could approach the challenge of the code – did not Nirenberg knit a poly-phe on the ribosome with the help of poly-U? Ochoa's group was able to repeat and confirm Nirenberg & Matthaei's experiment in one single day (see Lengyel, 1976). And then an avalanche followed: Ochoa's group had not only poly-U, but poly-U with some C, poly-C with some U, or A, or G, and so on... But Nirenberg & Matthaei had Heppel upstairs, who provided them with similar compounds... A hectic phase ensued: scrutinizing all these artificial RNAs, composed of random base sequences, as templates in Nirenberg's coli homogenates. When the raised dust finally settled, it was clear that:

The codon (or better, one of the possible codons) for phenylalanine was made up exclusively of Us, and corresponded – assuming the three-base code suggested by the experiments performed by Crick et al. (1961) – to the triplet UUU; the codon for serine had, besides U, also a C; the codon for tyrosine had U and A; methionine was codified by U, G, and A, etc.: a huge series of data... [for details see Matthaei et al. (1962); Speyer et al. (1962, 1963); Nirenberg et al. (1963); Crick (1963)].

But it was more than a mere detail that the sequence of bases within a codon could not be ascertained. The method utilizing random sequences of bases allowed exclusively the determination of what bases constituted the codon, but not their precise sequence in it.

Nirenberg's laboratory, however, had already maneuvered itself out of this impasse: the same Leon Heppel had developed a chemical method for synthesizing exactly defined trinucleotides. With that, Nirenberg and Matthaei achieved the feat of assembling a collection of all 64 possible trinucleotides, which they then employed in their protein-synthesizing system, containing: ribosomes, ATP, and all activated amino acids; in each sample, a different one was radioactively labeled. Noteworthy was the fact that the code words for the amino acids, essential for protein synthesis, were basically clarified, without any protein synthesis at all occurring, without one single peptide bond being formed. The ribosomes were able (an ability that could not be known beforehand) to bind the trinucleotides; and one of Crick's adaptors, i.e., a tRNA, charged with its corresponding amino acid, could then bind to its cognate ribosome-trinucleotide complex. Apparently, the tRNA presented the so-called anticodon, the complementary counterpart of the codon made up of the trinucleotide. All these processes normally preceded the actual peptide synthesis which, in this case, was unable to proceed because the template consisted exclusively of one single codon, the one being tested.

The whole complex consisting of one ribosome, one trinucleotide, and one molecule of its cognate tRNA charged with its corresponding $^{14}$C-amino acid, had to be reliably separated, using a simple method, from the excess unbound radioactivity. Nirenberg's group worked out an excellent, and simple way of achieving this: passing the assay solution through nitrocellulose filters with pores of 450 nm of diameter. The pores were actually large enough to allow the ribosomes through, but these were retained electrostatically, whereas the unbound charged tRNAs passed unhindered, being washed away in the process.

The first results from the binding experiments were published immediately (Leder & Nirenberg, 1964; Nirenberg & Leder, 1964).

Merely some few doubts and inconsistencies remained to be set straight (Fig. 10.2).

But then, a more elegant and convincing variation of the method with synthetic templates was developed.

Gobind Khorana, an American citizen of Indian origin, organic chemist at the University of Wisconsin, had kept a large group of co-workers busy producing synthetic DNA and RNA molecules for a long time. As an organic chemist, his original intent was to unearth increasingly complex synthesizing procedures leading to ever more complicated organic molecules (see for instance, Gilham & Khorana, 1958; Tener et al., 1958). His synthetic RNAs were not monotonous and also not mere random sequences of bases; they were short repeating sequences of bases. To obtain them, di-, tri-, or tetradeoxynucleotides with given base sequences were first synthesized, employing organic chemical methods. In a second step, these short blocks of deoxynucleotides were linked together, producing longer molecules, chemically similar to DNA, but with repeating sequences; these molecules could be further elongated (with the help of the enzyme DNA polymerase – see Chapter 17)

Figure 10.2 (right): Facsimile of an excerpt from the Proc. Natl. Acad. Sci. USA 53, 1161-1168 (1965).

*RNA CODEWORDS AND PROTEIN SYNTHESIS, VII.*
*ON THE GENERAL NATURE OF THE RNA CODE*

By M. Nirenberg, P. Leder, M. Bernfield, R. Brimacombe,
J. Trupin,* F. Rottman†, and C. O'Neal

NATIONAL HEART INSTITUTE, NATIONAL INSTITUTES OF HEALTH, BETHESDA, MARYLAND

*Communicated by Robert J. Huebner, March 26, 1965*

Nucleotide sequences of RNA codons have been investigated recently by directing the binding of $C^{14}$-AA-sRNA to ribosomes with trinucleotides of defined base sequence. The template activities of 19 trinucleotides‡ have been described and nucleotide sequences have been suggested for RNA codons corresponding to 10 amino acids.[1–5] In this report, the template activities of 26 additional trinucleotides are described and are related to the general nature of the RNA code.

*Materials and Methods.*—*Components of reactions:* E. coli W3100 ribosomes and sRNA were prepared by modifications of methods described previously.[6–8] Each $C^{14}$-aminoacyl-sRNA was

- - - - -

*The General Nature of the Code.*—Thus far, the template functions of 45 of the 64 trinucleotide sequences have been investigated in this system. A summary of the data and additional codon sequences which can be predicted from amino acid replacement data reported for E. coli[14] and TMV mutants,[19, 20] are shown in Table 4. Almost all of our earlier predictions were confirmed when the appropriate trinucleotide was tested.[2–5] Nevertheless, the summary shown in Table 4 should not be thought of as an invariant codon dictionary, since it is clear that codon recognition can be modified.

Previous studies with randomly ordered polynucleotides and cell-free protein synthesizing systems showed that synonym codons often differ in composition by only one base.[22, 23] This suggested that bases common to synonym codons occupy identical positions and, that either 2 out of 3 bases in a triplet sometimes may be recognized, or a base may be recognized correctly in 2 or more ways.[21, 24] On the

TABLE 4

NUCLEOTIDE SEQUENCES OF RNA CODONS

| | | | | | | | |
|---|---|---|---|---|---|---|---|
| UpUpU UpUpC | Phe | UpCpU UpCpC | Ser | UpGpU UpGpC | Cys | UpApU UpApC | Tyr |
| UpUpA UpUpG | Leu | UpCpA UpCpG | Ser | UpGpA UpGpG | Nonsense* or Trypt | UpApA UpApG | Nonsense† |
| CpUpU CpUpC | Leu or Nonsense* | CpCpU CpCpC | Pro | CpGpU CpGpC | Arg | CpApU CpApC | His |
| CpUpA CpUpG | Leu | CpCpA CpCpG | Pro | CpGpA CpGpG | Arg | CpApA CpApG | Glu-NH₂ |
| ApUpU ApUpC | Ileu | ApCpU ApCpC | Thr | ApGpU ApGpC | Ser | ApApU ApApC | Asp-NH₂ |
| ApUpA ApUpG | Met | ApCpA ApCpG | Thr | ApGpA ApGpG | Arg. or Nonsense* | ApApA ApApG | Lys |
| GpUpU GpUpC | Val | GpCpU GpCpC | Ala | GpGpU GpGpC | Gly | GpApU GpApC | Asp |
| GpUpA GpUpG | Val | GpCpA GpCpG | Ala | GpGpA GpGpG | Gly | GpApA GpApG | Glu |

* It is possible that these sequences are readable internal-, but nonreadable terminal-, codons.
† UpApA and UpApG may correspond to Terminator-, or Ser-codons in different strains of E. coli (see text or refs. 17 and 18).
*Summary and predictions:* The template activities of trinucleotides in BOLDFACE have been studied experimentally in this system. Other sequences are predicted. Although trinucleotides are arranged in pairs, one member of a pair may have greater template activity than the other. Estimates of relative template efficiencies are not indicated.
Amino acid replacement data used for these predictions were obtained with E. coli by Yanofsky,[14] or were induced by HNO₂ in TMV by Wittman and Wittman-Liebold[19] or Tsugita.[20]

Figure 10.3: Gobind Khorana (born in 1922).

and finally, again enzymatically, by means of the RNA polymerase (see Chapter 16), corresponding RNA copies could be produced.

When such RNAs were added to the cell-free system, very special polypeptides appeared, depending on the polymer used: [UC]n $\rightarrow$ 5'...UCUCUCUCUCUC...3', for instance, promoted the assemblage of polypeptides containing only serine and leucine. The trinucleotide binding tests performed by Nirenberg and Leder had already shown that linking seryl-tRNA or leucyl-tRNA to ribosomes was only achieved if, respectively, UCU, or CUC was present. Assuming a triplet code, one could then infer that a peptide with the alternating sequence ...ser-leu-ser-leu-ser... had been assembled.

If poly-UAC was employed, i.e., 5' ...UACUACUAC...3', then from all amino acids offered to the system, only three (tyrosine, threonine and leucine) were bound in polypeptides, which meant – again assuming a triplet code – that the formation of three different, repetitious polypeptides had been accomplished:

poly-tyrosine (if the reading-frame UAC was used),
poly-threonine (reading-frame: ACU), and
poly-leucine (reading-frame: CUA).

The synthesis of polypeptides based on RNA templates gained from the polymerization of tetranucleotides (see Fig. 10.4) was specially informative. Besides promoting the incorporation of 4 amino acids in repeating sequences, it evidenced that the presence of the triplets UAG or UAA was incompatible with the formation of long peptides (Khorana et al., 1966). These codons were of no use for promoting the

TABLE 7. AMINO ACID INCORPORATIONS STIMULATED BY MESSENGERS CONTAINING REPEATING NUCLEOTIDE SEQUENCES

(System, *E. coli* B)

| Messenger | Amino acids incorporated | Messenger | Amino acids incorporated |
|---|---|---|---|
| *Repeating Dinucleotides* | | Poly GUA | val, ser |
| Poly UC | ser-leu | Poly UAC | tyr, thr, leu |
| Poly AG | arg-glu | Poly AUC | ileu, ser, his |
| Poly UG | val-cys | Poly GAU | met, asp |
| Poly AC | thr-his | *Repeating Tetranucleotides* | |
| *Repeating Trinucleotides* | | Poly UAUC | tyr, leu, ileu, ser |
| Poly UUC | phe, ser, leu | Poly GAUA | *none* |
| Poly AAG | lys, glu, arg | Poly UUAC | leu, thr, tyr |
| Poly UUG | cys, leu, val | Poly GUAA | *none* |
| Poly CAA | gln, thr, asn*? | | |

\* The expected incorporation of asparagine has not been realized so far because of the presence of a powerful enzyme which deaminates asparagine in the amino acid incorporating system (cf. Schwartz, 1965).

CHART 2. CELL-FREE POLYPEPTIDE SYNTHESIS USING POLYMERS CONTAINING REPEATING TETRANUCLEOTIDE SEQUENCES

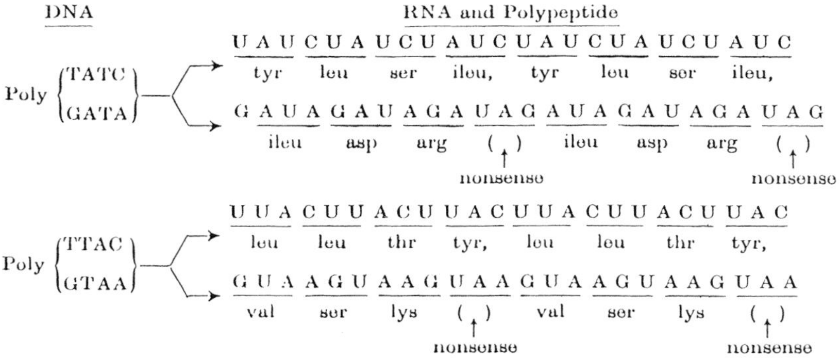

Figure 10.4: An example of codon identification by Khorana's method, using repetitive synthetic templates (Khorana et al. 1966). The symbols

$$\text{Poly} \left\{ \begin{array}{l} \text{TATC} \\ \text{GATA} \end{array} \right\} \quad \text{and} \quad \text{Poly} \left\{ \begin{array}{l} \text{TTAC} \\ \text{GTAA} \end{array} \right\}$$

correspond to the chemically and enzymatically produced double strands of DNA represented, respectively, by

5' ...TATCTATC...3'     and     5'...TTACTTAC...3'
3' ...ATAGATAG...5'             3' AATGAATG...5'

These DNA double strands, after being separated, were transcribed into RNA counterparts by RNA polymerase (see Chapter 16). The individual synthesis of each of the complementary RNA strands was achieved from the same DNA, by selecting the nucleotide building blocks to be offered to the system: A, U, and C, or A, U, and G.

growth of the peptide chain: they represented no amino acid. Accordingly, they were designated nonsense codons. Soon afterwards it became clear that these condons had indeed a sense, and quite an essential one for that matter. UAG, UGA, and UAA turned out to be the codons responsible for bringing polypeptide synthesis to a halt; they are now called terminator codons.

All codons had been indisputably assigned (Fig. 10.5)! M. Nirenberg and G. Khorana were bestowed with the honor of the Nobel Prize in 1968 for their contribution to this monumental feat.

Figure 10.5: The code "sun" : the codons are to be read from the inside (5') to the outside (3'); they symbolize the base sequence of mRNA codons, which correspond to the amino acids depicted around the circular scheme. [Scheme devised (here slightly altered) by C. Bresch – see Bresch & Hausmann, 1972.] "amber", "ochre", and "opal" are the nonsense, or terminator codons. Their designations derive from situations involving Anglo-Saxon humor. The amino acids accorded 6 codons (consequently shown twice), are marked by an asterisk. Dark triangles indicate the start codons, which at the beginning of the polypeptide synthesis bring about the incorporation of methionine, but otherwise codify methionine or valine.

Khorana's method, so elegant as it was – and, in contrast to the triplet-binding method, it actually promoted polypeptide synthesis –, nevertheless, gave occasion to fresh doubts. It was inconceivable that this chaotic system truly mirrored the events going on in the living cell. The experiments with the frameshift mutants performed by Crick, Barnett, Brenner & Watts-Tobin (1961) (Chapter 8) rapidly fed the notion that the genetic information started to be read from a fixed point. Since this notion also made sense from a biological point of view, it easily turned into a conviction. Alas, Khorana's ribosomes sprang indiscriminately onto any site of the templates, reading in any of the three potential frames. Of course, *in vivo*, merely one of them could be meaningful.

Did the cell possess a mechanism meant to curb this blind attraction between the RNA and the ribosomes, that appeared in the *in vitro* system? Could it be that the binding of ribosomes to RNA and the consequent initiation of protein synthesis was only allowed to occur at special, meaningful sites – the initiation sites?

The existence of such *in vivo* initiation sites was suspected not only by geneticists like Crick or Jacob & Monod (see Chapter 15), but also by biochemists. For instance, Waller (1963) noticed that, of all coli proteins, a disproportionately large tally had methionine as its initial amino acid. Was this observation to be related to a special role for methionine? (see Excursus 10-1).

The third type of RNA, identified in the *in vitro* system as template RNA, alongside tRNA and rRNA, and initially barely noticed by the biochemists, had been, for sometime, ghostly haunting a few *in vivo* experiments. Its initially shadowy appearance increasingly gained sharpness; paralleling the biochemical analyses, crucial observations were being made on living cells. A concrete profile was finally attained: that of messenger RNA, mRNA, the crucial carrier of the genetic information, migrating from the DNA to the ribosomes, the site of protein synthesis. This development will be covered in subsequent chapters.

Before going on, though, let us consider the melting together of the terms template RNA, as biochemically characterized in the *in vitro* systems, and messenger RNA, as derived from the *in vivo* experiments. This unification actually signaled the closing of the philosophical and academic rift till then existing between biochemistry and molecular genetics. It marked the turning point which caused biochemists to start considering molecular genetics, and molecular geneticists to stop fearing the embrace of the new biochemistry. This novel direction was conspicuously expressed at the Cold Spring Harbor meeting of 1963. There, leading biochemists and molecular biologists presented their contributions to the subject of the synthesis and structure of macromolecules, listened to each other patiently, and for five days, from the 5th to the 10th of June, debated their concerns with each other.

The 1963 Cold Spring Harbor meeting was attended by some 350 participants; among them were Mahlon Hoagland and Paul Zamecnik – but only as part of the audience. They were the ones who had launched the study of *in vitro* protein synthesis, who had described the tRNA and recognized its role in the system, and

who had demonstrated unmistakably and accurately that the ribosomes were the seat
of protein synthesis. Alas, all that had been relegated to the archives of history.
Crucial at the moment was exactly what they had overlooked: mRNA!

How did mRNA betray its secret existence? How and where was it built up?
What was its role and its fate in cell metabolism? The crude machinery of protein
synthesis had been unravelled, but not yet the way to control it. Would it be possible
to put it to work meaningfully? to accelerate it? to slow it down? to regulate it?

**Excursus 10-1**
**HOW DOES POLYPEPTIDE SYNTHESIS OCCUR ON THE RIBOSOME?**

Kjeld Marcker, working in Sanger's laboratory in Cambridge, U.K., made a chance
discovery of a tRNA charged with N-formyl-methionine, in *E. coli* (Marcker &
Sanger, 1964). [Referring to this event, Sanger declared later that it had been the
most successful digression from his actual work (Sanger, 1988).] It was then
postulated that this compound was the one responsible for the very first step in the
synthesis of a polypeptide, since the blocked amino group was prevented form
forming a peptide bond with a carboxy group; consequently, N-formyl-methionine
(Fig. 10.6) could only be conceived as being the first link in the peptide chain.
Marcker (1965) could show that the formylation of methionine took place only after it
became linked to a particular, methionine-specific, tRNA, the initiator-tRNA (besides

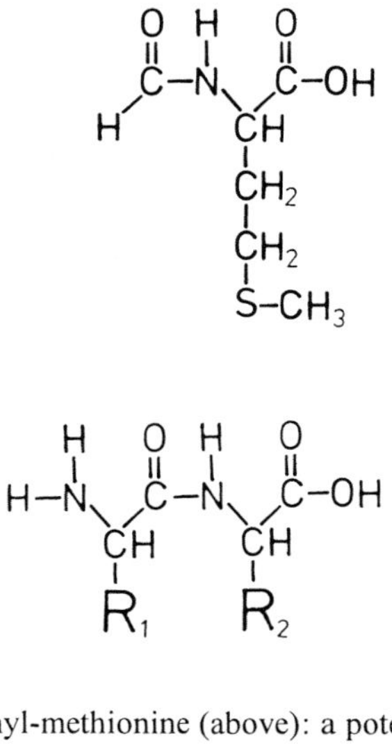

Figure 10.6: N-formyl-methionine (above): a potential analogue of a peptide.

this one, there is a "regular" met-tRNA), and that formyl-methionine was indeed incorporated exclusively at the N-terminal (that means, at the beginning) of a peptide (Clark & Marcker, 1966). (Formyl-methionine is often released from the peptide chain shortly after having been incorporated; if it were not so, all peptide sequences would necessarily start with met.)

It was to be expected that, in order for a peptide bond to be formed, two structures had to be located on the ribosome: either two amino acyl-tRNAs (for establishing the first link) or a peptidyl-tRNA followed by the charged tRNA carrying the next amino acid to be coupled with the nascent polypeptide (Bretscher, 1966). This concept required the existence of at least two sites on the ribosome capable of binding the charged tRNA molecules (see Fig. 10.7): one for the peptidyl-tRNA (P-site) and a second, receiving the newly arriving amino acyl-tRNA (A-site). The initial step of the process had to be, inevitably, the positioning of the first charged amino acyl-tRNA directly on the P-site. Was this privilege granted to the formyl-methionyl-tRNA? (The formylated amino group resembled structurally a peptide bond – see Fig. 10.6.) A straightforward and elegant experiment, conceived by Mark Bretscher, also in Cambridge, confirmed that assumption. The antibiotic puromycin blocks protein synthesis; this happens because, as an analogue of a charged tRNA, puromycin is able to react with a peptidyl-tRNA placed on the ribosome without, however, being able to go into a further peptide bond (Fig. 10.8). What Bretscher demonstrated was that among all charged tRNAs, solely N-formyl-metionyl-tRNA reacted with puromycin, an observation indicating that, on the ribosome, this particular tRNA was located where peptidyl-tRNAs were to be found: at the P-site.

Alas, the codon AUG could not be the complete signal responsible for the initiation of synthesis of a polypeptide, since many other methionines were to be found dispersed within its primary structure. Besides that, in the Nirenberg-Leder-binding-test the charged initiator-tRNA was not as specific as other tRNAs, recognizing two codons: AUG and GUG. The whole context involving the initiator codon AUG (or GUG) turned out to be influential; under regular physiological conditions, certain base sequences on the template RNA played an important role in determining where the binding of the ribosome was to occur. The mechanism of protein synthesis, then roughly understood, started making sense.

The innumerable details, the many additional factors, the dozens of proteins involved, the energy sources ATP and GTP, which regulate and maintain the machinery of the ribosomes both in prokaryotes and eukaryotes (see, for instance, Lipmann, 1969; Noller, 1991), were, and still are, the subjects of a multitude of intensive studies. But a complete understanding of the mechanism of translation at the molecular level seems now to be in sight, after decades of investigation have recently led two groups of crystallographers, one in Cambridge, U.K., and one in California, to arrive at exquisitely precise insights into the structure of ribosomal components and their detailed molecular functions (Wimbery et al., 2001; Yusupov et al., 2001).

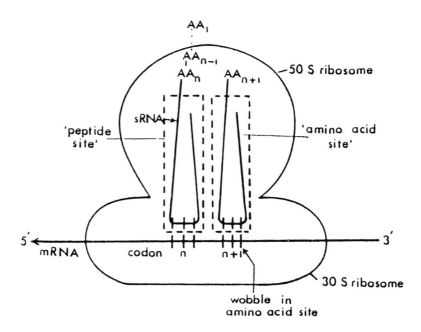

Figure 10.7: Protein synthesis on the ribosome (Bretscher's scheme, 1966). The ribosome itself is composed of two loosely linked subunits: a smaller one, designated the 30S subunit, which binds first to the mRNA, and a 50S one which subsequently joins the complex. [Both subunits are complex quaternary structures, compounded of ribosomal RNA (rRNA) and many proteins (in coli, for instance: the 30S subunit contains 21 proteins and one rRNA molecule with roughly 1500 nucleotides; in its 50S subunit 34 proteins are to be found along with 2 rRNAs with 3000 and 120 nucleotides, respectively).] "mRNA" is the template which presides the specific linear arrangement of amino acids in the polypeptide; it moves along the ribosome in the direction indicated by the arrow. The codons "n" and "n+1" are here, for more clarity, depicted as separated from each other, whereas in reality the distance between two neighboring bases is always constant, independent of their being positioned in the same codon, or in two different ones. "sRNA" stands for the tRNA linked to the peptide "AA1...AAn-1, AAn" at the P-site ("peptide site"); the tRNA charged with the amino acid AAn+1, sketched at the A-site ("amino acid site"), is not specifically indicated. The three vertical short lines on the tRNAs represent the so-called anticodon bases, complementary to the codons on the mRNA; the formation of hydrogen bonds between them leads to specific pairing. This complementary pairing – as all pairings of nucleic acids – rests on the principle of antiparallel disposition of the two single strand molecules (in this case: mRNA and tRNA). The implication is, for instance, that the codon 5'-CGA-3' has as counterpart the anti-codon 5'-UCG-3'; pairing scheme:

$$3'...GCU...5'$$
$$5'...CGA...3'$$

The term "wobble in amino acid site" refers to Crick's "wobble hypothesis", which accords to the third base in many codons a total or relative unspecificity in what concerns the mode of pairing to the corresponding first anticodon base. The wobble phenomenon makes sense, bearing in mind the fact of code degeneration (more than one codon for the same amino acid; Fig. 10.5). When the peptide bond of amino acid AAn+1 to amino acid AAn is established, the new peptidyl-tRNA moves on from the A-site to the P-site, after the tRNA on the P-site has delivered its peptide and diffused away from the ribosome into the cytoplasm; there it becomes charged anew. The template-RNA (mRNA) slides along one codon's length, allowing further codon-anticodon pairing. This move brings along the next codon (codon n+2) to the A-site; the next charged tRNA joins in. Thus, the same procedure is going to be repeated till a terminator codon reaches the A-site, prompting the hydrolysis of the peptidyl-tRNA on the P-site, liberating the completed polypeptide and tRNA.

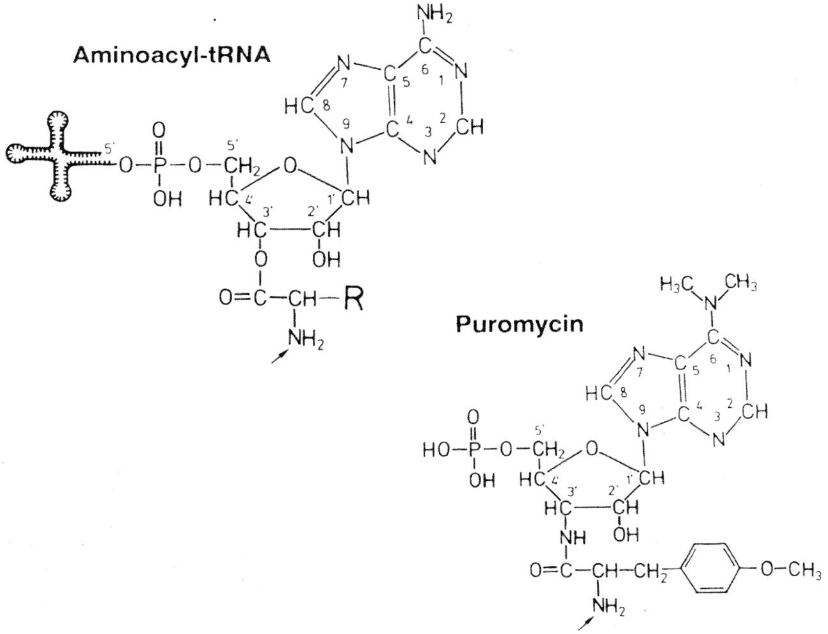

Figure 10.8: The antibiotic puromycin as an analogue of amino acyl-tRNA. Since in puromycin an amino group is located at a position analogous to that in an amino acid (arrow), this antibiotic can be incorporated into a growing peptide; however, the 3'-nitrogen atom of its sugar cannot undergo a carboxy-ester binding, thus blocking any further amino acid incorporation. (Only the last nucleotide of the depicted tRNA is represented in detail by its chemical formula.)

# CHAPTER 11

# INSIDE THE INSTITUT PASTEUR – PART I
# PROPHAGE AND THE FERTILITY FACTOR F

The Institut Pasteur in Paris, located at the very same premises since it was founded by its namesake in 1888, was originally meant to produce immunizing sera and consequently developed a long and sound tradition in medical bacteriology and virology. Through the latter, the institute was a venue for phage research of the pre-Delbrück era. It was Felix d'Hérelle, a French-Canadian researcher at the Institut Pasteur, who described what he called in 1917 bacteriophages which he isolated from the stools of patients convalescing from shigellosis disentery. Accordingly, d'Hérelle is considered to be one of the discoverers of phages. Although the British Frederick Twort had, already in 1915, observed a bacteria-killing, self-reproducing agent in culture plates with staphyloccoci, this was not known to d'Hérelle. Expert opinions diverge on to whom the merit of priority should go. While Twort's work ended abruptly with his death as a soldier during the First World War, d'Hérelle left a row of followers who proceeded to describe a series of different phenomena related to phages, but never succeeded in really interpreting their observations. Years later, the husband-and-wife team of Eugène and Elisabeth Wollman, also working at the Institut Pasteur, noticed a remarkable phenomenon: under certain circumstances, some phages would infect bacteria without causing their disintegration. Such phages faded from immediate detection, but each cell from the resulting clone stayed under the constant threat, albeit with low probability, of lysing after some generations, liberating progeny phages in the process. This was the phenomenon called lysogeny. Aside from the virulent phages which caused, unconditionally, the infected bacteria to lyse, there existed, apparently, another category of phages called temperate, which were able to enter the host, and then establish and maintain with it a state of non-lethal equilibrium for many generations. These viral entities – nowadays called provirus or prophages – divided synchronously with the host for many cell cycles, retaining nevertheless the capability of initiating a full-blown lytic cycle. Lysogenic bacteria, i.e. those harboring a prophage, live under the Damoclean sword of the all-time imminent, although rarely spontaneously occurring, death by lysis.

Delbrück's phage school resisted, for many years, accepting the very existence of lysogeny. Nothing similar had ever been noticed by its followers, they reasoned. No wonder: all of them – having the comparability of results in mind – accepted Delbrück's fettering dictate to exclusively work with phages of the so-called T-series, although these phages had been originally selected exactly because of their virulence (Demerec & Fano, 1945).

In 1943, Eugène and Elisabeth Wollman were arrested by the Gestapo in their laboratory – she first; he one week later – and deported to Auschwitz; no one ever heard again from them. Their son, Elie, took up the material worked on by Eugène and Elisabeth. After the war he held a scholarship at Caltech. There, while doing a survey of the literature at the library, he came across an index-card referring to his parent's lysogeny, with a remark scribbled across: nonsense!

André Lwoff, working at the Institut Pasteur since 1921, first as a student and then as a life-long research scientist, was convinced – this was in 1949 – that lysogeny, far from being nonsense, was rather an utterly interesting, even crucially significant phenomenon. His commitment to lysogeny was reinforced by the discovery of induction by his group: phage production could be triggered, in all lysogenic bacteria, by irradiation with ultraviolet light (Lwoff, Siminovitch & Kjeldgard, 1950). Induction was first demonstrated to occur in *Bacillus megaterium*, a bacterial giant (5 x 10 μm), microbiologically speaking, a soil microbe already known to the Wollman couple. The huge dimensions of *Bacillus megaterium* allowed Lwoff to perform manipulations under the microscope: placing one cell per droplet, he could observe the cells individually, catching eventual glimpses of single cells bursting apart, liberating phages in the process (Lwoff & Gutmann, 1950). Elie Wollman, in 1948, enthused about such observations, could not resist joining Lwoff's working group.

Soon after, Lwoff was blessed with another co-worker, François Jacob, a physician firmly decided to dedicate himself to biological research after the war – his body riddled with dozens of inoperable grenade shrapnels. This legacy from the allied Normandy landing was one of the reasons why he had to abandon his original dream of becoming a surgeon, and to embrace scientific research instead.

The trio reached the crucial decision of veering from *Bacillus megaterium* to *Escherichia coli* K12. They were prompted to make that move by Esther Lederberg, Joshua's wife, who had shown that this bacterial strain was lysogenic for a phage she named λ (Lederberg, 1951a), and Jean Weigle, a Swiss physicist at Caltech, who had demonstrated that *E. coli* K12 (λ) also could be induced by ulraviolet radiation.

*E. coli* K12(λ) was endowed with the decisive advantage of being able to undergo genetic recombination. The prophage λ could, thus, be assigned a genomic site on a linkage group near the gene coding for the fermentation of galactose (Lederberg & Lederberg, 1953). Apparently, the prophage did not swim freely in the cytoplasm of the host cell, but it was inserted into its chromosome; all preconditions for a genetic analysis of lysogeny were there. However, since recombination rates

Figure 11.1: Elie Wollman (born in 1917)

were exceedingly low, yielding merely a few recombinants for millions of parental cells, crosses remained somewhat intractable. Soon, though, this handicap was to be overcome.

By 1952, Wollman had heard of novel observations in *E. coli* by Hayes in London. Wollman arranged to visit him:

Williams Hayes was an utmost congenial and outgoing Irishman, a medical bacteriologist at Hammersmith Hospital in western London, a scientific maverick scattering an aura of friendliness. He had read about Lederberg's findings (Lederberg, 1947) and deemed that it could not hurt to check out the possibility of genetic recombination occurring in his bacterial system. Thus, in his small, scantly and humbly equipped laboratory (as petri dishes, for instance, he ingeniously used bottoms cut out from bottles), he started to play around with some strains of *E. coli* K12. For instance, on plates containing minimal medium he spread a mixture of two strains, one *met str$^r$* (methionine-dependent, streptomycin-resistant), and the other *thr leu str$^s$* (threonine- and leucine-dependent, streptomycin-sensitive); subsequently he added streptomycin to the plates at different times and checked for the emergence of the first prototrophic *str$^r$* recombinants. This happened when the time interval between mixing the strains and adding streptomycin was approximately 2 hours. Whenever the time lapse between the two events was shorter, no recombinants appeared;

Figure 11.2: William Hayes (1913-1994)

apparently the bacteria were being eliminated by the streptomycin treatment before recombinants were formed. If such recombinants arose from zygote formation, i.e., fusion of both parental cells – as assumed by Lederberg –, then the reciprocal cross (*met str* x *thr leu str*) should bring forth exactly the same outcome. In this reciprocal cross, though, the moment of adding streptomycin did not influence the results; whether streptomycin was present from the onset of the experiment, or whether it was added after hours had elapsed, turned out to be irrelevant in this case.

The reciprocal crosses were clearly asymmetrical, implying that the survival of parents from only one strain, but not from the other, played a role in the formation of recombinants. Apparently, the cells whose survival defined the outcome acquired the genes from bacteria of the other strain. These latter, acting merely as gene donors, were superfluous for the ensuing recombination process and could be eliminated without any consequences. The observed phenomenon apparently resulted from the existence of a donor strain, able to transfer its genetic material, even if further cell growth was hampered by streptomycin, and of a recipient strain, a receptor, which incorporated the donated genes into its genome and necessarily had to survive in order to yield the recombinant colonies (Hayes, 1952). The novel concept of an unidirectional gene transfer was presented by Hayes, in 1952, at a congress held in Pallanza, Italy; Jim Watson, present at this congress, got enthused to the point of introducing Hayes to the phage school; what followed was an invitation by Delbrück for him to spend half a year at Caltech.

The experiments then would get even murkier. The new observations seemed even more paradoxical and inscrutable, although Hayes, at Caltech, was not as isolated with his thoughts and experiments as in Hammersmith, counting as discussion partners all Caltech researchers, Delbrück included. In the meantime, it had become evident (Lederberg et al., 1952; Hayes, 1953a) that upon establishing contact with donor cells, almost all bacteria of the receptor strain turned into donor cells themselves, without having received any other genetic markers (which would show up in recombinants). [Recombination – as already demonstrated by Lederberg (1947) – was an extremely rare event (rate of occurrence in the order of $10^{-6}$ per parental cell.)] Clearly, a fertility or sex factor (F), transferable from donor to receptor, had to exist, able to transform receptors into donors, stated Hayes (1953a). Nevertheless, Lederberg held on to his conventional view of the formation of zygotes, i.e., cell fusion (see, for instance Lederberg, 1955). The controversy got even hotter when Hayes (1953b), by chance, isolated a donor strain able to transfer some genes – albeit not others – with a 1000-fold increased rate, as compared to normal donor strains ($F^+$ strains). These strains, called Hfr (High frequency of recombination), on the other hand, had lost the ability to transfer the F factor itself: i.e., after coming into contact with cells of Hfr strains, most receptor cells kept their status as such; merely very few recombinants were rendered Hfr themselves. No trace remained of the $F^+$ cells, able to transfer the $F^+$ factor. This state of affairs got even uncannier as the Hfr strain, itself streptomycin sensitive, transferred its genes even in the presence of streptomycin, as the conventional donors did. How could all that be possibly interpreted? Would any enlightenment on this matter be ever achieved? Watson and Hayes (1953) tried hard without succeeding, despite Watson having set great expectations in this subject regarding his future career (these events took place months before the discovery of the double helix). Some years were to elapse before a correct interpretation was reached. These observations, as unanalyzable as they seemed, encompassed, though, the foundations for the spectacular development of the crossing techniques for coli K12: mix together two strains, a donor Hfr and a receptor, allowing, thus, recombination to take place; bring about the separation of the cell pairs at different moments and look for what happened... Endlessly debating and theoretizing over that matter were the order of the day... and after Wollman's visit, Hayes made available his Hfr strain to the Paris laboratory. One of the first experiments involving this strain provided Jacob & Wollman with a decisive surprise. The obvious thing to be done first – for someone familiar with lysogeny, that is! – was to follow the fate of the prophage during and after a cross. If a non-lysogenic Hfr strain was crossed with a lysogenic F⁻ (without F factor), nothing happened – i.e., nothing excitingly new; everything went on as expected, as already described by Hayes. And then, the reciprocal cross: Hfr(λ) x F⁻, i.e., the donor was lysogenic, but not the receptor. The receptor cells lysed "en masse"! They were induced to lyse as though irradiated by ultraviolet. "Erotic induction" is what the phenomenon should have been called, later Wollman (1966) quipped; in print it became "zygotic induction" (Jacob & Wollman, 1954). (This was a curiously improper denotation, taking into account that Hayes had already convincingly shown that no zygotes were formed in the process.) Why did the receptor cells lyse after receiving the provirus? Or, in other words: what prevented donor cells, or for that matter, lysogenic bacteria in general, from lysing? Whatever it was, it was not transferred to receptor cells together with the prophage; consequently it was not located on the chromosome, but probably in the cytoplasm, and because of that it was not transferable. (Was this not a further vindication for the absence of cell fusion during bacterial crosses?) It was apparent that an inhibiting factor existed, which prevented the initiation of the lytic cycle. The concept of a cytoplasmic repressor was at hand – however, some time was still to pass until... (see Chapters 12 to 15).

**Excursus 11-1**
**PARASITES AT THE GENETIC LEVEL**

The concept of genetic parasite was elaborated by Luria in the 1950s. Back then, impressed by the survival strategies deployed by viruses, he described how their genomes could get rid of all superfluous genes, keeping merely the most essential information for self-replication, and how they appropriated the host cell genes, which then provided all necessary means needed for viral replication: the protein synthesizing apparatus, many other functions, and energy. Employing that strategy, the genetic parasite could dispense with all "housekeeping" genes, making do with only a handful of special genes indispensable for veering the cell metabolism into the direction of producing virus progeny. Bacteriophages also followed this scheme and could be deemed to be the ideal model organisms for more detailed studies. For instance, from the phage's point of view, two basically distinct strategies were available: virulence or temperance. The virulent phage followed strategy n° 1: it devoured the host cell in a matter of a few minutes, yielding a progeny of approximately 100 descendants. These, in their turn, depended on fresh bacterial victims to further reproduce. Not so the temperate phage, with strategy n° 2: its DNA stealthed into the host cell without destroying it, and without producing immediate progeny. Instead, it divided itself as a prophage, synchronously with the host's genome. This way, through one sole infection, thousands, even millions of descendants would be granted. The decision to follow strategy n°1 or n° 2 finds a parallel in the speculative stock market: in the case of virulent reproduction the phages speculate on a bear market (early host demise taken as probable) whereas lysogenisation would be the bullish alternative (many future cell division cycles likely to occur). The temperate phage is also familiar with "hedging": in case the host cell is injured (as, for instance, by UV radiation, it escapes by "inducing" a virulent cycle, thus rapidly switching to strategy n° 1 (also see Chapter 15).

A real optimist among the genetic parasites would repeal strategy n° 1 altogether in order to confidently associate itself with the host cell, come what may. It could then get rid of the last functions needed for self-replication, trusting on the host cell to take care of the perpetuation of its DNA. An appreciable part of the coli genome, it seems, consists of such parasitic DNA. A genetic burden – that is the way parasitic DNA relates to and is carried by the host cell. The larger this genetic burden is, the higher the risk for the host cell of being eliminated by selective disadvantage, together with all its parasites. This is an obvious liability, because replication of superfluous DNA demands an increased energy consumption. Keeping the DNA of parasites in check is crucial, especially in the case of bacteria; at stake is their long-range survival, which ultimately depends on their capability of dividing quickly. This selective pressure limits the proportion of DNA parasites that can intrude in prokaryotic genomes. In the case of higher organisms, the energy employed in the replication of DNA is relatively small, even negligible, as compared to the amount of nutrient energy devoted to other purposes. This situation brings about another scenario in higher organisms: apparently, the largest portion of their DNA sequences (compare to Fig 21.1) consists of functionless, parasitic DNA: "junk DNA" (see, for instance Doolittle & Sapienza, 1980; Orgel & Crick, 1980).

The functionless parasitic DNA is under the constant threat of being excluded by deletion from the main genome. Although such an eventuality is rather unlikely, many genetic parasites have developed different strategies to counteract such a menace.

For instance, the F factor, one example of many such parasites, has perfected a mechanism which, following contact of its host with another cell, allows it to move into, and propagate in, the receptor (the donor, transferring solely a copy of the factor, does not lose it in this process). It looks as though, in nature, the eventual loss of an F factor is compensated by the possibility of conquering new hosts. [Its role as a sex factor seems to be a laboratory artifact – so it can be argued – because in nature it does not play a significant role with regard to sexual exchanges involving the main chromosome (see, for instance, Ochman & Selander, 1984; Maynard Smith et al., 1993).]

The F factor encompasses a few dozen genes that guide its transfer from cell to cell (for comparison, the host deals with more than 4000 genes). Despite its modest genetic dimension, the F parasites are themselves subject to parasitism. There is a series of self-replicating parasitic mini-chromosomes which count on the F factor transfer mechanism for their own propagation to new host bacteria.

The game of parasitizing the parasite proved to be astonishingly versatile. Peter Starling and his co-workers, Elke Jordan & Heinz Saedler, in Cologne, Germany, were haunted for years by weird mutations occurring in E. coli, leading to the incapacity to ferment galactose (see, for instance, Jordan et al., 1967; Saedler &

Starlinger, 1967). Weird, because the synthesis of many of the enzymes involved in this metabolic process failed simultaneously. Deletions, as an explanation for this phenomenon, had to be excluded, since back-mutations to normal galactose fermentation happened to occur with the same probability as the forward mutation. (In the case of deletions, i.e., the loss of large DNA segments, there is no possibility of back-mutation.) Finally, the group in Cologne (see, for instance, Jordan et al., 1968; Saedler & Starlinger, 1992), and, almost at the same time, a young Cambridge graduate student (Shapiro, 1969) reached the insight that the observed mutations actually originated from insertions of extraneous DNA segments into the galactose genes (other genes are also susceptible to such insertions, but were not analyzed at the time). The inserting sequences wreaking havoc on the expression of galactose genes were always the same: one sequence of about 800 base pairs and another one with roughly 1400 base pairs, denoted, respectively, IS-1 (insertion element 1) and IS-2 (Schmidt et al., 1976). These insertion elements emerged at a rate of about $10^{-7}$ per cell division, only to disappear at the same rate. Where did they come from? Where did they go? What was the basis for their mobility?

A novel branch of research was born: the genetics of the IS-elements, "the jumping genes", as they were soon dubbed. The new scientific area turned into a competing ground for many rival groups (see, for instance, Bukhari et al., 1977). When the dust settled, it became clear that the insertion elements represented a new category of parasitic DNA, codifying for 3 or 4 genes, essential for their own occasional replication – if this came about too often, the death of host cells would be the consequence. After replication, the insertion element's copy intruded into another region of the host's genome or into one of its plasmids (Excursus 12-2), for example the F factor, mostly leading to deleterious mutations. It was shown that many bacterial strains carry along dozens of copies of diverse IS elements dispersed among different sites of their genomes, or inserted in their plasmids, (see, for instance, Matsutani & Ohtsubo, 1993).

Replication of IS elements within the host genome, does not, however, entail propagation to other bacterial clones: a precondition for this to occur is "sexual" contact. This occurrence offers the IS element the opportunity of invading a new host, carried along as a passenger by the transferred DNA segment.

IS elements may deploy other still more sophisticated strategies. For example, through flanking an essential gene, two such IS elements, acting as a unity (as the so-called transposons; see, for instance, Starlinger, 1980), replicate and invade other genomes, granting a selective advantage to the new host: the emergence of resistance factors (R factors) soon after antibiotic therapies became routinely used, can be so understood. R factors are nothing else but F factors, in which one or more genes conferring antibiotic resistance (from unknown origin) were incorporated by means of flanking IS elements. Such uncanny and rare events would remain undetected in nature, were it not for the overwhelming selective advantage conferred to the respective host in an antibiotic-treated patient. Through this mechanism, F factor-like parasites, in the form of R factors, came to exert an important role in bacterial sexual processes. (They also are responsible for the serious medical problem of multi-drug resistant pathogenic bacteria.)

This way, deleterious parasites turned themselves into symbionts, useful to the host. However, arguing about a possible usefulness or function of a structure endowed with the capability of self-replication is basically inappropriate because, if the structure replicates itself, this is sufficient to account for its existence, independently of its harmfulness or usefulness. (In which way is the flea useful to the dog? – although, considering the case of monkies, lice do fulfill the useful task of stabilizing the social ties among the members of a group, brought about by reciprocal lousing!)

These discoveries, so fascinating for molecular biologists, suddenly brought back to mind similar cases detected in maize, more than 20 years before. Barbara McClintock, in the 1950s, also working in Cold Spring Harbor but barely noticed by her colleagues, had spotted eerie genetic elements which, in the course of generations, changed their genome position, at measurably constant rates, and in this process induced mutations similar in character to those of Starlinger. McClintock's were somatic mutations, made noticeable by variegation of the corn grains. Moreover, also in the course of her studies, parasites of parasites came to light: "jumping genes" which were actually immobile themselves but could be made motile by the presence of autonomously jumping sequences. Thus the *Ds* element (from dissociation, because this element also caused chromosome breaks), for instance, could only change its genome location if another element, *Ac* (activator), this one autonomously mobile, was also present in the genome (see, for instance, McClintock, 1951, 1956; see also Starlinger, 1993). McClintock gained status as a cult figure. Justly so, because she had discovered and amply documented with dazzling analytical clarity and many years in advance, a phenomenon deemed noteworthy by microbial geneticists and molecular biologists (see Fig. 11.3); some decades went by till, in 1983, Mc Clintock, then 80-years-old, was tardily bestowed with the honor of the Nobel Prize for the merit of having discovered the mobile genetic elements. However, the fact that – even before the discovery of the mobile genetic elements – she had already enriched traditional classical genetics with a series of important contributions, remained almost unnoticed. For instance, in the 1930s she had identified genetic linkage groups in maize and assigned them to their respective chromosomes; she had recognized some cytological images as the morphological correlate of cross-overs, only detectable by the analysis of the progeny of crosses; she had defined the nucleolus-organizer as a chromosome structure and, in general, together with Morgan, she had strengthened the chromosome theory of inheritance.

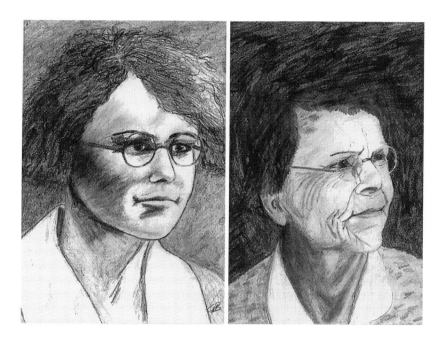

Figure 11.3: Barbara McClintock (1902-1992). Barbara McClintock received her doctoral degree from Cornell University, Ithaka, New York State, for her studies in genetics and cytology of maize, two disciplines that, at the time, had almost no common ground. (At that time, women were not allowed to pursue genetics as a doctoral theme at Cornell.) After some years of wandering, she found in 1941 an ideal position for her at Cold Spring Harbor. There she could dedicate herself exclusively to the genetics of corn till the end of her life. Scientific research was basically her one and only interest; albeit she enjoyed meeting and discussing with colleagues and students, she led an almost eremitic life; she also pursued her research preferentially alone. Evelyn Keller (1983), her biographer, described the young student Barbara as socializing, and lively extrovert; she was held in high esteem by her peers and was due to become a member of an elitist student sorority. When she learnt that Jewish students were excluded from that society, solidarity with a friend made her decline the membership. She could not accept the attitude of discriminating against anyone for any reason whatsoever. Disenchanted with people, she turned to herself, to nature, to maize. At least that is the impression that her biographer tries to mediate – it is difficult to judge if this representation is tailored to specially suit a feminist image ("women are the more sensitive beings"); but... se non e vero e ben trovato. For a more sober, comprehensive biography, see Comfort (2001).

# CHAPTER 12


## INSIDE THE INSTITUT PASTEUR – PART II
## INTERRUPTED MATING


Zygotic induction acquired an additional significance: it hinted at the possibility of analyzing the kinetics of the conjugational events. If Hershey, in a strike of ingenuity, took advantage of a kitchen blender with its shearing power to detach adsorbed phages from their host cells, why not try achieving separation of donor and receptor cells by a similar procedure?

Jacob & Wollman sure tried it out: liquid cultures of the pairs – Hayes' Hfr as a donor and F⁻ as a receptor – were mixed in a conjugation suspension, from which samples were taken at different times to be subsequently shaken in a blender. The outcome was strikingly clear: if the conjugation pairs were forced apart immediately after the beginning of the process, no recombinants emerged; if the conjugation was allowed to proceed for three minutes, some recombinants for the markers *leu* and *thr* appeared, indicating that solely the loci *leu* and *thr*, but no others, had already been transferred; the marker *tonA* (resistance to phage T1) followed after 10 minutes; and after 17 minutes, the *lac* genes (lactose fermentation) were to be transferred; it took 25 minutes for the *gal* genes (galactose fermentation) to reach the receptor, and so on... (Wollman & Jacob, 1955; Wollman, Jacob & Hayes, 1956) (Fig. 12.1). If the conjugation mixture was subjected to the blender procedure after about two hours, only then the last marker got to be transferred: the ability of functioning as Hfr (Jacob & Wollman, 1957). The individual markers migrated sequentially into the receptor cells as though bound to a chain which was gradually pulled into the receptor [or, alternatively, bound to a stick pushed into the receptor – see, for instance, Clark & Adelberg (1962)]. The kinetics of transfer revealed the gene sequence, allowing a linear chromosome map to be established, with all bacterial genes aligned on solely one linkage group.

Things would get even more fascinating. It had become evident that the serendipitous Hayes' Hfr strain had originated from a mutation-like event which had taken place in an F⁺ bacterium. Such Hfr strains could be specifically looked for, for

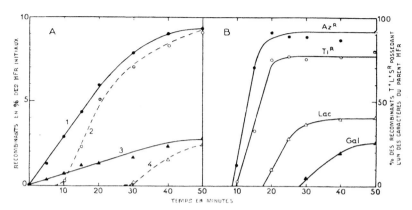

Un mélange en bouillon de bactéries Hfr($10^7$/ml) et F⁻($5.10^8$/ml) en voie de croissance exponentielle est préparé à temps o et agité à 37°. Des échantillons sont prélevés à temps variable, dilués, une fraction étant soumise au traitement mécanique, l'autre conservée comme témoin. A partir de chaque fraction des étalements sont faits sur milieux sélectifs.

Fig. A. — Fréquence, en fonction du temps, des recombinants T⁺L⁺Sʳ (avant traitement, courbe 1; après traitement, courbe 2) et Gal⁺Sʳ (avant traitement, courbe 3; après traitement, courbe 4). .

Fig. B. — Analyse génétique des recombinants T⁺L⁺Sʳ obtenus à partir des échantillons soumis au traitement mécanique. A chacun des temps indiqués, 120 recombinants ont été examinés. La distribution des caractères issus du parent Hfr est exprimée en fonction du temps auquel les échantillons ont été prélevés.

Figure 12.1: Transfer kinetics of genetic markers during the pairing of a streptomycin-sensitive donor (Hfr) with a streptomycin-resistant receptor.

A)     At time 0, two growing cultures, an Hfr and an F⁻, were mixed and further incubated at 37 °C under aeration. Then, samples were removed at different times (abscissae), and after appropriate dilution, spread on plates containing streptomycin, so that only receptor cells were able to originate colonies. These were subsequently tested for the presence of different markers received from the donors. The tally of recombinants is represented on the ordinates as a percentage of the Hfr cells of the original mixture. Curve 1 depicts the appearance of recombinants of the type *thr⁺ leu⁺ strʳ*; curve 3, type *gal⁺ strʳ*. Curves 2 and 4 relate to the same types as curves 1 and 3, respectively, with the difference that, before plating, the conjugation pairs were forced apart by the shearing force of a kitchen blender.

B)     The experimental procedure here was the same as in A), curves 2 and 4, i.e., the conjugation pairs were separated by a kitchen blender before plating. Samples from the conjugation mixture, appropriately diluted, were spread on streptomycin-containing minimal-medium plates, so that only *thr⁺ leu⁺ strʳ* recombinants could grow. These were further tested for T1ʳ (nowadays *tonA*, resistance to phage T1), Azʳ (now *azi*, azide resistance), *lac⁺* and *gal⁺* (fermentation of lactose or galactose, respectively) (Wollman & Jacob, 1955, © Editions Gauthier-Villars).

instance, using a method developed by Cavalli-Sforza in Milan, which originally aimed to demonstrate that a mutation to antibiotic resistance came about without the mutant cell or its ancestors ever having come into contact with the toxic substance in question (Excursus 12-1).

Experiments of interrupted matings, using Hfr donors of different origins, confirmed the one-dimensional arrangement of the *E. coli* K12 chromosome map – however, the order in which the genes were transferred varied according to the Hfr strain employed as a donor. In some crosses, the sequence of transfer was: *thr, leu, pro, lac, gal, trp, mal, xyl, B1* (this was typical for the original Hayes' Hfr strain); in others: *pro, leu, thr, B1, xyl, mal, trp, gal, lac* (this was observed when the donor was strain Hfr n° 2); or else: *xyl, mal, trp, gal, lac, pro, leu, thr, B1* (this was the case of strain Hfr AB313); and so on...

The detected gene sequences, albeit variable from strain to strain, were by no means haphazard, as though the genes to be transferred were chosen by chance in a genetic dice game. Much to the contrary, in all instances the sequence consisted of a so-called circular permutation of a basic sequence, suggesting a circular chromosome for coli (Fig. 12.2 & 12.3). The hunch was that the F factor (whose presence was the essential precondition for the emergence of Hfr strains), which in regular F$^+$ strains remained independent from the supposedly circular bacterial chromosome, would get incorporated (at a rate of about $1:10^{-4}$ per cell division) into this chromosome – there was already the precedent of phage λ! – and that, during conjugation, this integrated F factor would break open the circle at its incorporation site, prompting a copy of the chromosome to translocate towards the receptor, starting at this breaking point. And further, there was also the observation that the F factor itself, or at least an essential portion of it, trailing as the last marker on the chromosome, would only rarely be passed into the receptor – most conjugation pairs became separated before having the opportunity of transferring all markers (Jacob & Wollman, 1957, 1958b, 1961). [Lederberg (1959) kept protesting against such an abstruse interpretation, without any precedent in nature...]

Considering that the F factor could be inserted into the chromosome, the question arose of whether it also could, eventually, detach itself from the chromosome again. Jacob & Adelberg (1959) believed to have a good reason to affirm that. The marker *lac$^+$* (fermentation of lactose), usually transferred by a special Hfr strain as the last marker, preceding only the Hfr marker itself, could also be transferred, albeit at a very low rate, at the very beginning of the conjugation process. Clones derived from the so obtained *lac$^+$* cells showed the peculiar behavior of transferring the *lac$^+$* marker with a very high efficiency and very soon, after a few minutes already; moreover, cells from these clones proved to be F$^+$, transferring the F factor together with the *lac$^+$* marker. It looked as though the F factor had become autonomous again, having set itself free from the main chromosome, and, in the process, dragging away one or more neighboring genes, those flanking it on the main chromosome. Alan

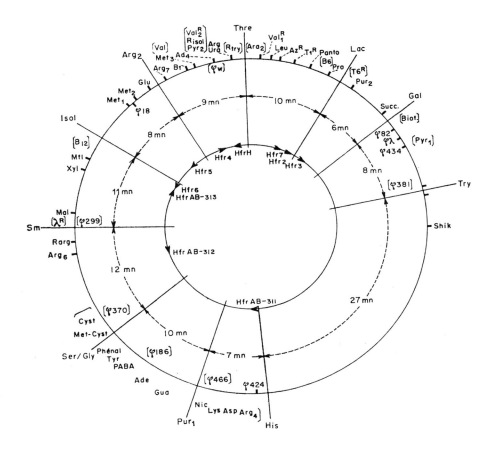

Figure 12.2: An early schematic representation of the *E. coli* K12 chromosome map. On the outer circle, individual genetic markers are displayed; in the middle, the approximate transfer times, by an Hfr donor, of the corresponding segments are represented (the total genome transfer time adds up to 108 min); the arrows at the inner circle indicate the starting points and directions of chromosome transfer for different Hfr strains (Jacob & Wollman, 1961).

Campbell (1962), in Rochester, New York, conceived a versatile model to explain this eventuality (Excursus 12-2). Such freed, autonomous F factors which, as a consequence of their having been once integrated in the main chromosome, carry along with them certain bacterial genes, are designated F' (F prime) factors. What Jacob & Adelberg had identified was a very special F', an F' *lac* factor, one that would come to play a crucial role in the further development of molecular biology.

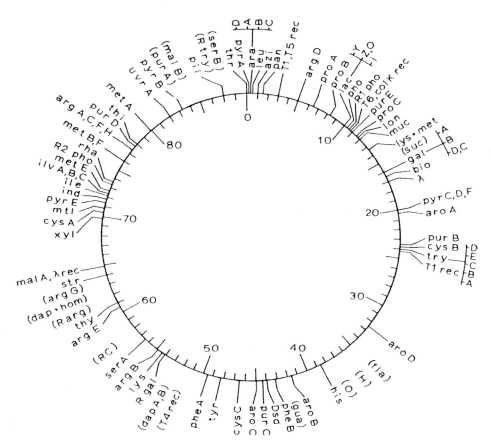

Figure 12.3: One of the first extensive chromosome maps of *E. coli* K12. This map was standardized to the starting point 0, corresponding to the transfer of *thr*+ by the strain Hfr H, and a total transfer time of 89 minutes (Taylor & Thoman, 1964). Later, 90 minutes and then 100 minutes became the norm (see Bachmann, 1987); nowadays about 4300 *E. coli* genes – although still many of unknown function – are identified and localized on the map, whose DNA sequence of about 4.6 million bases has been elucidated (Blattner et al., 1997; Riley & Serres, 2000).

Meanwhile, Wollman moved temporarily to Berkeley as a visiting scientist. Jacob kept his mind focused on bacterial conjugation, specially on zygotic induction. "Induction", was that not the very same expression earlier used by Monod for many years, while working on the other end of the corridor in the attic of the Institut Pasteur?

**Excursus 12-1**
**A METHOD FOR ISOLATING DIFFERENT Hfr STRAINS**

Luca Cavalli-Sforza, nowadays mainly interested in human population genetics, was back then one of the pioneers of bacterial genetics. He, in Milan, and Lederberg at the University of Wisconsin, conceived a straightforward, elegant, and informative experiment: Ten years after the fluctuation test of Luria & Delbrück (Excursus 4-2), it was designed, once and for all, to finally falsify the convictions of the last defenders of the theory that mutations derive from an adaptive process, to show that their Lamarckian concepts were totally ill-conceived. These stubborn Lamarckists argued that the fluctuation test was inherently unreliable since each one of its many test tubes offered the bacteria subtly different and uncontrollable environmental conditions. The bacteria in separate test tubes were, consequently, differentially influenced, so that their capability to react to the toxic agent, tested later on, could also fluctuate. This would very well explain the high variability of mutation rates of bacteria coming from those different test tubes. Truly, in the fluctuation test, each mutant came into contact with the toxic agent before it could be identified as a mutant. And according to the critics, this contact was decisive for triggering the mutation. Cavalli-Sforza & Lederberg's (1956) quest was to isolate mutants resistant to a toxic agent without their ever having been in contact with it. Their strategy was as follows: first, check the ratio of streptomycin-resistant ($str^r$) mutants contained in a coli streptomycin-sensitive ($str^s$) population. This was achieved by plating samples on plates containing streptomycin; the average tally observed was of one $str^r$ colony per 100 million $str^s$ cells plated (1 mutant : $10^8$ total number of cells). The original culture, never exposed to the toxic agent, was subsequently subdivided into portions, each containing about one million bacteria ($10^6$). According to the hypothesis of mutations emerging independently of the presence of the agent, one $str^r$ mutant should be hidden in one of these portions. If such was the case, then one of the many portions would have a ratio of $str^r$ mutants that was effectively 1:1,000,000 ($10^{-6}$), 100-fold higher than in the original culture. All populations were allowed to grow. The specific, enriched population was revealed by plating out samples from all the tubes on streptomycin plates.

Using this enriched culture, the procedure was repeated once more. This time, the portions were so prepared as to encompass 10,000 cells each. Again, one portion in about 100 should contain one mutant, so that its ratio in the corresponding population was about $1:10^4$ cells. After detecting the newly enriched population, the procedure was repeated once again, this time with about 100 bacteria per portion. The logical prediction: about every hundredth portion contained one mutant, the ratio being of $1:10^2$ cells. This could be confirmed by individually testing colonies grown on agar plates deprived of streptomycin, or by performing one further enrichment cycle. Employing this strategy, $str^r$ mutant clones could be isolated, whose cells never had come into contact with streptomycin. Irrational fantasies became the only escape left for the Lamarckian fundamentalists. This same tactic of successive enrichment – in many variations – can be deployed to pick out Hfr strains from F$^+$ cultures. In this case, parallel cultures of an F$^+$ strain are tested on the basis of their ability to efficiently transfer specific genes to F$^-$ receptors.

**Excursus 12-2**
**EPISOMES AND PLASMIDS: THE CAMPBELL MODEL**

Jacob & Wollman (1958b) created the expression "episome" as a general term defining all genetic factors of bacteria which could exist in two forms: either as self-replicating mini-chromosomes, or, alternatively, integrated into the cell's main chromosome. The phage λ stood model, as the prototype for all episomes; in the lytic cycle its DNA replicated autonomously in the cytoplasm, whereas in the form of a prophage it occupied a determined site on the coli's genetic map. The F factor also passed Jacob & Wollman's strict criteria to qualify as an episome: it could replicate autonomously in its original F$^+$ host, otherwise it could also be integrated into the host chromosome, as in Hfr strains. On the other hand, in the case of many R factors (Excursus 11-1), no such integration was observed, despite the many similarities between these factors and the classic F factor. Moreover, if the F factor used *Shigella* as a host instead of coli, no incorporation into the main chromosome came about. The F factor behaved as an episome in coli but not in *Shigella*. Allan Campbell, then at the University of Rochester and later at Stanford, reflected on this state of affairs, concluding that unifying all bacterial mini-chromosomes by taking into account their common characteristics, instead of pointing out their differences, would be a good no-nonsense approach. The expression "episome" was then extended to include all autonomously replicating DNA structures (Campbell, 1962). Later, the term "plasmid", an Anglo-Saxon synonym to the French "episome", gained ground, so that nowadays – defying priority – it is used nearly exclusively, especially in the field of gene technology (see Chapter 20).

Concerning those episomes – sorry, plasmids – able to be integrated into the main chromosome, what mechanism allowed them to perform this feat? A model – later named after his creator – occurred to Campbell (1962): the secondary chromosome to be integrated is present in the host cell cytoplasm as a ring; a region of this circle pairs with a DNA segment of the main chromosome; a crossover in this pairing region is sufficient to transform the DNA circle into a linear segment on the main chromosome (Fig. 12.4). According to the usual concept, the pairing segments, subject to crossover, are homologous, displaying identical or at least very similar DNA base sequences; alternatively, there is the possibility of a so-called "illegitimate" crossover; in this scenario, the crossover would take place between non-homologous

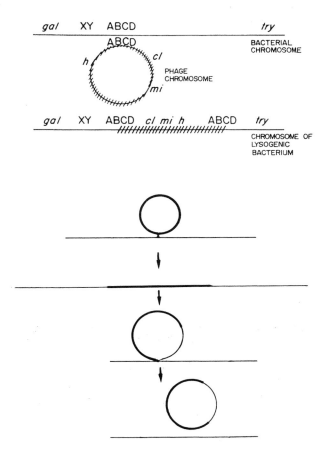

Figure 12.4: Top: original scheme representing the Campbell model. A homology between the sequences ABCD from the host DNA (straight line) and the episome (circle form) is postulated here; the integrating crossover is supposed to occur in this region (Campbell, 1962). For the episome to detach itself from the main chromosome, in other words, for it to regain the autonomous status, the process must run in the opposite direction. If the crossover is brought about at a point other than that of the integration event, some genetic material from the main chromosome may be inserted into the free episome, or some episomal DNA may remain bound to the main chromosome; through this mechanism it is possible to explain the emergence of, for instance, F' factors.

segments, and, as in the example of the phage λ, it was shown to be dependent on a set of specific episome-directed enzymes. [One should point out here that Campbell had postulated the ring closure of the λ chromosome on a purely theoretical basis; confirming the ligation of both λ DNA ends into a ring-like structure came about only years later (Wu & Kaiser, 1968).]

The Campbell model turned into a success story: its inherent concept of a recombinational interchange between the main chromosome and autonomous DNA molecules, although in innumerable variations, proved to be ultimately valid. Aside from that, it stimulated the fantasy of quite a few scientists in other research fields, as for instance, in immunology and animal virology. Even though Campbell's model does not apply to the chromosomal insertion of, for instance, retrovirus DNA copies, it triggered a way of thinking crucial to resolving intricate events involved in the integration of animal DNA viruses into their host chromosomes (see, for instance, Campbell, 1993).

# CHAPTER 13

## INSIDE THE INSTITUT PASTEUR – PART III
## THE FERMENTATION OF LACTOSE

𝒜t the other end of the corridor where Jacob's laboratory was located, Jacques Monod, encouraged by Lwoff, had kept himself busy for some years with ill-defined metabolic processes in coli. It had been known since the beginning of the 20th century that many enzymes of microorganisms were detectable only if their corresponding substrates were available in the culture medium. Microorganisms, so it seemed, could adapt to their environment. Enzymatic adaptation was the name for those observations. For years, the mechanisms for this supposed adaptation offered grounds for a series of diverse interpretations. At the experimental level, though, there were only scarce contributions to the understanding of these phenomena; for instance, no attempt was made to differentiate between selection of a few mutants already present in the bacterial culture, and substrate-induced modifications of the enzymatic set of the original bacterial population. With these facts as background, Monod had observed during the Second World War that the presence of the enzyme β-galactosidase in coli cells depended on these cells having been cultured in a medium containing lactose (β-galactosidase is the enzyme responsible for splitting lactose into glucose and galactose, the first step for lactose utilization). Absence of glucose was an additional precondition.

(At that time, Monod was working at the Institut Pasteur essentially as a member of the underground; his official working place was actually the Sorbonne. There the Gestapo would probably look for him, since he was a leading member of the "Résistance" – as was Lwoff, though in another faction.)

The appearance of β-galactosidase in coli cells after lactose supplemention was one of the examples of "enzymatic adaptation" or, in other words, adjustment to the environment. In the absence of lactose, β-galactosidase was obviously superfluous; under this condition, from the cell's point of view, it would be much more economical not to produce it. Soon after the war, the consensus was that a precursor should exist, which, influenced by the presence of lactose, would transform itself into β-galactosidase; this assumption was deemed absolutely logical...

But, in 1949, a young American immunologist, holder of a scholarship to work with Monod for one year (he would end up staying for six years), arrived at the Institute: Melvin Cohn. Cohn purified the β-galactosidase and injected it to rabbits, obtaining a specific antiserum able to react with the enzyme and to yield an immune precipitate. Bacteria which were not "adapted" did not react at all with this antiserum. The conviction of there being a precursor was shattered; if it truly existed, then it would have to be, structurally, totally different from β-galactosidase. Besides this piece of evidence, the fact that availability of a series of amino acids was a precondition essential for the appearance of β-galactosidase inflicted a second heavy blow on the notion of a precursor. For instance, if bacteria were starved, no new protein synthesis could possibly take place; and no accumulation of β-galactosidase after addition of lactose was demonstrable in this case; the starved bacteria had lost their "adaptability". If a precursor was present, one would expect that the addition of lactose would trigger a relatively quick response, turning the precursor promptly into the active enzymatic form. However, experiments to ascertain this point revealed a progressive increase of the enzyme concentration in the cell, starting three minutes after the supplementation with lactose, instead of the expected sudden surge. Rounding up the arguments against the precursor theory, it was also demonstrated that proteins reactive with the anti-β-galactosidase rabbit serum were not synthesized till the moment lactose (or a lactose analogue) was added; to reach this insight, radioactive sulfur (in the form of sulfate) – incorporated by the cells into methionine and cysteine – was monitored in immuno-precipitated material: no radioactivity showed up in this precipitate if the radioactive sulfate was only present until the moment of adding lactose [the radioactive sulfate was removed immediately before adding the lactose by centrifuging the cells and resuspending them in fresh non-radioactive medium containing lactose (Hogness, Cohn & Monod, 1955; Cohn, 1957)].

Meanwhile, Melvin Cohn had shown that the synthesis of β-galactosidase could also be triggered by lactose analogues – even better than by lactose itself –, although such analogues, as, for instance, methyl-thiogalactoside (MTG) were not accepted as substrates by the cell; indeed, the analogues subsisted unaltered inside the cells. Such an effect, exerted by non-degradable lactose analogues, could not possibly bring any advantage to the cell. Much to the contrary, it could be a disadvantage (synthesizing unneeded enzymes squanders energy). Therefore, the expression "adaptation" was inherently unsuitable to describe the process and was replaced by the term "induction" (Cohn et al., 1953).

"Induction", however, was the very same term that Lwoff, Jacob and Wollman used to define the experimental triggering of the lysis of lysogenic bacteria. Did such apparently so distinct phenomena, the "induction" of phage production on the one hand, and the "induction" of enzyme synthesis on the other, have more in common than solely their names?

Georges Cohen, a Frenchman, was a newcomer given the task of investigating the fate of lactose analogues, i.e., of non-degradable inducers, inside the cell, and determining, for instance, the cell structures to which they were bound. For that, he employed radioactive analogues. An unexpected result came about: if the cells were already induced, or if so-called constitutive mutant cells were used which produced β-galactosidase independently of the presence of an inducer [Lederberg (1951b) had already described such mutants], then the radioactive inducer was promptly and intensively taken up by the cells (centrifugation or filtration of the cells and subsequent monitoring of the radioactivity retained by them showed this effect); if, otherwise, the cells were not previously induced, their initial uptake of radioactivity was reduced to a mere trickle that only became more intensive after a few minutes (Rickenberg et al., 1956). These results conveyed the impression that the inducer, besides prompting the already known production of the enzyme β-galactosidase, also affected the cells in another way, namely by rendering them permeable to the inducer itself.

To account for the inducer's newly demonstrated effect, an additional enzymatic activity, also controlled by this inducer, was postulated, namely that of a permease. This putative enzyme, working like a pump, would be able to promote the active transport of the inducer – certainly of lactose itself as well – from the outside environment to the inside of the cell, working against a concentration gradient. Thence, the substrate could accumulate in the interior of the cell, even when its concentration in the surrounding medium was very low (Cohen & Monod, 1957).

Quite astounding was the fact that all mutants with constitutive synthesis of β-galactosidase also displayed a constitutive behavior in what concerned the permease activity [the enzyme itself would only be isolated years later by another group (see Kennedy, 1970; Büchel et al., 1980)]. However, individual mutational defects affected specifically one of the two functions, either the β-galactosidase or the permease; locating these mutations on the *E. coli* genetic map corroborated their independence: they were located clearly apart from each other, though in the same neighborhood (Fig. 13.1). Obviously, one was dealing with two separate genes which, nevertheless, could be induced together, regulated in common, by the same inducer. This fact would eventually provide the central pillar of the operon model which, some years later, would be proposed by Jacob & Monod (1961).

The gene with the allelic pairs constitutive/non-constitutive (i.e., normally inducible), denoted, respectively, $i$ and $i^+$ (now *lacI*), was located on the coli genetic map very close to the genes for β-galactosidase ($z$, now *lacZ*) and permease ($y$, now *lacY*); so close, in fact, as to cause Jacob & Monod to locate it erroneously between these two genes (Fig. 13.1).

What mechanism did enable gene $i$ to perform its function? Innumerable conjectures addressed this issue. Initially, Monod theorized that the presence of an inducer was a condition for enzyme synthesis to occur; constitutive mutants had this

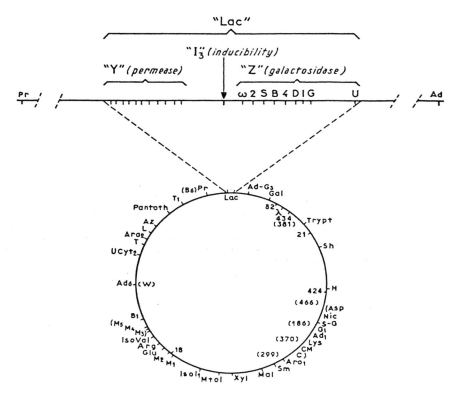

Figure 13.1: One of the first maps of the *lac* region of *E. coli* K12 (Pardee, Jacob & Monod, 1959). The definitive position of gene *i* (repressor) to the right of *z* was established only later (see also Fig. 12.3).

inducer, somehow, delivered by the cells themselves, so that no additional inducer coming from the outside was needed.

Some verifiable predictions could be inferred from this hypothesis of an internal inducer; for instance, according to it, the allele *i* (constitutiveness) should be dominant over $i^+$, since *i* would internally provide the inducer. Crosses of the type Hfr $i^+$ $z^+$ x F⁻ *i* *z* in a medium without inducer should corroborate or refute such a prediction. If Monod's theory was correct, β-galactosidase, absent at the onset of the procedure, should increase in concentration as the conjugation proceeded, even if no external inducer was offered. This should be so because none of the crossing partners could alone produce the enzyme: the donor Hfr ($i^+$, non-constitutive), on the one hand, needed an external inducer to synthesize β-galactosidase, and the receptor F⁻ (*i*, constitutive, but also a $z$ mutant, with a defective β-galactosidase), on the other hand, could not produce it at all. However, after some time having elapsed, the conjugation process on its way, the action of the receptor $i$ gene (proposed internal inducer) on the transferred $z^+$ gene should trigger the synthesis of β-galactosidase [the two genes, $i^+$ and $z^+$, as already mentioned, are located very close to each other (Fig. 13.1), thus being transferred almost simultaneously 15 minutes after the mixing of the two cultures]. The expected outcome would be comparable to the zygotic induction of phage λ. And exactly that came about – with a caveat: the synthesis went on solely for the first 90 minutes; after that time interval, production of β-galactosidase was curtailed and the receptor cells became inducible. In other words, after an initial phase of constitutive β-galactosidase synthesis, $i^+$ came to prevail; in the end, $i^+$ proved to be dominant over $i$ (Fig. 13.2). The hypothetical internal inducer turned out to be nothing more than a fata morgana which dissipated with this so dubbed PaJaMo experiment. [In the meantime, the Institut Pasteur had welcomed one more American guest scientist, Arthur Pardee, who, together with Jacob and Monod, had conceived and performed the experiments described above (Pardee, Jacob & Monod, 1958, 1959).]

Since the theory of the internal inducer could be dismissed, what could count as an alternative?

Leo Szilard was a physicist of Hungarian origin who worked on the Manhattan project during the war (in fact, he was one of its initiators). After the war he, too, turned to molecular biology. He had already a name in this circle in connection with the "chemostat". This was actually a quite banal apparatus, merely controlling the input of fresh medium into a bacterial culture, while, at the same time, the same amount of culture flowed out. The bacterial concentration in the culture could thus be kept constant by limiting supplementation with a nutrient, such as phosphate, for instance (Novick & Szilard, 1951). The bacteria, maintained indefinitely in the exponential growth phase, could then be subjected to experiments involving population genetics, as, for instance, selection experiments, detection of mutation rates, etc. Complex studies on the regulation of the rate of enzyme synthesis by mutants – similar to Monod's, but in a different system – were also carried out by Szilard. He kept an attentive eye on other groups' accomplishments, for instance, on the work carried out by his friend Werner Maas in New York on arginine synthesis (Chapter 3). The synthetic pathway of arginine started out with ornithine and passed through the intermediate product citrulline. If arginine was supplemented to the medium in which coli was being cultured, the cells failed to produce the enzymes which otherwise would be involved in its synthesis. When the coli bacteria were cultured in minimal medium (no arginine present), these enzymes were present. Remarkable was, that mutants unable to synthesize ornithine, if cultured in minimal medium, also displayed a highly increased concentration of the enzyme ornithine-transcarbamilase (the one responsible for transforming ornithine into citrulline),

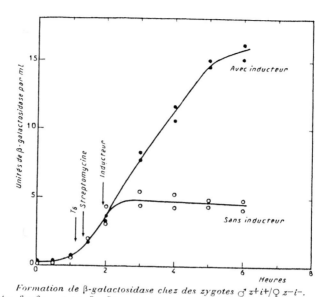

Formation de β-galactosidase chez des zygotes ♂ z+i+/♀ z-i-.

Les souches ♂ (Sm$^S$ T6$^S$) et ♀ (Sm$^R$ T6$^R$) sont cultivées sur milieu synthétique contenant du glycérol comme source de carbone. Au temps 0, les cultures sont mélangées (1 ♂ pour 5 ♀ : environ 2.10$^8$ bactéries/ml) et agitées doucement à 37°C. L'expérience est faite en quadruple. Du phage T6 (10$^4$/ml) et de la streptomycine (1 000 μg/ml) sont ajoutés aux temps indiqués pour arrêter la conjugaison et éliminer les ♂. A intervalles appropriés, des échantillons sont prélevés pour : 1° dénombrer les recombinants z+Sm$^R$ par étalement sur gélose lactose-streptomycine; 2° déterminer l'activité de la β-galactosidase. A 115 mn, on ajoute dans deux des suspensions du méthyl-β-D-thiogalactoside (10$^{-3}$ M).

Figure 13.2: The PaJaMo experiment. At time 0, two cultures of *E. coli*, strains HfrH *str$^s$ T6$^s$ z$^+$ i$^+$* and F$^-$ *str$^r$ T6$^r$ z$^-$i$^-$*, were mixed in a liquid culture medium deprived of lactose. This mixture was divided into two parallel cultures, which were further incubated at 37 °C and slightly shaken (for aeration). Phages T6 and streptomycin were added at the indicated times in order to thwart the conjugation process by eliminating the sensitive donor cells (the receptors were resistant to both agents). The β-galactosidase activity in both cultures was monitored at different times; after two hours, an inducer (methyl-thiogalactoside) was added to one of the parallel cultures. The other culture was not altered. Crucial was the observation that, during the first two hours, even without the addition of an inducer, the receptors, which in the course of the experiment had become *z$^-$i$^-$/z$^+$ i$^+$* (the genes *z$^+$* and *i$^+$* were transferred to the receptor during conjugation), actively synthesized β-galactosidase – an outcome comparable to the zygotic induction of prophage λ. After two hours, though, the synthetic activity was fully suppressed; apparently the *i$^+$* allele needed some time to become expressed, but it was, ultimately, dominant over *i$^-$* (Pardee, Jacob & Monod, 1958, © Editions Gauthier-Villars).

although, as pointed out above, no ornithine was being produced. But, also in this case, enzyme synthesis was curbed to an almost imperceptible level, as soon as arginine was supplemented to the cells. These facts paralleled those in Monod's lactose system, with the difference that the inducer in the arginine case was "nothing", namely the absence of arginine. Szilard reasoned that "nothing" can scarcely be an agent, so that the regulating factor had to be arginine itself; arginine had to have a negative effect! Apparently, its presence triggered a mechanism leading to repression of its cognate genes, which, if uncurbed, would spontaneously reach their maximal expression rate. Possibly, the synthesis of β-galactosidase could be similarly regulated by a repressor, and its inducers (lactose or MTG) would be needed to inactivate it. Minus times minus is plus! When Szilard, arguing along these lines, visited the Institut Pasteur in 1958, he and Monod nearly came to blows (see Szilard, 1960); alas, the theory of the inducer as an anti-repressor could not be so easily dismissed. The facts brought to light by the PaJaMo experiments implied that the $i^+$ allele was responsible for a cytoplasmatic agent able to thwart the synthesis of β-galactosidase. Thus, this agent was a repressor which, to be sure, became active only after some time (90 minutes), when it reached a threshold concentration in the cytoplasm; the $z^+$ gene, on the other hand, expressed its full action within a few minutes.

How could the repressor hypothesis be tested further? Jacob & Adelberg (1959) had already demonstrated that the autonomous F factor could be inserted into the main chromosome, so paving the way for the formation of Hfr strains. They reached that insight after having isolated strains in which the F factor carried, incorporated in its mini-chromosome, one or more genes from the main chromosome (an F' factor), which implied that the F factor in the process of becoming autonomous again had plucked these genes away from the host chromosome (Fig 12.4). Jacob & Adelberg's observations and consequent theorizing were grounded on a special F' factor, F'$lac$, which in addition to its own genes carried the three genes for β-galactosidase synthesis, $i$, $z$ and $y$. If F⁻ cells were infected with this F'$lac$ factor, they could, as a matter of fact, be deemed diploid for those three genes. Consequently, the infection with F'$lac$ rendered these cells partial zygotes – merozygotes. They could be either homozygotes or heterozygotes for the double set of genes in question (the homozygotes were called homogenotes, whereas the heterozygotes were the heterogenotes*. All imaginable heterogenotes for the $lac$-system were then concocted. It became an easy task to confirm that $i^+$ was indeed dominant over $i$ (see examples in Fig. 13.3). The hypothetical internal inducer was definitely dead. Still, the actual functioning of the system had yet to be figured out. But...

---

*After Crick (1988), the French had a strong preference for word constructs with roots in ancient languages, whereas the Anglo-Saxons chose instead regular words from everyday vocabulary to describe their phenomena and structures, for instance, adaptor, nonsense, overlapping, etc.

| Génotype | | Bactéries non induites. | | | Bactéries induites. | | |
|---|---|---|---|---|---|---|---|
| Chromosome. | F-Lac. | Galactosidase. | Protéine. Cz. | Perméase. | Galactosidase. | Protéine Cz. | Perméase. |
| $i^+o^+z^+y^+$............... | | $<1$ | – | nd | 100 | – | 100 |
| $i_3^-o^+z_4^-y^+/F\ i^+o^+z^+y^+$.... | | $<1$ | nd | nd | 320 | 100 | 100 |
| $i_3^-o^+z_4^-y^+/F\ i^+o^c z^+y^+$.... | | 36 | nd | 33 | 270 | 100 | 100 |
| $i^+o^+z_1^-y^+/F\ i^+o^c z^+y^+$.... | | 110 | nd | 50 | 330 | 100 | 100 |
| $i^+o^+z^+y_{II}^-/F\ i^+o^c z_1^-y^+$.... | | $<1$ | 30 | – | 100 | 400 | – |
| $i^+o^+z_1^-y^+/F\ i^+o^c z^+y_R^-$.... | | 60 | – | nd | 300 | – | 100 |

Figure 13.3: Expression of genes $z$ and $y$ in E. coli K12 heterogenotes with diverse F'lac factors. The constitutive $O^C$ mutants were originally isolated from the heterogenote $i^+ z^-/$ F $i^+ z^+$; the $z^-$ on the main chromosome was meant to render the heterogenote to be isolated easily recognizable on lactose plates. Left of the bar are the alleles of the main chromosome, to the right, the alleles on the F' factor; nd means "non-décelable" (non-detectable). The proteins Cz are the product of the $z$ alleles, $z_1$ or $z_4$; they do not show any β-galactosidase activity, but they are precipitable by the β-galactosidase antiserum – therefore called CRM, cross reacting material (Jacob et al., 1960, © Editions Gauthier-Villars).

One of Jacob's reasonings: if the repressor truly existed, it should necessarily have, somewhere, a site of action, a specific structure to which it would bind. This structure should be located near its target, i.e., the genes it represses... This hypothetical repressor's specific site of action should be mutable. Such types of defective mutants should turn the cells constitutive: if the repressor did not recognize any longer its mutated site of action, this repressor could, as a matter of fact, be deemed as non-existent. Well, not really, as we will see. First: yes, the defective mutants of the hypothetical site of action of the repressor would have to be constitutive. But, second, the expected behavior of these mutants in heterogenotes was clearly different from that of the already known constitutive repressor mutants. These, as described, were recessive, whereas the hypothetical operator mutants – "operator" was the new name accorded to the site of action of the repressor (Jacob & Monod, 1961) – would behave as dominants, at least if in cis-position. Therefore, the mutation, in order to express itself as dominant, had to be located on the same chromosome as the genes it affected (cis-dominant). The operator mutants were not expected to affect genes located on a different chromosome (depending on the construct, that meant either on the main chromosome or on the F'lac factor mini-chromosome.)

Merozygotes were the ideal subjects for searching for operator mutants – which are less frequent than the repressor mutants; we will soon see why this is so. If in a merodiploid one of the repressor genes lost its function (mutation $i^+ \rightarrow i$), this event would remain undetectable because of its recessiveness; however if the operator mutated, then the genes controlled by it would be expressed constitutively – quite indifferent to the type of genes placed on the other chromosome, i.e., the genes in *trans*-position.

The search for mutants of the operator was right away successful (Jacob et al., 1960), and the $O^c$ (operator constitutive) mutants were mapped in a short region preceding the β-galactosidase gene, $z$. The operator was very small in comparison to the extension of the repressor gene, $i$, and this was one of the reasons why operator mutations were relatively rare (one further explanation for the scarcity of operator mutations was the fact that not all point mutations in it lead to functional failure; most detectable $O^c$ mutations were deletions).

The consensus was that the genes $z$ and $y$ were regulated in conjunction. Either both were expressed concomitantly, with their gene products being intensely synthesized, or both were simultaneously repressed. The repressor gene, $i$, by means of its product, the repressor, was responsible for this inhibition. The repressor would bind to the operator, thus blocking gene expression. The suspension of the blockade, i.e., the synthesis of the corresponding gene products, came about by means of an inducer (lactose or MTG, for instance), which apparently interacted with the repressor, thus abolishing its affinity to the operator.

Cohen & Jacob (1959) observed a similar situation in an anabolic system, namely the synthesis of tryptophan. The genes for the enzymes involved in this process were also mapped close together and were coordinately regulated by another gene – let us call it $R$, for regulator. $R^+$ was also dominant over $R$, implying that it also coded for a diffusible cytoplasmatic product. The tryptophan system was only different from the lactose system in that the repressor of the lactose system was inactivated by the inducer (lactose or MTG), whereas in the *trp* system the repressor, inactive by itself, was activated by tryptophan. The implication is that tryptophan, if available, interacted with the repressor in such a way as to create an affinity for the operator. Tryptophan, in its system of synthesis, acted actually as a co-repressor rather than an inducer; either inducer or co-repressor, both being small molecules, had the common characteristic of being able to control the function of the repressor, be it by their absence or presence. Thus, a common expression was coined for them: both are effectors.

# CHAPTER 14

## INSIDE THE INSTITUT PASTEUR – PART IV
## mRNA

Indeed, all of that was very satisfying – but then came the queries. The PaJaMo experiment brought to light that, after the $z^+$ allele entered the $z^-$ receptor cell, less than three minutes elapsed till the rate of β-galactosidase synthesis attained its peak. And yet, Jacob and Monod knew well – Mahlon Hoagland, who in 1959 had been working at the Institute, had told them personally – that protein synthesis, most certainly, took place on the ribosomes. The notion – fed by various previous hints – that RNA was the template for the alignment of the amino acids during the process of protein synthesis was well established. Since in eukaryotes the nucleus englobed practically all cell DNA whereas the cytoplasmatic ribosomes garnered the bulk (80%) of the cell RNA, it seemed only too logical to conclude that the ribosomal RNA was that template – and there was no reason whatsoever to assume that in bacteria things should be different. Nevertheless, rRNA was an extremely stable molecule, not prone to be broken down once assembled – Davern & Meselson (1960) had demonstrated that fact (Excursus 14-1). Let us reason: would a gene, the allele $z^+$ of coli, be apt to assemble the amount of ribosomes needed for the β-galactosidase synthesis, which totalled a full 5% of all cell protein – and this in a mere few minutes? ... and further, after accomplishing that feat, would it be able to halt this process abruptly in order to impede the synthesis rate from going up continuously (which surely would happen if more and more specific ribosomes kept accumulating as time went by)? [That a swift increase and a sudden stop of β-galactosidase synthesis occurred had been clearly shown by the PaJaMo experiment (Fig 13.2).]

One easy way out of these difficulties would have been to admit one exception for the case of β-galactosidase and to assume that its synthesis took place directly on the respective gene, as soon as it migrated into the receptor cell, thus escaping the repressive action imposed by the donor cytoplasm. This scenario would also explain the immediate synthesis with a constant rate and the later inhibition, when enough repressor had been produced by the gene $i^+$.

Enter François Gros. He, a younger co-worker of the Jacob-Monod team, fed β-galactosidase-producing bacteria with 5-fluorouracil – an uracil analogue, taken up by the RNA as uracil. All of a sudden the enzymatic production came to a halt (Bussard et al., 1960). And, upon removing the 5-fluorouracil, the regular rate of enzyme synthesis was fully restored, and this, in a matter of a few minutes. RNA, after all, had to be involved in the process: the 5-fluoroU-RNA was not functional, probably because it was not suitable as a template. However, the RNA in question could not possibly be ribosomal RNA; unlike this latter, it was unstable, disappearing almost as swiftly as it was formed. An RNA with such an attribute had at that time never been described!?

Francis Crick, Sydney Brenner, François Jacob and some others hotly debated those issues – PaJaMo, etc. – at Cambridge in the spring of 1960. [Crick repeatedly stated (Judson, 1980; Crick, 1988) that these discussions took place on a Good Friday. With these assertions he apparently wished to emphasize his anticlericalism.] An unstable RNA? It came to Brenner's mind that such an RNA species had indeed been described some years ago, and again recently. Only, no one had paid any attention to it; nobody had tried to convert these reports into a testable hypothesis. [Volkin & Astrachan (1956, 1957); Nomura, Hall & Spiegelman (1960); Yčas & Vincent (1960) – see Excursus 14-2]. This fleeting, transitory, almost ungraspable type of RNA seemed to exist after all; one was even tempted to say: it had to exist. Everything would start making sense. An RNA molecule would be assembled by

Figure14.1: François Jacob (born in 1920)

Figure 14.2: Jacques Monod (1910-1976)

copying one of the strands of the DNA of a gene (through base pairing between free nucleotides and the DNA strand complementary to the RNA to be synthesized); the newly assembled template RNA would migrate from the nucleus to the cytoplasm, where ribosomes would attach to it. (Failing to recognize this detail was the main flaw of Spiegelman and his group; they were not alert enough to discern between the elusive template RNA, bound transitorily to the ribosomes, and the structural ribosomal RNA, a stable architectural feature of the ribosomes – see Excursus 14-2.) The ribosome would represent the apparatus able to translate the 4-symbol-script of the template RNA into the 20-symbol-language of the proteins; its own, ribosomal RNA, packed between the ribosomal proteins, would solely serve as a scaffold, but not as a template. This third species of RNA – besides rRNA and tRNA (Excursus 14-3) – had been detected by Volkin & Astrachan (1956, 1957) and by Yčas & Vincent (1960), but its meaning remained undisclosed; it had been observed by Nomura, Hall & Spiegelman (1960), but erroneously interpreted as newly assembled rRNA (Excursus 14-2). This third species of RNA had to decay rapidly (at least in *E. coli*), after serving as template for the synthesis of a few protein molecules. This instability was required to fit the PaJaMo and the 5-fluorouracil-incorporation experiments. No wonder that this RNA remained unrecognized for so long. New experiments, meant to verify this novel concept, were hastily planned; then the crucial small gathering in Cambridge was over.

Some weeks later, Brenner and Jacob met again at Caltech, in Meselson's laboratory (it coincided that both, though for different reasons, were in California at the same time; Brenner had been invited by Delbrück, Jacob by Stent). Matthew Meselson, as the unchallenged world expert in density gradient centrifugation (Excursus 6-2 and 14-1), was an indispensable consultant on technical matters. But Brenner and Jacob knew already perfectly well what they were after. They aimed to demonstrate the existence of a newly synthesized, unstable RNA, bound to previously existent ribosomes. A phage infection with T4 should clarify the problem that they were tackling. [Seymor Cohen (1947) had shown that infection with T4 imposed on the host bacteria an immediate disruption of its net RNA synthesis.] $^{32}$P-phosphate would be offered to coli cells immediately after phage infection and all newly synthesized RNA would be identified through its radioactivity. Demonstrating that this radioactivity was linked to the pre-existing "old" ribosomes was the goal. In order to tell apart the "old" ribosomes from the possibly newly formed ones, a trick was employed: colis would first be cultured in "heavy" medium, i.e., a medium containing nutrients labeled with the heavy isotopes $^{15}$N and $^{13}$C (see Excursus 14-1) and subsequently transferred to regular medium (with the normal isotopes $^{14}$N and $^{12}$C). From this moment on, all newly assembled ribosomes would have normal density, in contrast to those from the first phase (with heavy isotopes). The "old" and the newly synthesized ribosomes should become distinguishable by their different densities and could, thus, be eventually separated by CsCl density gradient centrifugation. The so prepared cells (first grown in heavy medium and then transferred to a normal one) were infected with phage T4 and, after a few minutes, collected by centrifugation, broken open and mixed with CsCl for the ultracentrifugation.

Because merely a scant amount of the heavy isotopes $^{15}$N and $^{13}$C was available to Brenner and Jacob, the amount of heavy ribosomes would not suffice to grant a visible ultraviolet absorption band in the CsCl density gradient ultracentrifuge tube. That is why, before ultracentrifugation, the "heavy" cells were mixed with a large surplus of "light" (normal) ones. This would guarantee, at least, to pinpoint the exact position occupied by the light ribosomes. The result: in a control experiment with solely light ribosomes (from a normal, "light", coli culture infected with the phage T4), the distribution of radioactivity (radioactivity emitted by newly synthesized RNA) overlapped with that of the ribosomes. In the actual experiment ("heavy" colis infected with T4 in normal medium), the radioactivity was clearly placed at a position different from that of the light ribosomes and this position corresponded to a higher density (Brenner, Jacob & Meselson. 1961). The only viable interpretation for this outcome was that, after T4 infection, newly synthesized RNA was attached to heavy ribosomes (too few to be seen), which were present before the infection with T4 took place. That was the first close encounter with the RNA of the third kind.

In the meantime, François Gros landed on the continent's east coast, heading for Harvard, more precisely for Watson's laboratory. He, too, was focusing on the same issue. Even more convincing than the phage infection experiment would be a demonstration that the novel, fleeting type of RNA existed in normally growing coli cells; the postulated template RNA should also exist in these. The following procedure was attempted (Gros et al., 1961): $^{32}$P-phosphate was offered to coli cells for a brief time interval (for instance, one minute). After that short period, the uptake of radioactive phosphate was curtailed to a negligible level by adding a massive excess of regular, non-radioactive phosphate. This procedure allowed the identification of a quickly disintegrating RNA species which was bound to the ribosomes and whose specific base composition was different from that of the coli ribosomal RNA, and corresponded to that of coli DNA (equating uracil with thymine, of course). During the short time interval – a so-called pulse – in which the cells were given $^{32}$P-phosphate, a few rRNAs incorporated it too; this rRNA was, as we know, stable; however, a large proportion of the total RNA being radioactively labeled during the short pulse was of an unstable type. This observation was not essentially different from that made a year before by Yčas & Vincent in yeasts (Excursus 14-2). Alas, essentially different was the prestige of the protagonists, and the glamor of an experiment elegantly corroborating an exciting novel hypothesis.

Jacob and Monod deemed the moment opportune for compiling all available data in an extensive overview (Jacob & Monod, 1961). There it was exposed, explicitly and for the first time, the concept of mRNA – to the chagrin of Crick and Brenner (see Judson, 1980) who themselves had actively contributed to all these novel, fundamental insights.

Except Elie Wollman, the central figures of the Institut Pasteur of that era – André Lwoff, François Jacob and Jacques Monod – became the Nobel Prize laureates for Physiology or Medicine in the year of 1965.

## Excursus 14-1
## RIBOSOMAL RNA IS STABLE (DAVERN & MESELSON, 1960)

Matthew Meselson and his graduate student at Caltech, C. I. Davern, fed coli cells with the heavy isotopes $^{15}$N and $^{13}$C. Labeling with $^{15}$N was an easy matter; it sufficed to offer $^{15}$NH$_4$Cl as a nitrogen source to the growing bacteria (Excursus 2-1). Labeling with $^{13}$C, on the other hand, required some tricks. [The isotope itself was courteously made available to them by Linus Pauling. Pauling, for his part, had been given it, courteously, by the Academy of Sciences of the USSR, of which he had been elected a member – no banal feat at the Cold War's apogee.] The $^{13}$C, first in the form of $^{13}$CO$_2$, was then photosynthetically assimilated by single-celled green algae (*Ankistrodesmus*) in a closed system in which $^{15}$NH$_4$Cl was the nitrogen source. A hydrolysate from these algae, added to the medium, provided the organic building blocks for a coli culture. A sample from this culture with high density-labeled coli cells was then removed, to be used as a control, and its RNA prepared for ultracentrifugation (80 hours at 50,000 rotations/min) in a CsCl gradient (Excursus 6-2); within this gradient, heavy RNA could be monitored. A large excess of nutrients with normal isotopes ($^{14}$N and $^{12}$C) was given to the remaining culture which was further incubated. From this moment on, all new RNA was synthesized using building blocks of normal density. After many cell divisions, the RNA of this culture was equally subjected to a CsCl density gradient centrifugation. Just two RNA bands, corresponding to two different densities, were seen: one with "old" heavy RNA and another one with newly synthesized RNA of normal density. No bands of intermediate density showed up; i.e., the ribosomal RNAs were stable structures: their nucleotides, once having been assembled, were enduringly inseparable and non-dissociable, no exchange of building blocks was taking place.

**Excursus 14-2**
**A NOVEL, UNSTABLE SPECIES OF RNA**

Elliot Volkin and Lazarus Astrachan, from the biology divison of the Oak Ridge National Laboratory in provincial Tennessee – better known for its nuclear technology –, had treated T2-infected coli cells with radioactive phosphate and noticed that a remarkable RNA fraction of the infected cells became radioactively labeled. What made this RNA so remarkable was its base composition: it mirrored that of the infecting T2 DNA (equating hydroxymethylcytosine with cytosine and thymine with uracil). The fact that the A + T fraction of T2 DNA was only half that of coli left no room for doubt. Doubts arose, however, when an interpretation of this observation was attempted. Was this RNA, endowed with a relatively high turnover rate (quickly synthesized and equally swiftly broken down), perhaps, a precursor of T2 DNA? I.e., would it be converted to DNA (Volkin & Astrachan, 1956)? The researchers kept their experiments going (Volkin & Astrachan, 1957), extending them to include some with T7-infected colis (Volkin et al., 1958). The results were basically the same. However, the author's way of interpreting them had radically changed. The previous assumption, that the RNA was a precursor of DNA, turned out to be unsustainable: the rate of synthesis of T2 DNA proved to be many-fold higher than that of the RNA. They stated: "It is possible that the specific kind of RNA, synthesized by the host under the direction of the phage, functions as unity (template?) responsible for the synthesis of phage-specific proteins" (Volkin, Astrachan & Countryman, 1958). But, who would possibly be interested in statements made by outsiders from Tennessee? The authors were never cited.

And then, Sol Spiegelman and his group at the University of Illinois, Urbana, exactly as Volkin and Astrachan had done some years before, labeled T2-infected bacteria with $^{32}$P-phosphate; the labeled RNA, however, was analyzed somewhat more thoroughly and more refinedly than it had been by their colleagues in Tennessee: not only was its base composition gauged, but also its sedimentation rate and its mobility in an electric field. All characteristics were clearly distinct from that of previously existing coli RNA. Besides, they observed that a large proportion of that RNA was bound to the ribosomes, although less firmly than the regular ribosomal RNA. Unfortunately, though, they came to recognize the meaning of this detail only later – too late. What they originally believed was – and here was the flaw – that RNA bound to ribosomes could be nothing but ribosomal RNA (Nomura, Hall & Spiegelman, 1960). Spiegelman went even a step further in the characterization of the T2-specific RNA in infected coli by demonstrating that this RNA hybridized with T2 DNA, which meant that its base sequence must be complementary to one of the T2 DNA strands (Hall & Spiegelman, 1961). Besides that, one of Spiegelman's graduate students, Masaki Hayashi, radioactively labeled a transient RNA species of uninfected coli cells, and, on top of that, showed that this RNA hybridized with denatured coli DNA (Hayashi & Spiegelman, 1961; see also Excursus 6-3).

[Open to debate is the question as to what extent the idea of searching for hybrid double helixes was influenced by a brief experiment, carried out by the X-ray crystallographer Alexander Rich at MIT. Rich had mixed synthetically made polyriboadenylic acid (poly-A) with polydeoxyribothymidylic acid (poly-dT), demonstrating that they intertwined spontaneously (one strand RNA-like, the other DNA-like). Rich, for his part, made it quite clear that he was aware of the crucial meaning of the transfer of genetic information from DNA to RNA; and he even suggested that the evidence he had gathered with regard to DNA-RNA hybridization could help to address the issue of genetic information transfer (Rich, 1960).]

Spiegelman (1961) quickly coined the expression "informational RNA" (as opposed to rRNA and tRNA – but see also Excursus 14-3) meaning the "informed mediator" between DNA and protein. In the meantime, he had painfully grasped that a third kind of RNA, hitherto not clearly defined, had to exist in the cell – but to no avail – it was too late, Jacob and Monod (1961) had already stolen the show.

The one participant in this competitive endeavor who did not stand any chance was Martynas Yčas, Gamow's friend and theoretical code breaker (see Chapter 8) in Syracuse, N.Y. To start with, the aims of his queries were somewhat unintelligible: he tried to elucidate a synthesis mechanism for RNA and, at the same time, its nucleotide composition. He labeled the RNA of yeast cells with radioactive phosphate and proceeded to hydrolyze this material, using either snake venom or alkali; snake venom diesterase cleaved the 3'-sugar-phosphate-ester bonds of the RNA molecule; alkali split its 5'-bonds. The $^{32}$P-phosphate was offered to yeast cells for variable periods of time; and considering that phosphate incorporation into the different RNA building blocks, the nucleotides, occurs at correspondingly different rates, it was to be expected that, upon hydrolysis of the puls-labeled RNA by the two methods, eight different nucleotides resulted, (each one of the four bases as a 3' or a 5' phosphate), which would display different levels of specific radioactivity. Based on such differences in RNA fractions obtained after varying labeling times, Yčas & Vincent (1960) could show, for their own astonishment, that in normally growing yeast cells different RNA species existed: one of them was synthesized relatively quickly and then broken down; and this RNA fraction had a base composition that paralleled that of the yeast DNA. What Volkin and Astrachan had demonstrated for phage-infected coli had its analogue in normally growing yeast cells. Yčas and Vincent's reasoning also matched that of Volkin et al. (1958): "The function of such an RNA fraction is not clear. In view of its composition, it might be a primary gene product, acting as an agent for transmission of genetic information from DNA to protein. Alternatively, it could be storing information for the replication of DNA itself, if such a process is, as has been suggested, of an indirect nature." [This was a reference to one of Stent's wild speculation; see Delbrück & Stent (1957)]... Had they only dropped this last sentence, they would seem to us, today, quite a lot smarter.

**Excursus 14-3**
**rRNA-DNA & tRNA-DNA HYBRID HELIXES**

After Spiegelman had succeeded in demonstrating the formation of hybrid double helixes between coli DNA and its counterpart "informational" RNA (today mRNA), his laboratory developed a taste for this type of work and decided to expound the origin of the other, regular, RNA species, namely rRNA and tRNA (back then still called sRNA; see Chapter 9). This issue had yet to be addressed experimentally and was still involved in many uncertainties. Whereas the cell's total mRNA had at its disposition innumerable complementary sequences spread all over the genome – this RNA was essentially a collection of copies from all cellular genes –, one had to expect that the rRNA, with its sequence of just around a few thousand bases, and the tRNAs, would hybridize, if at all, with merely a minimal fraction of the cellular DNA. That these RNA species were indeed copies of matching DNA base sequences was no settled matter; especially the tRNAs, with their relatively small and compact molecules displaying many internal hydrogen bonds, lent room for assuming an autonomous replication or an enzymatic synthesis without templates. Again, two of Spiegelman's graduate students, Saul Yankofsky for rRNA, and Dario Giacomoni for tRNA, succeeded in demonstrating that these RNAs were also clearly hybridizable with the DNA of the organism of origin, although, as already surmised, with less than 1% of the total amount of DNA. The fact that pancreatic RNAase was not able to digest hybridized RNA was an important asset for the demonstration of such scant quantities of DNA-RNA hybrid helixes. After subjecting the material to be analyzed to RNAase treatment, it became feasible to detect even the faintest radioactivity emitted by RNA-DNA hybrid helixes as acid precipitable material (Yankofsky & Spiegelman, 1962,a,b; Giacomoni & Spiegelman, 1962; see also Giacomoni, 1993).

# CHAPTER 15

# THE OPERON

The cornerstone of that review (Jacob & Monod, 1961) lay in the novel, seminal concept of a coordinated regulation mechanism affecting neighboring genes, all of them involved in the same metabolic pathway, each of them commanding one of its steps. The genes in question were to be expressed or switched off conjunctly. This coordinated regulation depended primarily on the action of a specific regulator gene, whose product affected the genes to be controlled by repressing their expression: it was a repressor. The repressor acted directly on its target structure situated on the chromosome (or on a copy of this structure; see Fig. 15.1), immediately adjacent to the genes to be regulated. Those, the genes responsible for the structure of the respective enzymes, were denominated structural genes (in contrast to the regulator gene, the one commanding the synthesis of the repressor). The repressor's target structure on the chromosome was the so-called operator. If no repressor was available, the expression of the regulated genes would start from this operator site. A repressor molecule, by occupying the operator, would block this process. All the structural genes, together with their operator, aligned in a row, made up a functional unity: the operon. The expression of the operon consisted in the production of the so-called transcript of the respective genes, starting from the operator site. This transcript was assumed to be an RNA molecule, built as a sequence of bases complementary to the base sequence of one of the DNA strands. The postulated transcript, once produced, moved away from the DNA, its place of origin, into the cytoplasm, the venue where the ribosomes were to be found. The hypothetical transcripts, representing the blueprint for the synthesis of their respective proteins, attached to the ribosomes, the protein synthesis apparatus, providing them with the specific piece of information these transcripts carried. These revolutionary new ideas stood as a veritable breakthrough in the field of molecular biology: ribosomes were solely translation machines, adept at processing any genetic information supplied to them by the RNA transcripts, themselves copies of the informational sequences stored by the genes. One was dealing, thus, with a totally new kind of RNA, the messenger RNA (mRNA). The mRNA was unidirectionally assembled along the operon, the process having the operator as a point of departure; this scenario would account for the coordinated expression of the operon's structural genes. The putative mRNA would also explain, for instance, why the *lac* operon's expression (production of β-galactosidase) reached its peak level with such astonishing speed, as demonstrated by the PaJaMo experiment (Fig. 13.2): the already existent ribosome would promptly be engaged in the assemblage of new proteins as soon as the freshly made mRNA, carrying the specific information, became available. And beyond that, the mRNA – at least in coli – would be transient, as required by the observed immediate curtailment of protein synthesis as soon as production of mRNA is halted. And further, the mRNA would stop being synthesized as the operator became blocked by the repressor. The repressor was, thus, an agent of negative control; namely, its action was to prevent the otherwise spontaneously and intensively occurring synthesis of mRNA, the mediator between gene and gene product.

In the case of catabolic operons, like the well documented example of the *lac* operon, the repressor, by itself, was endowed with affinity for the operator. That affinity would however be lost upon interaction with the inducer. This inducer, a small organic molecule, was, in most cases, the starting substrate in the catabolic pathway, to be worked upon by the gene products resulting from the expression of the operon. The inducer for the *lac* operon system was, thus, lactose. The situation faced by the anabolic operons was slightly different; in the example of the *trp* operon, the repressor was actually inactive, as such, till it interacted with tryptophan, the end product of the synthetic pathway; this interaction with tryptophan was essential for establishing the affinity of the repressor for the operator.

The insights previously gained from studies with lysogenic bacteria and their temperate phages were essential for Jacob and Monod's coming to terms with the new data; phage λ, for instance, expressed a series of genes in the course of a lytic growth cycle. However, if the same virus was trapped in a prophage stage, these genes were not expressed. This state of affairs could be accounted for by a repressor, coded for by a λ regulator gene. This assumption was supported by observations made with some λ mutants which failed to establish lysogeny upon single infection, but could overcome this incapacity if infecting in conjunction with wild type λ. Such a mutant could only be maintained as a prophage in doubly lysogenic cells, when it could rely on the wild type λ as its partner. None of the two phages – neither the wild type nor the mutant – could, under this circumstance, overcome the host's immunity. This example of λ mutants was a strict parallel to the repressor-defective, recessive, constitutive mutants of the *lac* system. Other, exclusively virulent, λ mutants were also discovered. Apparently, defective or deleted operators stood behind such phenotype. These mutants would always entail a lytic cycle even if the infected cells were already lysogenic (the virulent mutant's DNA was, apparently, impervious to the prophage λ repressor); even a double infection, together with the wild type, failed to impose the prophage status on these virulent λ mutants. Comparing them to the *cis*-dominant operator mutants of the *lac* operon ($O^c$ mutants) lay at hand.

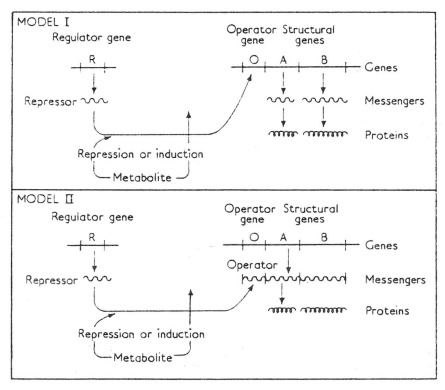

Figure 15.1: The first model for the regulation of an operon (Jacob & Monod, 1961). The regulator gene *R* (*i*, in the case of the *lac* operon), produces a repressor (first believed to be an RNA molecule, later identified as a protein) which, depending on whether the operon is anabolic or catabolic (as, for instance, those for tryptophan or lactose, respectively), will be activated or inactivated (in the figure described as "Repression or induction". The active repressor occupies the operator. In model I, the operator is represented by a DNA segment at one end – actually the beginning – of the operon; here, this segment is named the "Operator gene". Nowadays, the operator is not considered as a gene (it does not code for a gene product), but as a "signal sequence". If the operator is occupied by a repressor, no mRNA can be formed; this mRNA would represent a copy of the base sequences of one of the DNA strands of the structural genes (in the scheme, A and B). In model II, the possibility is raised that an RNA, copied from the operator, binds to the repressor; in this instance, mRNA would always be churned out, but could only be translated in the absence of the repressor; Model II was thought of as a theoretical possibility, which could not be confirmed in this case, although today – specially in eukaryotes – many examples are known in which the translational step depends on specific regulation mechanisms: translational control (see, for instance, Kozak, 1992; Richter & Theurkauf, 2001).

It looked as though all the well analyzed systems were based on the same fundamental principle of gene regulation: negative control! And these systems were essentially different: the one referred to a catabolic metabolism (the case of the *lac* operon), the other to an anabolic synthetic pathway (the *trp* operon); and the strategies of virus infection also fitted the model. This state of affairs enticed Jacob and Monod to speculate that negative control was the universal principle behind all mechanisms of gene regulation.

According to them, that same principle would reign over the most diverse models of gene expression in cells of higher organisms, affecting, for instance, cell differentiation in various tissues, tumor development, and even embryonic growth. The similarities between cancer cells and constitutive *lac* mutants seemed only too cunning not to be true.

It turned out that resisting the hypnotic power wielded by the original operon model required an enormous effort. Nevertheless, innumerable observations contradictory to the pristine model started to accumulate. Not to consider the possibility of a positive control became gradually untenable. Regulation by positive control implies that the product of the regulator gene stimulates the expression of structural genes, instead of repressing them, as in all cases documented till that time. In a positive control scenario, the regulator gene product acts as an activator – not as a repressor (see, for instance, Reznikoff et al., 1985). Ellis Englesberg at the University of California in Santa Barbara described such a situation for the arabinose operon of *E. coli* (see, for instance, Englesberg et al., 1965, 1969). However, it was only at the cost of intense energy and talent of persuasion that his results were finally granted general recognition (see, for instance, Beckwith, 1987). Later, other types of regulation mechanisms intrinsically different from the ones already known were discovered. For instance, the principle of attenuation which underlies the activity of the operons for the synthesis of many amino acids. There are no regulatory proteins for these operons, neither repressor nor activator exists. The synthesis of their respective mRNAs always starts out with constant rates; when the pool of newly synthesized amino acids reaches an adequate level, their respective charged tRNAs, consequently, also attain such a critical level. This high supply of amino acyl-tRNA leads to a change in the pattern of distribution of ribosomes on the corresponding nascent mRNAs; the resulting secondary structures of these mRNAs lead to their cleavage and thus to an abrupt halt of the transcription process, before the structural genes are transcribed (see, for instance, Kolter & Yanofsky, 1982; Yanofsky, 1988).

Even the classical *lac* operon model was far from totally correct and complete in its original form:

The chemical nature of the repressor was unknown, although Jacob and Monod strongly believed it was an RNA molecule. This assumption was based on the erroneous interpretation of experiments with antibiotics that had an inhibitory action on protein synthesis. However, the proteic nature of the repressor was later clearly established (see, for instance, Gilbert & Müller-Hill, 1967).

The mechanism by which mRNA was formed was not given a thought. The concept of an RNA polymerase (see Chapter 16) acting as the RNA-synthesizing enzyme was missing; mRNA just popped up, somehow. Consequently, the notion of a promoter, as a binding element for the RNA polymerase, independent of the operator, was only later conceived (see Ippen et al., 1968).

The fact that the lactose operon could also be subjected to a positive control mechanism was not realized by Jacob and Monod; this was no surprise, because from the beginning of their review, they emphasized that they were not taking the glucose effect into account. This effect – no β-galactosidase production occurs if glucose is present, not even in constitutive $O^c$ mutants – seemed too uncomfortable to them. They summarily declared it as not particularly significant; their peremptory but flawed judgement led them to ignore and forget the glucose effect. Other authors thought differently and interesting findings resulted from their objections (see, for instance, Kolb et al., 1993).

The interaction mechanism between repressor and effector (effector: the general term for inducer and co-repressor) was another issue not yet grasped. This state of affairs would not change as long as the repressor was seen as an RNA molecule which supposedly bound to the operator through base pairing. When its proteinic nature was finally recognized, only then, could an appropriate mechanism for its mode of action be postulated; and this was exposed in an extensive essay. The repressor's tertiary structure would be slightly altered by binding to the effector, so that its affinity to the operator would be lost (in the case of catabolic operons) or brought about (in the case of anabolic operons). At the beginning of the 20th century the physiologist Christian Bohr (father of the quantum physicist Niels Bohr) had already described how proteins could have their characteristics influenced by specific small molecules (as verified later, without altering the topology of the folded peptid chain). Bohr's observations referred to the hemoglobin molecule. But, in the time span between these observations and the publication of that extensive essay (Monod, Changeux & Jacob, 1963), many other good examples of protein suppleness had become known. With their essay, the authors earned themselves the merit of putting all these examples on a common ground and, last but not least, finding an all alluring name for the phenomenon they described: allostery! The interaction of an allosteric effector with the allosteric center of a protein caused a – subtle – allosteric change in its tertiary structure, leading to an essential modification of its affinity for other structures.

This was a crucial mechanism underlying not only the activation or inactivation of repressors, but also such other phenomena as feed-back inhibition, cooperation between subunits of quaternary structures, like the hemoglobins, or still more intricate systems... It was an all encompassing principle, its scope covering the whole of molecular biology.

Today, the mechanisms of gene regulation are recognized to be a lot more complex and variable than Jacob and Monod's original notion stipulated. This applies not only to regulation at the level of mRNA synthesis, i.e., transcription, where scores of variations are now known to exist, but also at the level of protein synthesis on the ribosomes, the so-called translation (see, for instance, Kozak, 1992). Yet, that pristine model stands as the essence of the generally applicable principle of gene regulation; it recognizes gene regulation as dependent on specific interactions between genetically informative macromolecules. The revolutionary concept of specific protein-nucleic acid interaction, indirectly brought forth by the Jacob and Monod model, emerged as an unheard of novelty for molecular biologists and biochemists. The last decades have witnessed the validation of that concept as the basis of genetic specificity, so that, nowadays, in the field of molecular biology, no issue can possibly be addressed without taking it into account.

# CHAPTER 16

# RNA POLYMERASE

The RNAs were an undeniable reality: rRNA, tRNA, and finally the ephemeral mRNA; they had moved to the center stage of molecular biology. However, the process of their synthesis was still awaiting elucidation. Jacob and Monod did not even care to admit the existence of a problem. This issue was taken up by the classical biochemists, typifying an era whose pinnacle had already passed. Many of these scientists were quite accessible to new challenges coming from molecular biology – they were, actually, too smart to react otherwise, contradicting insinuations concocted by some of the new heroes of molecular genetics. The biochemists' main field of action, where their visceral main interest lay, were the classical problems of metabolism (where does the energy come from?), the enzymes. Epitomizing that school of thought, R.M.S. Smellie at the University of Glasgow, one of the pioneers in RNA synthesis, published a review which starts out with a never ending reflection on the synthesis of nucleotide building blocks: "The reaction sequence... involves the reaction of carbamyl phosphate with $CO_2$ and glutamate to yield ureidosuccinic acid, ring closure to form dihydroorotic acid, and oxidation to form orotic acid, which then reacts with phosphoribosyl pyrophosphate to give orotidine 5'-monophosphate. 5'-UMP is formed from orotidine 5'-monophosphate by decarboxylation and the cytidine nucleotides are derived from the uridine nucleotides by amination...". And beyond all these details, there were systems in which one or two nucleotides were attached to RNA molecules. All these themes spoke directly to biochemists and warmed their hearts....but where were the concerns for the problems of information transfer hidden? Politely mentioned, yes – but merely at the end of the 30-pages text: ...it is postulated that during the assemblage of a protein molecule, amino acids, coupled to sRNA, are transferred to a template, most probably an RNA molecule associated to a ribosome (Smellie, 1963).

Very murky for biochemists seemed to be a variety of enzymatic activities already known to incite, in manifold ways, the attachment of individual nucleotides to RNA: the one enzyme managed to attach an uridyl residue; another one, an adenyl residue; further, there was one known to attach pCpCpA (three nucleotides with the

bases cytosine, cytosine and adenine) to the RNA, and still another one could
polymerize any nucleotide, but only if a primer was present [this could be an RNA
(see for instance, Chung et al., 1960; Straus & Goldwasser, 1961; Hurwitz &
Bresler, 1961), or else a DNA (Weiss & Nakamoto, 1961,a,b; Hurwitz et al., 1961)];
and that not to mention the polynucleotide phosphorylase which, without any primer
or template, polymerized diphosphonucleosides to long chains of random sequences,
liberating inorganig phosphate (see Chapter 10).

Of the multitude of these reactions, which ones were those to be seen as crucial?
Since the discovery of the DNA double helix it was accepted as an uncontestable fact
that its base sequence was responsible for ultimately determining the expression of the
respective genetic traits. And RNA was accorded the role of an intermediary in the
process. As a corollary, its base sequence was assumed to be a direct reflection of the
DNA sequence. A supposition fraught with such a fundamental meaning did not,
however, suffice to impel neither molecular biologists nor biochemists to the quest for
the biochemical mechanisms of information transfer. Some minds among the latter,
though, keen to unearth ever new or ever more exciting enzyme activities, had
already come across the polymerization of triphosphonucleosides into RNA in cell-
free extracts from either coli (Hurwitz et al., 1960) or liver (Weiss & Gladstone,
1959; Weiss, 1960), with liberation of pyrophosphate – as long as DNA was present!
In the wake of these observations, two insights led to the discovery of RNA
polymerase, a crucial enzyme responsible for catalyzing these polymerizations: the
first was to offer DNA as template (the terms template and primer were
interchangeable at the time – we will come back to this point in the following
chapter); the second was to provide 5'-triphosphonucleosides as substrate building
blocks. Because RNA (of course also DNA) was built up of nucleoside
monophosphates, which, depending on the method of hydrolysis, were liberated as 5'-
monophosphates or 3'-monophosphates (see, for instance, Excursus 14-2, last
paragraph), it was not evident from the beginning that 5'-triphosphoribonucleosides
had to be supplied as substrates for the polymerization reaction. But Arthur Kornberg
– as we will soon witness – had already tackled the issue of DNA replication. He had
noticed that for *in vitro* synthesis of DNA 5'-triphosphodeoxynucleosides were
required as energy-bearing building blocks; and the template dependence of the DNA
polymerase reaction, a sensational novelty, had not escaped his attention. Not wanting
to demean the merit of the participants in the enzymological clarification of RNA
synthesis, the analogy to DNA synthesis was only too evident as not to have served as
an inspiring model.

Biochemists fell under the spell of the cell extracts endowed with the enzymatic
activity of promoting the template-dependent polymerization of nucleotide building
blocks. It became an irresistible challenge for them to purify and enrich such an
activity by means of traditional biochemical fractionation methods. Their undertakings
led to the identification of a relatively large protein, isolated from *E. coli*, with
molecular weight of 400,000, today known as RNA polymerase but at the time also

named RNA-nucleotidyl-transferase, which directed RNA synthesis. Once available, the purified enzyme helped to unravel intricacies of the reaction: it started in the presence of DNA and all four triphospho-nucleosides ("triphosphates"), ATP, CTP, UTP, GTP; the essential conditions were: presence of magnesium or manganese ions, temperature optimum of 38 °C and pH 8 (Chamberlin & Berg, 1962). The most effective template turned out to be double-stranded DNA, although single-stranded DNA would also do. With double-stranded DNA as the template for the reaction, free molecules of RNA were swiftly garnered in the test tube, whereas if the template was single-stranded DNA, free RNA was only slowly churned out, starting at the second cycle of template use. This could conceivably be due to initial RNA molecules remaining bound to the single-stranded DNA template, forming hybrid double helixes, thus suggesting base complementarity between the template and the produced RNA. In any case, the base composition of the RNA produced was strictly template-dependent (Hurwitz et al.,1962).

The biochemical demonstration of the enzymatic activity, as well as the characterization of the enzyme, certainly brought the researchers involved a great deal of satisfaction and recognition. Nevertheless, these first accomplishments had raised more new problems than answers to old questions.

One example: in the early *in vitro* experiments both DNA strands were used as templates. Nevertheless, from the two RNA copies obtained, only one could presumably be endowed with a physiological meaning, as for instance, that of an mRNA. Would Nature be that wasteful? If yes, what was the fate of the anti-mRNA inside the cell? And, after Jacob & Monod (1961) had launched the notion of mRNA, the lines of reasoning had to change, even for biochemists. The acceptance of the idea that the synthesis of specific mRNAs, as, for instance, that of the lactose operon, was individually triggered or prevented, led scientists to consider when and where, in the cell, such RNA synthesis would start; and how would it be brought to a halt? These matters, implying a totally different category of biochemical problems, could not possibly be approached by such rough *in vitro* RNA polymerase experiments. Or could they...?

The more carefully controlled and elaborate the assay conditions for the RNA polymerase reaction became, the clearer it was that a special mechanism guaranteed the choice of only one of the DNA strands as the true template; a breakdown of this mechanism made both strands available as templates, as was initially observed. Moreover, if the template DNA was denatured, or if short double strand segments or double strand DNA with many single strand breaks were offered as templates, then both strands would be transcribed; apparently, RNA polymerase jumped indiscriminately on single strand DNAs or free ends (Vogt, 1969). How the meaningful DNA strand was selected as a template was a question that remained unanswered.

Technical innovations are known to have, sometimes, more leverage than brave new ideas for the advancement of science.

Hitherto, a chromatographic column of phosphocellulose had been the method of choice for the purification of the coli RNA polymerase (see, for instance, Burgess et al., 1969). The process yielded a protein composed of four polypeptide subunits; two of them were identical (the α-subunits with a molecular weight of 40,000) the third and fourth were unequal (β- and β'-subunits with molecular weights of 155,000 and 165,000, respectively). This RNA polymerase composed of ααββ' was adept at binding to any site on double-strand DNA, so that, for instance, in a scenario where the enzyme was in excess, the DNA would be crammed with the enzyme molecules. In the presence of the four triphospho-ribonucleosides, the DNA was symmetrically transcribed, i.e., transcription went along both strands, but only – a critical point neglected for years – if single-strand breaks (the so-called nicks) were present in the template. Nicks always sneaked in when the procedure for DNA preparation had not been exceedingly and painstakingly cautious (Geiduscheck et al., 1964; Green, 1964). The ααββ' complex could still bind to a double-stranded DNA devoid of nicks, but without proceeding to transcription. The not so heedful manners of the first RNA polymerase researchers turned out to be their good fortune!

If less drastic methods for isolating RNA polymerase were used, as for instance, centrifugation in a glycerin gradient, then a protein with five subunits resulted (Burgess et al., 1969), the ααββ' complex being complemented by another polypeptide, the so-called σ-factor (sigma-factor, molecular weight 95,000). The relatively stable ααββ'-complex became known as "core polymerase" and the whole complex, including the σ-factor, was the "holoenzyme". The σ-factor, apparently, was attached only loosely on the ααββ' complex, being detached from it by the phosphocellulose. Thus, it had remained unnoticed at the beginning. At first glance, the σ-factor seemed to exert, primarily, a restrictive role on the process. For instance, it refrained the core polymerase by curbing its typically indiscriminate and intense mode of binding to the double-stranded DNA: if double-strand DNA had been handled cautiously, displaying neither single-stranded segments nor breaks of the sugar phosphate backbone, then merely a few holoenzyme molecules would bind to it (one holoenzyme for many thousand base pairs). The choosy holoenzyme, in contrast to the promiscuous core polymerase, limited the binding to special sites on the

Figure 16.1: The binding of RNA polymerase to promoters. Purified preparations of *E. coli* RNA polymerase and DNA of phage T7 were mixed and analyzed under the electron microscope. Genetic analysis had attested that three promoters for the coli RNA polymerase were localized at one end of the T7 DNA (E). Indeed, in the preparation one can see one RNA polymerase molecule (P) bound to each of these promoters; because of the absence of RNA building blocks, no RNA synthesis went on; this forced the RNA polymerase molecules to remain linked to the promoters; *b*, *c* and *d* are images similar to *a*, but enlarged (Bordier & Dubochet, 1974).

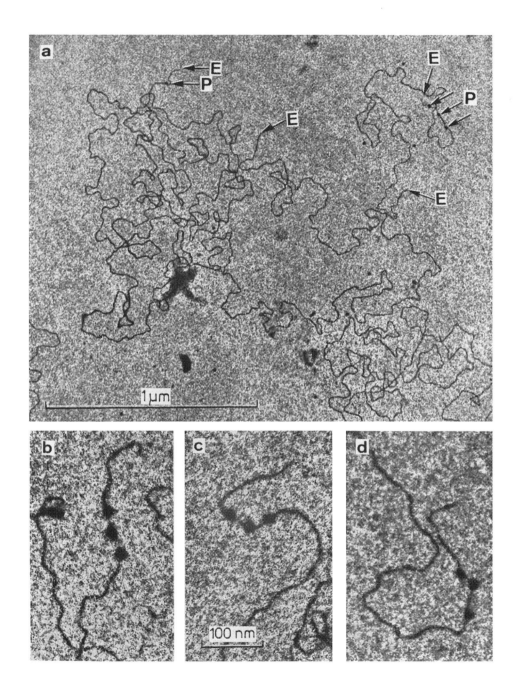

template, conceivably the natural starting points for the physiologically meaningful transcription. The holoenzyme's binding sites on the DNA could even be isolated and identified under the electron microscope (Fig.16.1); it fitted the image of the already postulated promoter (see Chapter 15). The specific binding of the integral RNA polymerase to the promoter caused, at that site, a slight distortion of the double helix – conditioned by the σ-factor; the hydrogen bonds of the base pairs at this point became transitorily disrupted; the bases on the strand to be transcribed became thus available to pair with complementary bases of free triphospho-nucleosides diffusing to the active center of the RNA polymerase holoenzyme. These were the initial steps leading to the coupling of the first two nucleotides into a dinucleotide (beginning of RNA synthesis), with liberation of pyrophosphate. The RNA polymerase, then freed from the σ-factor which was only crucial for the initiation of the process, wandered along the DNA, in the direction established by the promoter. Because the RNA polymerase could read the template exclusively in the $3' \rightarrow 5'$ direction (with RNA being churned out in the $5' \rightarrow 3'$ direction) the issue of which DNA strand was to be transcribed was automatically solved. What remained to be clarified was how the transcription mechanism came to a halt. A so-called terminator turned out to be the structure responsible for that function; the terminator, encompassing about 50 bases in a special sequence, forced the detachment of the RNA polymerase from the template (see, for instance, Chamberlin, 1970, 1974; Richardson, 1993).

The general scheme of regulated gene expression and the details of the subtle biochemistry of RNA synthesis – at least those aspects of interest to molecular biologists – fitted together. A veritably monumental perspective emerged: crucial life events –  cell differentiation, cancer, etc. – would be dependent on the activity of specific promoters and terminators! Disentangling the minutiae of this general scheme – certainly a fantastic one – remained a goal for the future.

The controllable action of the RNA polymerase viewed as one of the life-supporting pillars: what an astounding attribute for such a humble enzyme! This possibility had already crossed the mind of the ingrained biochemist Fritz Lipmann (who, in 1953, had earned the Nobel Prize for having discovered coenzyme A); in a bout of molecular biological insight – he had been involved in an intense discussion with Francis Crick; or was he, with hidden irony, simply mocking the molecular biologists? – he remarked: To emphasize this, I would prefer to call the replicating polymerases "replicases". The much discussed genetic information transfer is dependent upon an almost foolproof functioning of the sequence determination by the template. The name polymerases for the enzymes transacting this information transfer actually focuses too much attention on a relatively trivial function and not enough attention on the all-important replication function (Lipmann, 1963) ...exerted at the right moment, one would like to add.

# CHAPTER 17

# THE CORONATION: DNA REPLICATION

We have by now elaborated a panorama view over the paths leading from the genetic information stored in DNA to the meaningfully regulated transcription of one of its strands into RNA (rRNA, tRNA and mRNA), and through translation of mRNA into polypeptide chains and the folding of these chains into protein molecules with enzymatic activity. These enzymes, in their turn, are the instruments forging new building blocks for renewed macromolecular synthesis. At this point, we reach the peak event of a cell's life cycle, its coronation: the self-duplication of the genetic information, the replication of DNA.

DNA replication deemed the coronation of a cell's life cycle: why is that? Life on earth is characterized by the competition of the so-called replicators (Dawkins, 1976). Replicators are nucleic acids, mainly DNA molecules, endowed with the prerogative of self-replication. The faster a replicator arrives at reproducing itself, over and over again across a large time span, the more copies of itself will it manage to leave behind. The lazy replicators lose out, being overgrown and eliminated by the swifter ones: the molecular "survival of the fittest". In order to have at their disposal increasingly efficient strategies of self-perpetuation or "autocatalysis", the replicators elaborated, during their evolutionary history, the most refined tricks for garnering the essential building blocks and energy: they architectured, for their own use, cells with metabolism, and multicellular organisms, contraptions which allowed them to conquer ever more new ecological niches where they could, ultimately, flourish. [Monkeys are machines with whose help certain DNA molecules manage to replicate themselves in the trees (Dawkins, 1976).] Replicators which, through billions of years of evolution – mutation and selection –, developed ever more complex and cunning tactics, finally generated, beyond enzymes, cell structures, and organs, the most subtle of the emotions of the human soul, committing them to their cause...

This provocative theory of life and its meaning corresponds to how the sociobiologists of the 1970s contemplated the issue (see, for instance, Wilson, 1975; Dawkins, 1976). Their way of thinking was often condemned and resisted – vehemently and with penetrating moral revulsion – even by molecular biologists (see, for instance, Bateson & Dawkins, 1985; Segerstråle, 2000; Alcock, 2001). But, inherently, sociobiology simply took hold of Darwin's concepts, adjusting them to the evolution of behavior. Getting disturbed by such ideas means that the power of science to mold character and philosophical convictions is being taken far too seriously. These are exactly the same sins purportedly perpetrated by sociobiology, through its scientism and biologism, which its opponents claim to be fighting.

Thus, perhaps, the coronation indeed. Molecular biologists at first never really gave heed to the biological aspects of DNA replication. Watson & Crick (1953b) even imagined, in the beginning, that DNA replication would possibly proceed spontaneously, according to the rules of base pairing. Crick later commented on this flaw, admitting that there was "a gap in our comprehension of molecular biology". Closing this gap became mainly the task of biochemists rooted in the old school.

Arthur Kornberg was a young physician who, during World War Two, after serving several months in the navy, was finally stationed on land. [Referring to this episode in his reminiscences, he coquettishly insinuated that his transfer was due to his strong aversion to naval etiquette (Kornberg, 1976); later, though, he hinted his transfer to land had been provoked by the attention accorded to his first scientific publication: observations on his own rare form of jaundice, the Gilbert disease (Kornberg, 1989).] In 1942, at the National Institutes of Health in Bethesda, Maryland (back then only a handful of small laboratories; today rendered a working place for over 10,000 people), he undertook nutrition experiments on rats. However, to him, the vitamins became soon less interesting as food supplements than as biochemical substances whose role in cell metabolism he decided to explore. Vitamins, coenzymes, oxidative phosphorylation, ATP synthesis, energy metabolism, enzymes: biochemistry got hold of him, and would not let him escape ever again – till late in his life.

His interest in nucleotides especially concentrated on the issue of their synthesis: …were the bases and the nucleosides first synthesized and subsequently phosphorylated, or were perhaps phosphorylated precursors further processed enzymatically? Exactly, what reactions lead to the synthesis of the bases – the purines (adenine and guanine) and pyrimidines (cytosine, thymine and uracil)? All these were queries which deeply bored Jim Watson in 1951, during his stay at Herman Kalckar's laboratory in Copenhagen.

Working with nucleotides also logically implied that the issues linked to their polymers, RNA and DNA, had to be tackled. For example, what building blocks contributed to the assembly of RNA? Was the sugar-phosphate-backbone structured first, to be then appended with bases? Or would each nucleotide be fastened to a growing chain? As a 5'- or a 3'-phospho-nucleoside, or both, or maybe a cyclic one?

The part played by activated nucleotides in various metabolic reactions became ever clearer, and it was not such absurd an idea to speculate that nucleoside triphosphates could serve as substrates for polynucleotide synthesis. Kornberg – in the meantime settled at Washington University in St. Louis – mixed radioactive ATP to a coli extract and noticed that it could partially be retrieved – although in exceedingly low amounts – as acid-precipitable (i.e., macromolecular) material, apparently RNA; what else but RNA? Then, alarming news reached St. Louis, coming from New York: it was being claimed that nucleoside diphosphates were the building blocks of RNA. This was distinctly shown by Marianne Grunberg-Manago, a member of Severo Ochoa's group. It would just not be worth pursuing further the issue of RNA synthesis if the New Yorkers were so near to settling it. What Kornberg could not know is that Ochoa was on the wrong track, leading to a dead end (see Chapter 10).

In 1955, Kornberg moved on to tackle DNA synthesis, instead. And this he did not because Watson & Crick had published their model two years before – Kornberg did not even know about it! – but because of his intrinsic interest in biochemistry, in the enzymatic activities lurking in cell extracts, challenging him to purify and characterize them. From one colleague, he received radioactive thymidine, a substance which was incorporated in no other macromolecule than DNA. The outcome was not so exciting: a very low amount of radioactivity was incorporated into acid-precipitable material (some 100 cpm out of millions of cpm added to the cell extract). Nonetheless, this material, if treated by DNAase, turned acid soluble – it was DNA, indeed. A young graduate student, Robert Lehman, enthused by these results, decided to invest further efforts. Soon he had evidence that thymidine-monophosphate (dTMP) was a lot more efficiently used as substrate and that the triphosphate (dTTP) was even better.

Adding molecules of DNA to the coli extract, together with the radioactive building blocks, brought them a step further. Kornberg had decided to include DNA in the extract with the expectation that it would act as a so-called primer, a sort of initiator for the polymerization reaction. This scenario would find a parallel in the synthesis of glycogen or starch. There, the sugar molecules had to be attached to a primer in order for the polymerization to proceed. Moreover, the added DNA was expected to compete for the DNAase in the extract, preventing it from destroying the scarce newly synthesized material. Lehman and Kornberg did not immediately notice that the supplementation with DNA fulfilled two further essential functions.

First, because it was degraded by DNAase in the cell extract, it provided the other building blocks for DNA synthesis (exclusively dTTP was offered to the reaction); the freed deoxynucleoside monophosphates, after being activated by their respective kinases in the presence of ATP, yielded dTTP and the hitherto unkown deoxynucleosid triphosphates, dATP, dCTP, and dGTP.

The cell extract was fractionated and the enzyme activity responsible for the incorporation of dTTP enriched. That set the preconditions for a further characterization of the reaction, and caused the DNA offered as primer to be recognized as the agent of a still more remarkable role; namely, the base composition of the newly synthesized DNA mirrored precisely the primer's own. This fact gave the DNA the faculty not solely of a primer, but also of a template. The template function was an absolutely novel biochemical category, providing the information for the enzyme to polymerize the offered building blocks in a specified way (Kornberg, 1960). The ability of the enzyme – named DNA polymerase – to work according to the instructions conveyed by the template signified a revolutionary concept which led biochemists to react with astonished incredulity and rejection. Nonetheless, their negative attitudes were rendered ever less pronounced in the face of concrete observations: no arguments can belie facts. And Kornberg's group (Josse et al., 1961), after moving to Stanford in California, had demonstrated that the DNA polymerase accurately copied not only the base composition of the DNA template but also the precise sequence of its bases. The frequencies of the 16 possible pairs of neighboring bases of the template coincided exactly with those of the synthesis product. This method, the so-called nearest neighborhood analysis, produced the first evidence – beyond that exposed by the Watson-Crick model – that the two strands of the DNA double helix displayed opposite polarities. (In the meantime, Kornberg's group had come to realize that their subject of interest, DNA, was, simply, of crucial meaning for the issue of genetic inheritance – they had moved, unintentionally, to the center of interest for molecular biologists.)

Their system seemed so perfect that its failure to yield biologically active DNA as, for instance, transforming DNA, turned out to be an embarrassing and unfathomable situation. Obviously, something basic was still missing (although Kornberg, back in 1959, together with Ochoa – see Chapter 10 – had already been bestowed with the Nobel Prize).

The first breakthrough came only in 1967 in the form of a novel enzyme, ligase, which was able to covalently close interruptions in the sugar phosphate backbone of DNA strands: the single-stranded DNA of phage ΦX174, a closed molecular circle, was completed to a double strand by DNA polymerase; the so derived complementary DNA strand – which was still a linear molecule placed on a circular template – was converted into a covalently closed circle through the action of ligase; this circle, through the combined action of DNA polymerase and ligase, was copied anew to yield a circle which was fully infectious, being taken up by coli cells, inducing them to produce progeny phages! The news of "life out of the test tube" travelled quickly around the globe in 1967: an obvious piece of journalistic hype. ΦX174 embodied a special case of biologically meaningful DNA synthesis, but this could not, by any means, be generalized; as yet, no conventional transforming principle had been synthesized *in vitro*. Clearly, in addition to ligase, there had to exist other factors essential for supporting the action of the DNA polymerase. Nevertheless, the central role played by the DNA polymerase with regard to DNA synthesis was never disputed. On the contrary, the studies undertaken by Reiji Okazaki at Nagoya University in Japan seemed to remove one serious obstacle to the understanding of the DNA polymerase's way of working: Kornberg's enzyme could only bind 5'-phosphonucleosides to free 3'-OH ends of a growing chain; at a replication fork, however, only one of the two strands grew in the 5' $\longrightarrow$ 3' direction – no problem there for the DNA polymerase; the strand on the other side must, however, grow in the 3' $\longrightarrow$ 5' direction, and this stood in conflict with the known behavior of the DNA polymerase (Fig. 17.1). Okazaki et al. (1968), using radioactive thymidine for pulse experiments, showed that the newly synthetized DNA was present, for a very short time span (measured in seconds), in the form of fragments of only 1000 to 2000 nucleotides, before binding covalently to the already synthesized DNA. In this process, ligase apparently had to intervene. This insight came from the use of temperature-sensitive ligase mutants: at higher temperatures, these short DNA fragments – dubbed Okazaki pieces – accumulated inside the cells, which led to their deaths (Pauling & Hamm, 1969). All Okazaki pieces were synthesized in the 5' $\longrightarrow$ 3' direction (Okazaki & Okazaki, 1969). The interpretation of this experiments was at hand: DNA synthesis on the problematic strand (i.e., on the template requiring newly synthesized DNA to grow in 3' $\longrightarrow$ 5' direction) only gave the overall impression of proceeding in this 3' $\longrightarrow$ 5' direction; individually, the Okazaki pieces were synthesized in the 5' $\longrightarrow$ 3' direction and then bound together by the ligase. The combined action of DNA polymerase and ligase could, therefore, explain the simultaneous growth of both strands of a replication fork (Fig. 17.1).

And yet, the then apparent quiet was disturbed by a mutant produced by John Cairns' laboratory in Cold Spring Harbor (DeLucia & Cairns, 1969). This mutant, devoid of the Kornberg polymerase, grew apparently unhindered in normal conditions; its sole handicap was a slightly higher sensitivity to UV radiation. Arthur Kornberg was cornered, even ridiculed (Anonymus, 1971). Because he was absent from Stanford, spending a sabbatical in England, one of his three sons, Thomas, who was studying biology at Columbia University, N. Y., felt obliged to intervene. He, together with the young biochemist, Malcom Gefter, who had not, as yet, seriously occupied himself with the theme, started the search for new DNA polymerases in Cairns' defective mutants, as well as in normal coli cells. They found not only one, but two new enzymes endowed with activities similar to the discredited one of father Kornberg. The old Kornberg enzyme got the denomination of DNA polymerase I, whereas the new ones became number II and III. Soon it was verified that DNA polymerase II, just like I, was not essential for normal cell growth. But DNA polymerase III was different: mutants with temperature-sensitive DNA polymerase III stopped synthesizing DNA at higher temperature (Kornberg & Gefter, 1971).

Even if this new DNA polymerase could be deemed the veritable replication enzyme, the complexity of the events occurring at the replication fork was still grossly underestimated. In one of his last publications, Okazaki (he, 40-years-old, succumbed to leukemia – a consequence of a too liberal use of $^{32}$P?) pointed out that the short segments of newly synthetized DNA, in polymerase I-defective mutants, were only in a somewhat laggardly manner covalently bound to the growing strand (Okazaki et al., 1971). Polymerase I had, thus, regained a significant role at the replication fork – a role which, if necessary, would be taken up by another enzyme, although not with the same efficiency. Therefore, at least three distinct coli enzymes participated in the process of DNA synthesis; DNA polymerase III, apparently, fulfilled the task of producing rough Okazaki pieces, whereas the polymerase I filled in the empty gaps among these; and ligase was summoned to seal the last remaining single strand interruptions.

DNA polymerase I was largely rehabilitated, especially because the mode of action of both other coli DNA polymerases (and, as later confirmed, also of the DNA polymerases of higher organisms) were intrinsically the same as Kornberg and his co-workers had so convincingly revealed: elongation of a polynucleotide by successive attachment of 5'-phosphonucleosides to its 3'-OH end along a template and following the pairing rules of Watson and Crick.

But the pre-existence of a polynucleotide, namely the primer to be elongated, was a condition for the synthesis reaction to start. Apparently, the DNA offered to the reaction, not so heedfully treated, provided enough single strand breaks and free ends to stand as primers. One should also take into account that single-stranded DNAs, by haphazardly folding unto themselves, originated so-called hair-pin structures. Such hair-pin structures – totally senseless, biologically speaking –, supplied the DNA with a point of departure for its further elongation (though not replication, *sensu stricto*). (This was, probably, one of the reasons why DNA synthesized *in vitro* lacked any biological activity.) Not only DNA, but also small segments of RNA could act as primers for DNA synthesis, if they were bound to the strand of template DNA through complementary base-pairing. DNA polymerase would also be able to work on this RNA primer, elongating it by attaching deoxynucleotides and so creating a "chimeric" product: a small RNA head, with a long DNA tail.

Figure 17.1: Simplified scheme of the enzymes active at the DNA replication fork. Top: from Alberts & Sternglanz (1977); bottom: from Kornberg (1989), reproduced with permission from the Annual Reviews of Biochemistry, Vol. 58, © 1989 by Annual Review Inc. One branch of the template ("leading strand") is straightforwardly complementarily copied by DNA polymerase III. On the other side of the fork ("lagging strand"), an RNA polymerase ("primase") acts, producing short RNA primers; these are elongated, first by DNA polymerase III, and then by DNA polymerase I, with concomitant degradation of the RNA primer by this same enzyme; the ligase covalently closes the last gaps between the newly synthesized DNA segments. Directly at the fork site, a helicase carries out the untwisting of the parental double helix, after a gyrase, at a certain distance from the bifurcation point, had cleaved the sugar-phosphate backbone, rendering this short segment, to be replicated next, able to swirl independently from the rest of the DNA double helix (see also Fig. 6.4); renaturation of the single strands originated from helicase action is counteracted by a single-strand-binding protein (SSB)

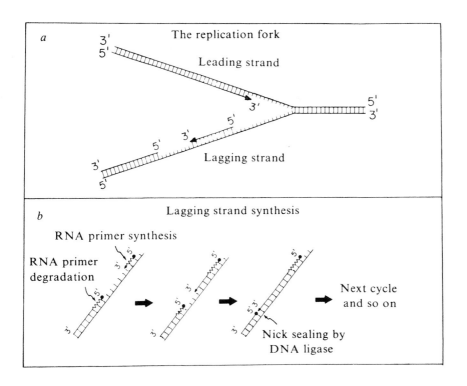

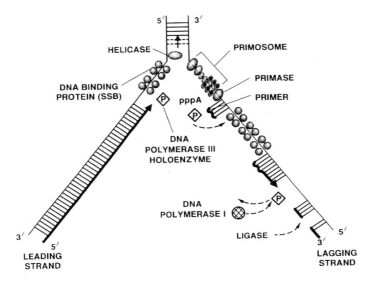

    This situation seemed to bear a significant role in the regular *in vivo* replication
of DNA – a veritable Columbus' egg for solving the primer issue, since the RNA
polymerase is not primer-dependent: at the replication fork, small segments of RNA
primers would be synthesized, which, in their turn, would be worked on by the DNA
polymerase III. Truly, at the 5' ends of short, still unfinished Okazaki pieces, the
presence of small pieces of RNA, covalently bound to them, could be detected (see,
for instance, Sugino & Okazaki, 1973). Soon after having performed their duty, the
RNA primers were degraded by action of an RNAase, no other than the Kornberg
polymerase proper: what a versatile enzyme! Not the regular RNA polymerase (the
one carrying out the transcription processes) was active in building the RNA primer,
but a further, long unnoticed enzyme with RNA polymerase activity (see, for
instance, Rowen & Kornberg, 1978). With this, the list of proteins directly involved
in the *in vivo* DNA replication was enriched by still one more enzyme: the primer
RNA polymerase or primase, the product of gene *dnaG*.
    Nonetheless, the list was far from complete. Still many other proteins performed
essential functions in the replication fork:
    DNA polymerase III, in the meantime, had been characterized as a complex
quaternary structure englobing ten different polypeptide subunits (see, for instance,

Figure 17.2: Arthur Kornberg (born in 1918).

Marians, 1992); the *dnaE* gene, whose mutation had originally hinted at the function of the DNA polymerase III (Nüsslein et al., 1971; Gefter et al., 1971) coded only for one polypeptide of the set, the so-called α-subunit.

And, in addition to all these enzymes, an unswirling protein, helicase, cared for the untwisting of the double strands of the DNA at the replication fork; and a single-strand-binding protein hindered the separated DNA strands from renaturing at the replication fork, averting, at the same time, the action of DNAases.

The DNA gyrase (or topoisomerase II, sometimes also referred to as swivelase) was also recognized as a relevant enzyme. This protein was first spotted through its noteworthy ability of adding so-called "supertwists" (extra twists of a covalently closed circle of double-stranded DNA) to a double helix circle, or, alternatively, dissolving such supertwists, without leaving behind a hint of single strand breaks. The meaning of DNA gyrase activity for DNA replication is revealed by its capability of introducing transitory breaks into the DNA double helix, just ahead of the replication fork, not too far from it, thus preventing the rotation (see Fig. 6.4) from being transmitted along the whole DNA molecule (the swirling just in front of the replication fork is an essential operation so that the plectonemic twisting of both DNA strands can be reversed; see Excursus 6-2.)

After Watson and Crick had suggested the possibility of DNA replication occurring spontaneously, without any intervention of enzymatic catalysis, and after the naive notion of the existence of solely one DNA duplication enzyme had turned out to be unsustainable, today no one dares to affirm that the problems of DNA replication are fully grasped. Suffice it to mention that, although today over 30 genes are known to participate directly in the process of DNA synthesis in coli, by no means all pertaining issues have been solved (see, for instance, Marians, 1992; Baker & Wickner, 1992).

Not to mention the situation in higher organisms: the exceedingly complex structures of their chromosomes and the limited access of genetic and biochemical analysis render the present understanding of DNA replication in eukaryotes even more inadequate than that in bacteria (see, for instance, Kelly & Brown, 2000).

# CHAPTER 18

# THAT WAS THE MOLECULAR BIOLOGY THAT WAS

𝒜 generally satisfying, all encompassing view was by then attained: the double helix was a template for DNA polymerase in order to create a copy of itself, as well as a template to be transcribed by the RNA polymerase into RNA which, on the ribosomes, directed the production of proteins according to an established code. The amino acid chains, then, folded themselves into characteristic tridimensional structures, exposing special active sites for manifold enzymatic reactions of the metabolism to occur. Beyond that, the capability of these events to be regulated, as exemplified by the operon model, rounded off the picture. The most complex processes of cell differentiation or embryological development were seen as mere variants of the basic principle of protein-nucleic acid interaction.

All processes carried out by the living cell, such as the shaping and replication of its most elaborate structures, which just a few years before appeared absolutely opaque to analysis, turned out to be comprehensible, although in general terms only. But the end of molecular biology, as a passionate quest for understanding the principles of life as a biological phenomenon, drew near; the main goal had been reached. This, anyway, was how Gunther Stent (1968) deemed the situation to be, as expounded in a highly controversial article titled: "That was the Molecular Biology that was". In it, the author explains how the avant-garde views of pioneering molecular biologists had given way in time to the tightly regulated propositions of the scientific business. Stent anticipated the imminent demise of the new science and its special flair. Molecular genetics, as seen by him, was subdivided into three epochs. The first, the romantic phase, was permeated by dreams about the novel laws of nature. Its principal representative was Max Delbrück. It was not a long-lasting dream, as the discovery of the DNA double helix dispensed it an early fatal blow. The philosophical background of the romantic phase of molecular biology had to be revised from its ground: the mystifying principles and forces turned out to be mainly plain hydrogen bonds...

The dogmatic phase followed; in it, the dogma of molecular biology (Crick, 1963; Olby, 1970), namely the statement that genetic information flows from DNA to RNA into proteins, was verified, confirmed, refined and extended. This phase, lasting from 1953 till approximately 1963, witnessed the deciphering of the genetic code, the elucidation of the mechanism of protein synthesis and the elaboration of the operon hypothesis for the regulation of genetic activity. While the romantic phase engaged a mere handful of idealistic researchers in the pursuit of the fundamental principle of biological self-replication, the dogmatic phase kept busy a flock of some hundreds of molecular biologists.

From the mid-1960s on, nothing else remained to do except the elaboration of the details based on the rules set by the dogmatic phase.

This meant the onset of the academic phase and its swelling army of many thousands of scientists, busily entrenched in their research institutes, at old and new universities (as, for instance, shown in Fig. 18.1) and intensively supported by expansion programs, many of them coming from government institutions. That was, at least partially, motivated by the fact that, in 1956, the Soviets had shocked the Western World with the Sputnik, provoking a surge of hysterical reactions in its wake. Such slogans as "the Russians will surpass us" played a pivotal role in the arousal of interest from responsible administrative bodies: research and teaching funds started flowing in as yet unknown proportions.

It is not surprising, therefore, that this routine mass endeavor no longer radiated the aura of intellectual adventure, the trademark of the pioneering molecular biologists. As a consequence, many of these reached out anew for other romantic fields of activity. A formidable task, apparently insoluble, comparable in scope to that of self-replication, presented itself: the enigma of the nervous system and of consciousness. Would this last challenge posed to biological science be approachable? Delbrück himself had already tried to solve the tangles of the sensory stimulus by means of the single-cell fungus *Phycomyces* (see, for instance, Cerdá-Olmedo & Lipson, 1987); however, his small school of followers practically dissipated after his death. Other researchers took up different paths in the same direction. Seymor Benzer, for instance, aimed at the same goal by analysing the genetics of behavior in *Drosophila*. One of his noteworthy observations referred to the *dunce* mutant (see, for instance, Davis, 1993) – nonetheless, his work had a muffled repercussion, remaining confined to a close circle of specialists. Gunther Stent, for his part, got involved with the nervous system of the leech (see, for instance, Stent & Weisblat, 1982). He was successful, although no overwhelming breakthrough was attained. And Francis Crick, of course; he, too, believed to have found in neurobiology an appropriate field for his inquisitive intellect. At the beginning of his scientific life, both this area of studies as well as molecular biology had stood as possible choices for his future research career. Crick, plunging into his new speciality, made some astonishing findings almost immediately, though not about neurobiology itself, but about established neurobiologists (Crick, 1988): these, according to him, did not really aim at disclosing the working mechanisms of the mind, but rather deemed neurobiology as a sort of playground or a vastly open field of action where manifold observations, any observations, could be produced. Alas, the fresh topic of his attention did not allow Crick a veritable comeback as a leading research personality.

The nervous system, the mind, consciousness – these topics seemed to have an effect similar to the astrophysicists' black holes: they swallowed all those who approached their gravitational field, no meaningful message being emitted to the outside world (see, for instance, Horgan, 1999).

Many scientists, including Crick, were never really concerned about defining the field containing the mass of phenomena conceivably analyzable by scientific methods: they simply and naively considered the boundaries of their field to be inexistent. [Gunther Stent is definitely one of the exceptions to that assertion – see, for instance, Stent (1978); Horgan (1998).] The ultimate failure was inevitable, a fact already foreseen by Niels Bohr, a fact that even today merely few thinkers recognize: that the universe ought to be subdivided into observer and the observed; the observed is apt to be scrutinized, the nature of the division between the two categories is not. Before trying to understand nature, the universe, by means of science, the scientist or philosopher ought to take for granted this fundamental tenet.

This state of affairs would prevail even if it were possible to actually perform what has to remain a thought experiment, namely to pin down the action of each neuron, each molecule in the brain, before and after a thought has traversed this brain, or a sensation felt by its owner (Fig. 18.2). The romantic endeavor to master this last obstacle, to try to comprehend consciousness, certainly is a quixotic insurgency against transcendence; the obstacle certainly will remain insurmountable. This fundamental truth is commonly disavowed because accepting something as unconquerable does not suit the idealized image of an omnicompetent scientific method. One final solace remains, though, as Sir Peter Medawar (1978) once remarked: "Fortunately one does not have to understand the brain in order to use it".

But, considering the approaching senility of molecular biology, things would turn out to be quite different from what Stent (1968) prophecied. The academic phase was not the end. Actually, it was just the beginning of an unexpected development, involving a new incremental order of magnitude with regard to the impact of molecular biology on society. This new phase, elaborating on Stent's categories, could be called that of the academic-governmental-industrial complex, characterized by massive financial rewards, involving billions of dollars, pounds, euros, or yens, which decisively shape the future of the field. Let us follow that development step by step.

Figure 18.1: Two examples among hundreds, of how, worldwide, in the 1960s, new architectures came to replace dignified, romantic old buildings. Concomitantly, research transfigured itself into a mass endeavor. Pictures at top: the new and the old Institutes for Molecular Biology of the Göttingen University (Gottschalk & Schlegel, 1982). Pictures at bottom: the new and the original buildings of the Biological Faculty of Freiburg University (photo from the archives of the Biological Faculty).

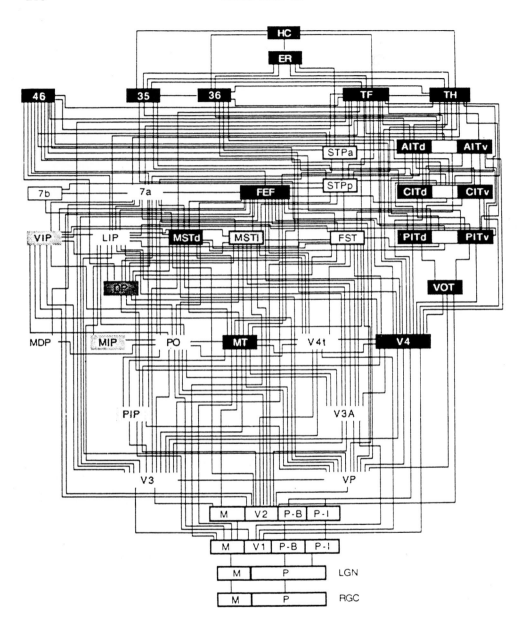

Figure 18.2: Representation of some intracortical links between visual regions of a monkey's brain (Felleman & Van Essen, 1991; see also Crick & Jones, 1993). Will the disentangling of the intricacies of brain structure bring us nearer to comprehending its higher faculties?

# CHAPTER 19

# R & M

𝒜t the apogee of the early phage school, Luria & Human (1952) made a remarkable observation: phage T2, grown on a strain of coli B/4, was unable to go through a second growth cycle on the same host (as shown much later, the critical step lay in traversing the host cell membrane). Only a special strain of *Shigella dysenteriae* would permit growth of these phages; however, the descendants from the growth cycle in *Shigella* could grow again on B/4. It looked as though the phages coming out of B/4 could remember their last host; they were somehow modified by it, marked. This example brought to light the phenomenon of host-dependent *modification* (M), which cried out for an explanation. This observation described by Luria & Human turned out to be quite a special case, based on the incapacity of the strain *E. coli* B/4 to synthesize uridine-diphosphoglucose (UDPG), a compound necessary for the typical glucosylation of T2 DNA. The glucosylation of T2 DNA was a prerequisite for its penetrating through the cell membrane of *E. coli*; this, because a DNAase located there (absent in *Shigella*), broke down the unglucosylated T2 DNA. The phage T2 was thus "restricted" in its growth; in other words, it was subjected to the phenomenon of *restriction* (R) by this particular strain of coli. This peculiar occurrence was – and still remains – a typical instance of the matters preoccupying researchers confined in their academic ivory towers. Many further restricting and modifying hosts, as well as restricted and modified phages, were discovered; one special case was scrutinized by Werner Arber and his co-workers at the University of Geneva (see, for instance, Arber, 1965a, 1968; Arber & Linn, 1969):

Phage λ, cultured in *E. coli* B (denoted λ·B), was able to replicate normally in this host, but not in *E. coli* K. Although the λ·B virus particles adsorbed to K cells, their proliferation inside this host was curtailed: most phage DNA was broken down in the cell. The few λ·B phages escaping this fate generated progeny (denoted λ·K) able to grow in K, but not in B (where their DNA would be degraded). The resulting situation was similar to the initial one: a few λ·K phages surviving in B yielded λ·B progeny, able to replicate in B. Those few λ phages escaping the adverse action of the inhospitable strain were not genetic mutants (as could be shown by the Luria-Delbrück fluctuation test, for instance); they were – as paradoxically as it may sound – genetically unchanged phages, whose DNA, nonetheless, was somehow modified, so that it became refractory to the otherwise hostile strain. This state of affairs was documented by Arber & Dussoix (1962) and Dussoix & Arber (1962) (Fig. 19.1).

What could possibly be the secret mechanism that protected the DNA from enzymatic degradation without affecting its base sequence?

Odd observations kept accumulating. For instance, coli B mutants, as well as coli K mutants, were found which were no longer able to break down the otherwise incompatible phage DNA; in other words, could no longer restrict those phages. These were the $r$ mutants (not to be mixed up with the $r$ mutants of phage T4, with which they had nothing in common, aside from the denotation). Several of these $r$ mutants concomitantly lost the ability to modify invading DNA; these were the $r$ $m$ type mutants. Other $r$ mutants, though, kept on modifying as usual; those were the $r$ $m^+$. (Mutants of type $r^+$ $m$ never emerged; these certainly would be lethal, as they would attack their own, non-modified DNA.) Conjugation experiments showed that the loci $r$ and $m$ were close neighbors on the genetic map of both B and K (near the marker *thr* "threonine"). Clearly, the genes $r$ and $m$ coded the restriction and modification enzymes, whose effects were, accordingly, the degrading of foreign DNA or the altering of that DNA so that it became protected from the disintegrating action. What could conceivably be their mode of action? At exactly what DNA site?

In addition to the host genes $r$ and $m$ (alleles $r_B$ and $m_B$ or $r_K$ and $m_K$), physiological conditions also played a part in the restriction and modification processes. Starvation did not influence normal cells insofar as these events were concerned. Similarly, in most cases, starving auxothrophic mutants of their essential building block also did not affect the occurrences referred to. This was the situation for *pro* mutants when starved of proline, *arg* mutants when starved of arginine, and so on. There was one exception, though, involving the *met* mutants. In this special case, absence of methionine led to a loss of the capability to restrict and modify (Arber, 1965b); what was the meaning behind this outcome?

Even before the discovery of the double helix, it had been known that animal as well as plant DNA displayed around 5% of their total cytosines as methylated at their 5 position, thence transformed into 5-methyl-cytosine (Wyatt, 1951). Later, a further methylated base was detected in the DNA of bacteria and phages; this was 6-methyl-aminopurine, a methylated adenine (Dunn & Smith, 1958). The unexpected DNA glucosylation of phages T2, T4 and T6 (Kornberg et al., 1961) and the methylation of tRNA (Borek, 1963) had already been characterized as steps which occurred after these nucleic acids had been synthesized. This knowledge helped Marvin Gold & Jerard Hurwitz, at the Albert Einstein College of Medicine in New York's Bronx, to realize that the methylation of some bacterial DNA adenosines was

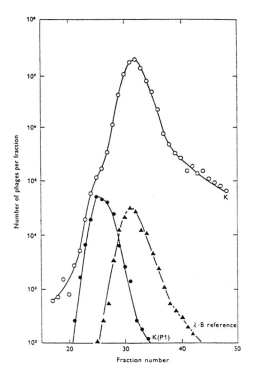

Figure 19.1: "Modification" is based on non-inheritable specific changes of the DNA (Arber & Dussoix, 1962). The experiment here depicted was performed with λ phages cultured in the modifying and restricting host *E. coli* K(P1) cultured in medium containing heavy isotopes: thence, "heavy" λ·K(P1). The so obtained "heavy" λ·K(P1) phages were used to infect *E. coli* K cells cultured in a medium of normal density (*E. coli* strains non-lysogenic for phage P1 inflict no P1-specific modification and restriction). The resulting progeny from λ·K(P1) in *E. coli* K, thus λ·K, was centrifuged in a CsCl gradient. The centrifuge tube was punctured at the bottom, its content dropped out in individual fractions. The phage titre in such fractions was monitored (phage titre in the ordinates, fraction number in the abscissae – higher density left) on three different bacterial lawns: in *E. coli* K [○], *E. coli* K(P1) [●] and *E. coli* B [▲]. On *E. coli* K(P1) exclusively "heavy" phages [with a density corresponding to that of normal protein coats combined to a "semi-heavy" DNA (formed by "semiconservative replication" modus of the parental phage – see Excursus 6-2)] gave rise to plaques. This means that solely phages maintaining at least one strand of the parental phage, λ·K(P1), could grow on K(P1). The phages λ·B were admixed as density markers to the λ·K lysate to be centrifuged; λ·B phages can only originate plaques on a coli B lawn, since they are restricted by both coli K and coli K(P1) (in the latter, even restricted two-fold: by the K system as well as by the P1 system); λ·K and λ·K(P1) were, for their part, restricted by coli B.

also an event taking place on the already assembled DNA molecule (Gold & Hurwitz, 1964). But how and what for?

Let us first address the how: S-adenosyl-methionine (SAM) was known to be an essential donor of methyl groups for a series of metabolic processes. Thus, it seemed quite a reasonable initial experimental step to incubate cell extracts and DNA with radioactive SAM (radioactively labeled in the methyl group) in the hope of obtaining labeled, therefore methylated, DNA, as acid-precipitable material. If the DNA, i.e., the potential receptor of the methyl group, and the cell extract came from the same organism, the results were meager or totally negligible. However, if the DNA was from a different source than the cell extract (for instance, salmonella DNA combined with cell extract from coli, or coli DNA combined with cell extract from staphylococci), then a clear and reproducible, although limited, transfer of radioactive methyl groups from SAM to the DNA was detected (Fig. 19.2); this effect was also discernible if the tests included two different strains of coli. And yet, the methylated bases, mainly 6-methyl-aminopurine and 5-methyl-cytosine, added up to a very minor proportion of the total DNA bases. This state of affairs also encouraged researchers to neglect this unimposing proportion of methylated bases as functionally negligible by assuming that they were nothing but insignificant flaws slightly tainting the beauty of the DNA double helix. Truly, the process of methylation did not affect the intrinsic DNA characteristic of base pairing. On the other hand, the species and strain specificity of the cell extract methylases (they were not yet purified) suggested a function for them: perhaps these methylated bases were markers for intervening regulation processes? One had no idea.

In any case, the constellation was engaging: on the one hand a biochemical effect, the methylation of DNA, looking for a plausible function, on the other hand, an odd function, restriction-modification, waiting for a conceivable biochemical explanation. Would it be possible to kill two birds with one stone by linking both issues together? A challenge was there to validate the supposition that DNA methylation underlay the phenomenon of host-dependent modification. By employing *Met⁻* mutants, it became feasible to certify that, indeed, a specific methylation of modified DNA took place. The DNA from such mutants, fed with $^{14}$C-methionine, was extracted, hydrolysed and subsequently subjected to chromatographic analysis. John Smith, at Cambridge, shuttled repeatedly between his laboratory and Geneva in order to work out this issue with Arber's group. Their favorite subject of study, the familiar phage λ, turned out to be rather unsuitable on account of a considerable proportion of methyl groups hanging on its DNA bases which, apparently, had nothing to do with the modification process; this unspecific methylation strongly interfered with the assessment of the few decisive bases putatively methylated as a result of the process of host-dependent modification. The solution to this deadlock came in the form of another phage, fd, isolated by Hoffmann-Berling et al. (1963) in Heidelberg. [fd was characterized by some exotic features: first, its filamentous

TABLE IV

METHYLATION OF DNA

| Enzyme source | Source of DNA | | | | | | | | |
|---|---|---|---|---|---|---|---|---|---|
| | Starved E. coli | Normal E. coli | M. lysodeikticus | C. diph-theriae | L. del-bruckii | Cl. past-eurianum | Mb. phlei | S. typhi-murium | S. aureus |
| | *C—CH₃ group incorporated in 30 min (mμmoles)* | | | | | | | | |
| (a) Cell extract from | | | | | | | | | |
| E. coli | 0.022 | <0.001 | 0.024 | 0.026 | <0.001 | 0.006 | 0.042 | 0.008 | 0.016 |
| M. lysodeikticus | <0.001 | <0.001 | <0.001 | 0.001 | 0.001 | 0.026 | 0.004 | 0.001 | 0.002 |
| C. diphtheriae | 0.006 | 0.015 | 0.013 | <0.001 | 0.006 | 0.031 | 0.015 | 0.003 | 0.008 |
| L. delbruckii | 0.017 | 0.004 | 0.015 | 0.017 | <0.001 | 0.007 | 0.034 | 0.009 | 0.013 |
| Cl. pasteurianum | 0.013 | <0.001 | 0.056 | 0.041 | 0.001 | <0.001 | 0.027 | 0.032 | 0.049 |
| (b) Purified E. coli enzyme | 0.200 | <0.001 | 0.503 | 0.428 | 0.013 | 0.032 | 0.705 | 0.149 | 0.212 |

The reaction mixture (0.25 ml) contained: 10 μmoles of Tris buffer, pH 8.0, 1 μmole of MgCl₂, 2 μmoles of mercaptoethanol, 10 mμmoles of ¹⁴C—CH₃-labeled S-adenosylmethionine (2.2 × 10⁷ cpm per μmole), 150 mμmoles of deoxynucleotides as DNA in each case, and 60 μg of protein from crude extracts. In the experiments with purified enzyme, 5 μg of protein was added, the DNA concentration was reduced to 75 mμmoles of deoxynucleotide, and 0.2 μg of heated RNAase was included in each reaction mixture. Cell extracts prepared from *Mb. phlei*, *S. typhimurium*, and *S. aureus* did not lead to detectable methylation of any DNA preparation and for this reason have not been included in the above table.

Figure 19.2: Methylation of DNA by cell-free extracts from different sources (a), or by purified methylation enzymes from *E. coli* (b). DNA was incubated in the presence of radioactive S-adenosyl-methionine, precipitated, and the DNA-associated radioactivity monitored (Gold et al., 1966).

shape; second, its infection mode: it adsorbed exclusively to the tip of the so-called F pili (Fig. 4.11, g), its progeny provoking no lysis of the infected cells, but oozing out continuously through the host cell wall; third, its DNA, comprising merely ca. 5000 nucleotides – a dwarf, even among phages – was single-stranded (though turning to a replicative double-stranded form inside the host).] Phage fd grown in *E. coli* K (thus, fd·K) was restricted by *E. coli* B; just about seven from 10⁴ fd·K phages overcame this blockade, originating plaques on a lawn of coli B. The phages picked from these plaques (then, fd·B), could grow on K (in contrast to λ, fd was not restricted by K), thus losing their specific B modification. These phages, repeatedly plated, first on B and then on K, allowed to select several mutants less intensively restricted by B. Starting with these, further mutants, totally unaffected by B restriction, could be isolated. The small fd genome apparently encompassed only two restriction sites, which were successively stripped away by the selection procedure. A decisive observation came in the wake of this insight: the B-modified DNA of the fd wild type displayed around four methylated adenines, fd·K DNA just one or two of them. In contrast, the non-restrictable (also non-modifiable) fd mutants displayed the same one or two methylated bases, independently of their having been cultured on B or K. The modification was, therefore, paralleled by the methylation of a few special adenines of fd DNA, converted to N-6-methyl-aminopurine (Arber & Kühnlein, 1967; Kühnlein & Arber, 1972; Smith, Arber & Kühnlein , 1972).

Figure 19.3: Werner Arber (born in 1929).

Matthew Meselson and Robert Yuan, working at MIT, took the next step; namely, they isolated and purified from *E. coli* the respective cutting enzyme which attacked predefined sites on the DNA, and whose activity was dependent on SAM and ATP. Everything fell into place, starting with the methionine dependence of the restriction process (SAM is synthesized from methionine), and including the details of the restriction substrate; for instance, partially modified double helixes (just one strand modified; the other unaltered) were not recognized by the enzyme (this outcome was actually a logical presupposition; since modification occurs only after DNA synthesis, the newly synthesized strand of a DNA double helix must be protected against restriction by the old, already modified strand, before getting modified itself – see Fig. 19.1).

It was a grandiose success story (Meselson & Yuan, 1968). The enzyme, however, was somewhat fussy in what concerned its requirement for SAM (not really a household chemical) and huge amounts of ATP; it recognized indeed the special sites already postulated by Arber, but it cut somewhere else, dozens of base pairs away. Nevertheless, it all amounted to a successful accomplishment, alas, one which still remained exclusively confined in the academic ivory tower. In this sense, a misfortune, too. The phenomenon of restriction-modification broke out from its confining academic isolation only after the third start – an unintentional one whose significance was at first unrecognized.

Hamilton Smith, Daniel Nathans and their co-workers at Johns Hopkins Medical School in Baltimore, Maryland, were actually kept busy by studying the phenomenon of transformation in *Hemophilus influenzae*, one of the few microorganisms allowing an easy approach to the subject. They noticed by chance that a *Hemophilus* (today spelled *Haemophilus*) cell-free extract cleaved the DNA of the salmonella phage P22 into well defined segments; more precisely, it cleaved the DNA of any source, with the exception of the DNA of *Hemophilus* itself: it was an event analogous to the enzymatic restriction process described by Meselson & Yuan. Purification of the endonuclease activity became an alluring task. Similarly to the restriction enzyme of coli, that of *Hemophilus* provoked a limited number of double strand breaks; as a co-factor it only needed magnesium ions, no SAM (Smith & Wilcox, 1970). Still more important was the assessment that the so derived fragments, all of them, displayed the same terminal sequences. The compelling inference was that the enzyme, then called endonuclease R (today denoted *Hind*II), cleaved a specific, symmetric sequence of 6 base pairs in the middle (Fig. 19.4). Wherever the critical sequence was to be found in the DNA, the novel, specific endonuclease would cut the double strand. Soon there was a first collection of precisely defined, electrophoretically characterized, restriction fragments (Fig.19.5); for instance, the DNA of the small tumor virus SV40 was cleaved to 11 typical fragments (Danna & Nathans, 1971).

Setting these fragments in their proper original arrangement was the next goal; with it emerged the first so-called restriction map (Fig. 19.6). And suddenly, several hitherto unknown restriction enzymes, aiming at other target sequences, were identified (Fig.19.4). The huge variety of specific restriction endonucleases – hundreds of them are known today – hints at the importance and extent of the phenomenon of DNA restriction in nature. Its biological function is, apparently, one of defense against penetrating foreign DNA, especially that of phages.

[However, phages threatened with curtailment by restriction evolved mechanisms to efficiently counteract it. A straightforward solution would involve mutating away from the target sequences recognized by the host's restriction enzymes; indeed, most phage genomes display fewer restriction sites than would be expected by chance. Beyond that, some phages developed enzyme systems for

| Denomination | Origin | Target sequence with cutting site (arrows) |
|---|---|---|
| AhaIII | *Aphanothece halophytica* | 5' TTT↓AAA 3'<br>3' AAA↑TTT 5' |
| BamHI | *Bacillus amyloliquefaciens* H | 5' G↓GATCC 3'<br>3' CCTAG↑G 5' |
| ClaI | *Caryophanon latum* | 5' AT↓CGAT 3'<br>3' TAGC↑TA 5' |
| EcoRI | *Escherichia coli* RY 13 | 5' G↓AATTC 3'<br>3' CTTAA↑G 5' |
| HindII | *Haemophilus influenzae* Rd | 5' GTPy↓PuAC 3'<br>3' CAPu↑PyTG 5' |
| HindIII | *Haemophilus influenzae* Rd | 5' A↓AGCTT 3'<br>3' TTCGA↑A 5' |
| HpaII | *Haemophilus parainfluenzae* | 5' C↓CGG 3'<br>3' GGC↑C 5' |
| KpnI | *Klebsiella pneumoniae* | 5' GGTAC↓C 3'<br>3' C↑CATGG 5' |
| Sau3AI | *Staphylococcus aureus* 3A | 5' ↓GATC 3'<br>3' CTAG↑ 5' |
| TaqI | *Thermus aquaticus* YTI | 5' T↓CGA 3'<br>3' AGC↑T 5' |
| XhoI | *Xanthomonas holicola* | 5' C↓TCGAG 3'<br>3' GAGCT↑C 5' |

Figure 19.4: Examples of restriction endonucleases, their respective target sequences and cutting sites.

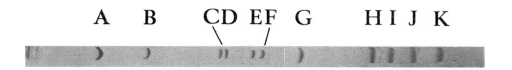

Figure 19.5: The first identification of defined restriction fragments from an individual DNA type (Danna & Nathans, 1971). SV40, a monkey virus with a small, circular, double stranded DNA (ca. 5500 base pairs), was cultured in cells in the presence of $^{14}$C-thymidine. Its labeled DNA was then extracted and incubated for 6 hours with the *H. influenzae* restriction endonuclease (today denoted *Hin*dII). The various fragments derived from such treatment of the DNA were subsequently separated by electrophoresis (running left to right, that means, the larger fragments are more to the left). Autoradiography of the gel followed.

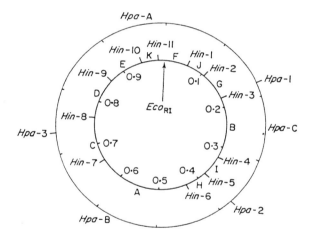

Figure 19.6: The first restriction map. The 11 fragments of the SV40 DNA, as described by Danna & Nathans (1971) (see Fig. 19.5), were arranged by Danna, Sack & Nathans (1973) in a physical circular genetic map. Two methods, still valid, could be employed: 1st: after incomplete digestion with the restriction enzyme in question, larger fragments remain whose size correspond to that of two or more combined smaller fragments, so that these latter are recognized as adjacent. 2nd: individual restriction fragments are subsequently treated with a different restriction enzyme, originating smaller pieces to be arranged, as a puzzle, when compared to the fragments derived from the digestion with single restriction enzymes. Cutting sites of three distinct enzymes are indicated: Hin (today *Hin*dII), Hpa (today *Hpa*I) and EcoR1. The latter cuts the SV40 genome at one site only, which was arbitrarily taken as point zero for the length scale (the total genome was normalized to length 1).

glucosylating their DNA or they substituted, for instance, hydroxymethyl-uracil for thymine; all that in order to escape the curtailing action of restriction enzymes. Another strategy was to perfect mechanisms to alter the restriction enzymes themselves, so that they lost their efficiency; but many host organisms responded with fresh restriction enzymes especially adapted to the particular situations (see, for instance, Bickle & Krüger, 1993): there appears to be a continuous molecular genetic arms race for still better weapons deployed to outmanoever the enemy – a battle without winner or loser.]

Werner Arber, Daniel Nathans and Hamilton Smith were honored in 1978 with the Nobel Prize for physiology or medicine, in recognition of their contribution to the unravelling of the restriction processes.

Independent of the role played by restriction enzymes in the tribal wars between microbial strains, their use for mankind, albeit in a totally different context, turned out to be of utmost importance...

# CHAPTER 20

## MOLECULAR CLONES

Stanley Cohen at Stanford, Herbert Boyer at the University of California in San Francisco, and their co-workers (Cohen et al., 1973) were the first to set into action the theoretical knowledge referring to restriction. With the help of ligase, they fused a piece of foreign DNA, in the form of a restriction fragment, with a small plasmid, cleaved by the same restriction enzyme (Excursus 20-1). Then, they tried transforming *E. coli* cells using this artificially combined DNA: it worked! The experiment marked the onset of a new era.

Bringing the new era to blossom was, nevertheless, no easy task; the first cloning experiments exposed a multitude of uncertainties. A main goal in subsequent experiments was the use of a defined, functional gene, rather than a meaningless DNA fragment, like the one in the seminal experiment.

How to pinpoint, isolate and prepare the gene destined to be cloned?

Cloning virus genes presented no apparent problem, at first – it was everyone's guess that, with the help of a restriction map, a fragment encompassing the desired gene could be recognized, isolated and cloned.

Alas, exactly that apparent ease was bound to turn into the problem: during a traditional cozy meeting in rural and idyllic New England (Gordon Conference, New Hampshire), in June 1973, it was debated how the DNA fragments of the monkey virus SV40, for instance, could be transferred into *E. coli* by means of the novel technique. Concerned warnings as, for example, "We are witnessing now a pre-Hiroshima situation!" lingered already in the air (see Wade, 1973); the participants were sensitized: was SV40 not a virus with the potential to trigger tumors in newborn hamsters? An absolute assurance was demanded that such colis, endowed with the SV40 genes, would never be able to contaminate the human population. Even an infinitesimally small chance had to be excluded that, due to carelessness, or by a malevolent plot, a dire lethal cancer epidemic could result, capable of decimating mankind. (It should be pointed out here that the original polio vaccines, distributed to many thousands of children, carried SV40 as an unrecognized contamination; fortunately, no consequences accompanied that flaw.) Concerns and uncertainty pervaded at the conference as its participants decided to inform the National Academy of Sciences, which, in its turn, called a commission into life (Berg et al., 1974) which, for its part, organized a meeting.

In February 1975, around hundred molecular biologists – all of them utterly proud of having been invited – met on the park-like premises of the Asilomar Conference Center. The venue lies not very far away from the beach, on the city limits of Monterey, California's colonial capital, about 100 miles south of San Francisco. Conferences, consultations and deliberations were, for four days, the order of the day: discussions involved the freedom of science, responsibility of scientists... Because of their openly displayed sense of responsibility and the fettering measures they had imposed on themselves even before Asilomar, the participating scientists entertained the naive belief of having the right to settle all possible problems for themselves. They accepted and ratified, though in a somewhat less stringent form, the recommendations of the Berg Comission: a sensible gradation of precautionary measures aimed at experiments conveying different degrees of risks. This agreement was the core resolution of the Asilomar "recommendations". It happened that the press was represented in Asilomar; it sniffed a sensation and informed its readers, with the effectiveness of its media resources. Horrifying news headlines are known to ensure high sales and/or ratings; the general public enjoys shuddering with fear raised by blood-curdling themes, especially if the unknown dangers and menacing threats are perceived as merely vague suspicions (see Fig. 20.1). Governmental institutions felt obliged to intervene. In the wake of all that followed investigative committees, genetic regulations, bureaucracy, courses for biologists held by bureaucrats, demonstrations organized by fundamentalist groups. Rightly so? Scores of questions certainly remain open for debate; violent protest, however, will not be very helpful. Molecular biology had lost its comfortable shadowy niche in society.

The momentary surge of excitement would eventually calm down, only to be raised time and again by new biotechnological breakthroughs, as for instance those in agriculture. People do get accustomed to novel life styles, and they eventually will stop arguing about the new breeds of animals or plants that are the result of genetic engineering rather than classical methods – the outcome is very much the same: better and increased yields of crops or domestic animals, suiting the ever growing world population's demands. Anyone for more colorful aquarium fishes? Manipulation of the human genome with the intent of attaining special human features? What is feasible will really be done? Well, breeding human types could conceivably have been done for a long time already. Indeed, the classical methods employed for the selective breeding of plants or animals could have brought forth variable human types,

Figure 20.1: The monster of the Haveberg forest (from Boaistuan's "Histoires prodigieuses", 1566). The literature of centuries past witnesses the importance of the alleged existence of horrific creatures for placating a deep–seated urge of the human soul. The fables of the present also provide us with such symbols: Dracula, snowman, zombies, Loch Ness, etc. Apparently, science destroyed humanity's belief in illusory gnomes, giants and monsters. As a countermeasure, a substitute had to be provided in the form of gene technology, which ideally conveyed the grounds for the sadomasochistic joy of fancying about the horrifying menaces lurking around humankind. With that, a psychological barrier was erected, preventing any rational debate on the theme. It does not matter that the experience gathered over 30 years has already vindicated the safety of gene technology, and that modern basic research, the pharmaceutical industry, breeding of crops and animal husbandry are all intimately dependent on it. The issue seemingly lies less in gene technology itself, but rather in society's lack of tolerance and emotional openmindedness for scientific innovations: a matter involving sociology, mass psychology and politics. Help! Who is in charge of all that?

emphasizing diverse characteristics: athletism, musicality, mathematical talent... Owing to the enormous variability of the human wild type, such endeavor would even have a better chance of succeeding than, say, the selection of the very diverse breeds of dogs. Will such scenarios become reality some centuries or millennia ahead? (750 after Ford – see Huxley, 1932)? Should we get involved with issues affecting humanity's very far away future? Individual considerations will certainly lead each one of us to take diverse positions on the subject, but whether present attitudes will come to affect future generations is another matter...

...and back to basics: the fettering rules imposed on gene technology were progressively loosened worldwide, but technical difficulties were still to be mastered. Concerning our initial query of how to pinpoint the gene to be cloned, the project of forging a gene synthetically was evolved in Herbert Boyers' laboratory (Khorana's work had already led to a suitable technique – see Chapter 10). The gene of choice was the one encoding somatostatin. Starting with the 14 known amino acids making up that small peptide, Itakura et al. (1977) picked the corresponding codons, aligned them, fused this sequence to a special plasmid and inserted it into coli cells. This was the first experiment using a synthetic gene to obtain its respective peptide. The method could however not be generalized. The amino acid sequences of many proteins were unknown; other proteins were so complex that synthesizing the corresponding sequence of nucleotides would have been a hard – even impossible – task, back then. Somatostatin was indeed an exceptionally small peptide; however, the same procedure employed for its synthesis was also successfully used for the therapeutically essential insulin (Crea et al., 1978; Goedell et al., 1979).

Once the surprisingly straightforward principle of cloning DNA sequences was established (Excursus 20-1), researchers quickly improved and extended basic techniques. One such pivotal improvement was based on the chance discovery of a remarkable enzyme, one year before, made by David Baltimore at the Massachusetts Institute of Technology, and concomitantly, but independently, by Howard Temin at the University of Wisconsin in Madison. By studying the reproductive cycle of RNA tumor viruses – retroviruses as they are named today – both researchers had observed that the genome of these viruses, after infecting a host cell, was integrated into its chromosomes in the form of a DNA copy of the viral RNA. This DNA was synthesized by means of an enzyme using RNA as a template (Baltimore, 1970; Temin & Mizutani, 1970), unlike the DNA polymerase which copies DNA from a DNA template, or RNA polymerase which forges RNA from a DNA template. This odd enzyme was the "reverse transcriptase". Its discovery as a component of the retrovirus particle brought Baltimore and Temin the honor of the 1978 Nobel Prize. Remarks that the reverse transcriptase was a sensational finding because it refuted the central dogma that "DNA makes RNA makes proteins" accompanied its discovery; Crick remained unperturbed: the essential tenet of the central dogma was that the information flows from nucleic acids to amino acid sequences and not in the opposite direction. (Crick, apparently, did not realize that Boyers' laboratory had achieved exactly that inverted information flow from peptide to polynucleotide, though by means of human manipulative inventiveness and skill.)

The reverse transcriptase proved indeed to be sensational, but in quite another context. By its action, a DNA gene (in the form of a copy-DNA, cDNA) could be prepared for cloning on the basis of its corresponding mRNA, providing that this latter could be found and isolated. For highly specialized cells, this goal was achieved with relative ease, since most of their synthesizing capacity is devoted to one particular protein, say, hemoglobin in the case of reticulocytes.

mRNA of less specialized systems was made accessible by, for instance, deploying specific antibodies to precipitate nascent proteins, still bound to their specific mRNA, directly on the ribosomes. The whole complex, including the desired RNA, was thus precipitated. And, of course, it is possible to simply isolate RNA molecules from cells or tissues and clone corresponding DNA copies, in the hope that important genes, or fragments of genes, might later be identified (see Chapter 23).

**Excursus 20-1**
**RESTRICTION ENZYMES AND DNA CLONING – THE BEGINNING OF GENE TECHNOLOGY**

The essential new feature introduced by gene technology to biological research is the possibility of the so-called cloning of DNA sequences (in short: DNA cloning). For this, a DNA segment of any source – generally about the size of a gene – is incorporated into the structure of a small plasmid (Fig. 20.2), which is then used to transform a bacterial host, such as, for instance, *E. coli* (see Excursus 7-1). The plasmid, together with the integrated foreign DNA, replicates inside the host, giving rise to many copies of itself. Restriction enzymes are the cornerstone behind this technique, bringing about the formation of accurately defined fragments. Some restriction enzymes cleave the single strands of the DNA double helix in a staggered way (see Fig. 19.4). From the action of such restriction enzymes, DNA fragments result which display short terminal, protruding, single-stranded regions of complementary sequences. This cutting mode – in contrast to precisely opposing cleavage sites on the single strands – is actually irrelevant for the natural function of the enzyme, because, under physiological conditions, the breaks inflicted on each strand, even if not rigorously opposed, will lead to an irreparable break of the double strand of DNA. This is so, because the relatively few hydrogen bonds between the single strands within the restriction site are not enough to hold the fragments together. However, at lower temperatures, when the molecular thermic movements are well reduced, even those few hydrogen bonds will suffice to maintain intact the double strand structure at the cleavage site and will even allow renaturation of small single strands of complementary sequences. The experimental implication of these facts is that the restriction fragments, simply by virtue of the cooling process, will be bound together again, since their terminal protruding single strand domains – the so-called sticky ends –, at the cutting sites, are bound to renature. Ligase then seals the single strand interruptions at the cleavage sites, so that the effect of the restriction enzyme can be thoroughly reverted. All types of DNA fragments originated from the same, specific restriction enzyme are, thus, potential renaturation partners. Biological research acquired thus a powerful new technique for covalently binding precisely defined DNA fragments from any source.

Such DNA fragments can also be integrated in, say, plasmids. For that purpose, the circular plasmid is first isolated from the host and subsequently opened up by a restriction enzyme (it is quite helpful if the plasmid has only one restriction site specific for that enzyme). The DNA fragment in question, obtained by the action of the same restriction enzyme, is incubated together with the opened plasmid, in the presence of ligase, giving rise to artificial structures similar to F' factors (see Chapter 12) in what concerns the mixed origin of their DNA. By various means (for instance, as described above, by its being transferred into a coli cell), the fragment of foreign DNA may then be replicated many times in a corresponding host cell culture, resulting in a DNA clone ( Fig. 20.2).

The efficiency with which DNA fragments (passenger DNA) are integrated into the plasmid (cloning vehicle), as well as the acceptance of the recombinant DNA circle by the host cells, may often be unsatisfactory. To get round this problem, different methods were developed. For instance, resistance factors (R factors, see Excursus 11-1) may be chosen as cloning vehicles in order to permit the selection of antibiotic resistant host cells, which likely will also carry the passenger DNA fragment. A second genetic marker of the plasmid may be helpful to differentiate between vehicles with or without passenger DNA. This second marker, conferring, say, a further antibiotic resistance, will have to encompass the restriction enzyme's cutting site. If no foreign passenger is inserted into the plasmid vehicle, renaturation of its terminal sites restores the function of the second resistance gene; if, however, the foreign DNA was incorporated into the ring, restoring that function will be impossible. The use of a selective medium will then allow only the growth of cell colonies transformed by the R factor, and the subsequent use of the replica plating

Figure 20.2: The cloning technique. The DNA of a plasmid, frequently an experimentally mutilated R factor (all non-essential DNA deleted), or an altered phage, such as λ, is cleaved by a suitable restriction enzyme, mixed with the foreign DNA cut by the same enzyme, and incubated to allow the annealing of the protruding single-stranded regions of complementary base pairs. A ligase binds both fragments covalently ("ligation"). (Ligase is also capable of promoting the ligation of DNA fragments devoid of terminal single-stranded regions – so-called blunt ends –, although less efficiently.) The so manipulated plasmids are inserted into a host cell – often coli – through a transformation procedure. The plasmid, together with the inserted foreign DNA, starts replicating synchronously with the host cell over an unlimited number of cell division cycles. In this synopsis, the main steps are summarized in three text boxes; 1$^{st}$) top, left: digestion with restriction enzymes; 2$^{nd}$) top, extreme right: cleaving of the cloning plasmid by the same type of restriciton enzyme, in the example of the plasmid pBR322, and, 3$^{rd}$) bottom, middle: ligation of the mixed fragments by ligase. This basic scheme may be complemented or altered in innumerable variants. For example, the DNA to be cloned may be obtained from mRNA by the action of reverse transcriptase – as so-called cDNA (complementary or copy DNA) – or it may be synthesized in the laboratory (left, second box from bottom). In some cases, a series of enzymatic treatments (middle) may be necessary to allow or to facilitate the cloning procedure. By adding λ-specific termini (bottom, right) to the DNA to be cloned, packing in λ phage coats and replication in phage-like fashion, becomes feasible. (From the catalogue "Biochemicals for Molecular Biology 1987" – Firma Boehringer Mannheim.)

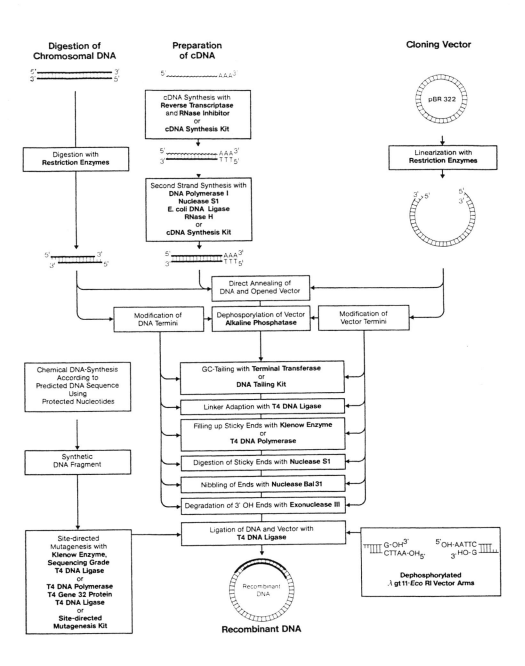

**Recombinant DNA**

technique will then provide an easy method for discarding the unwanted colonies with the intact second antibiotic resistance gene. Though this procedure retains its historic value, it has been followed by several still more efficient ways of pinpointing the desired clones.

Cloning fragments of more complex genomes, as those of higher organisms, is, accordingly, more laborious (see Excursus 21-3); some spectacular examples can be given, as the gene-technological production of human insulin and of human clotting factor for the treatment of hemophilics. Erythropoietin, the substance promoting the development of erythrocytes, is one further early example of a therapeutical compound (administered to patients with kidney failure, but also used as an illegal doping agent for athletes) obtained by means of gene technology.

In recent years, high expectations have been raised with regard to the possibility of countless further medical applications of gene technology; thus, in the wake of the Human Genome Project (see Chapter 23) and its development into proteomics, thousands of genes are being continuously cloned in a gigantic routine endeavor (whose value – as some argue – is still to be convincingly demonstrated).

**Excursus 20-2**
**HOW MOLECULAR CLONING ALMOST BECAME SUPERFLUOUS: PCR**

Why did this idea not occur first to me? That was the thought haunting many smart colleagues, when reflecting on the DNA polymerase chain reaction. Indeed, in retrospect, the concept is ludicrously simple. Through the PCR technique, churning out any desired amount of a certain DNA fragment, say, a gene, departing from minimal quantities, even from one single DNA molecule, became reality. The method of PCR circumvents the necessity of a host cell for cloning; it is, effectively, cloning *in vitro* (Saiki et al., 1985).

The first step involves heating the sample with the DNA fragment to be replicated, in order to separate the double strands into single ones; addition of suiting primers turns the DNA single strands into templates for the DNA polymerase. These primers, offered in a large excess, must be complementary to short segments on the 3'-terminals of the single strands of the DNA to be amplified. This condition being fulfilled, and nucleotide building blocks being available, DNA polymerase comes into action, complementing each DNA single strand so that full double stranded versions are formed. This cycle is repeated as many times as desired: first denaturation of the double strands by heat, followed by cooling down to allow the attachment of the primers – still in great excess. The polymerase will then lengthen these primers. After the second reaction round, there are four times as many of the desired pieces of DNA, as compared to the beginning. And after the third cycle, 8 times as many; then 16; then 32... Allowing the procedure to be repeated, say, 20 times – a question of a few hours – guarantees $2^{20}$ copies of the desired segment, that means, around a million!

To resist the heating steps between each round of PCR, temperature-resistant DNA polymerases are employed today. These heat-resistant polymerases, isolated from thermophilic microorganisms, like *Thermus aquaticus* (Taq-polymerase), inhabitants of hot springs, facilitate the procedure enormously.

For the outcome of the reaction, it does not even matter whether the DNA sequence to be amplified is pure or not; pivotal for the specificity of the process is that the primers be long enough to assure accurate homing in on the target sequence. If the critical sequences are known, synthesizing the suitable primers (comprising sequences of some tens of bases) presents no difficulties whatsoever; alas, taking advantage of the possibilities offered by PCR may be constrained exactly by the absence of such knowledge. To start the reaction, not more than the DNA from one cell, one chromosome only, or – crucial for criminal evidence – one isolated hair, or traces of blood or semen are required. Today, research laboratories take advantage of special variants of the technique to amplify single-stranded DNA or RNA... the range of possibilities opened up by the PCR revolution seems to be unlimited (see, for instance, Arnheim & Erlich, 1992).

The passionate surfer and excentric, Kary Mullis (the founder of a biotech company specializing in sequencing the DNA from rock stars – see Fig. 23.1 – and known to display nude fotos of his girl friends in his lectures to keep students from falling asleep, etc.), was already known in the biotech enterprise Cetus Corporation located in Emeryville, California, for his endless, crazy conceptions which never evolved to anything productive; that is why, at first, no one took him seriously when he came forward with his visionary fabulation about PCR. However, this one idea – a stroke of genius during a nightly car trip – guaranteed him the 1993 Nobel Prize for chemistry.

# CHAPTER 21

# BEYOND COLI

What Monod meant by saying: "What holds true for *E. coli*, holds true for elephants" was on the one hand correct. The elegant and straightforward notion of the genetic information being stored in DNA, then passed on to RNA in a controlled fashion (transcription), and further to proteins (translation), was soon validated in general as a process also occurring in higher organisms.

Those for whom the fascination of molecular biology lay hidden behind the mystifying self-replication ("We aim at understanding life..." but, of course, only in its epistemologically significant basic elements) had reached the goal of their yearnings: grasping the basic mechanisms underlying life processes, at least as much as they judged necessary.

And, concerning molecular biology of higher organisms? For instance, embryonic differentiation? "Differentiation is a bore!" – remarked a young and brilliant colleague in 1968, representing a view held by all romantics. It consisted, again and again, in endless examples of interactions between information-bearing macromolecules, mainly nucleic acids and proteins – no new principles! The operon model had forestalled any conceivable exciting novelty: the gene is no abstraction; it is directly and intrinsically involved with the action of proteins in cell metabolism. [This principle remained unchanged, even when an overwhelming number of additional gene-regulating proteins were progressively discovered – from activators to zink-finger proteins, enhancer-binding proteins, homeobox proteins, all sorts of modulators, the different RNA polymerases, of course, but also steroid receptors plus scores of further transcription factors (for reviews, see, for instance, Fry & Peterson, 2001; Gerasimova & Corces, 2001; Näär, Lemon & Tjian, 2001; Orphanides & Reinberg, 2002; Ptashne & Gann; 2002; Veenstra & Wolff, 2001; see also Fig. 21.13).] Molecular biology, as the science of the principles inherent to the transfer of genetic information, was finished – the romantics ought to seek other challenges.

And yet, this was not the overall prevailing opinion; much to the contrary: the majority of molecular biologists marched, in the 1970s, towards the analysis of higher organisms, the eukaryotes (Excursus 21-1).

And soon, it became recognizable that Monod was – on the other hand – deeply mistaken.

Beyond the concept of the central dogma, a huge array of newly discovered complex phenomena kept accumulating; working them out was a task far beyond conventional genetics, which drew upon the realms of biochemistry, cell biology, immunology, embryology, physiology, neurobiology, and medicine. Molecular biology was not dead (one remembers Mark Twain: "Rumors about my death are greatly exaggerated") – it had simply merged with all those disciplines with their manifold engaging aspects. Without aiming at the impossible task of giving an overview of all of today's biology, we will rove through some of the topics encompassing genetic research.

The mere comparison between the genomes of higher organisms and those of coli or other prokaryotes raised the suspicion that they, in effect, could not be that similar after all.

Starting with their sizes: the most simple of eukaryotic genomes, that of the yeasts, was five times as large as that of coli; ...and vertebrate genomes, as well as those from higher plants, comprised thousands of times as much DNA (Fig. 21.1).

The parcelling of the genome in – from 2 to over 100 – distinct chromosomes, with their typical, microscopically recognizable, individual shapes, was documented as early as the beginning of the 20th century. However, the knowledge that each of those chromosomes was composed of solely one linear DNA molecule, reaching a length of several centimeters, was novel (Fig. 21.7 & 21.8). And since the beginning of the 20th century, it was also clear that this DNA was intimately associated with basic proteins, the histones; the insight that these histones, in the form of the so-called nucleosomes, were enwrapped in the DNA, was also new (Fig. 21.6 & 21.7).

Certain was the fact that some gene-regulating events had to be peculiar to higher organisms, since typical features, such as, say, oncogeny or embryogenesis had no counterpart in prokaryotes. And yet, the fact that operons, the functional multi-gene structures, were absent from eukaryotic genomes was utterly unpredictable. Essential differences in the processes of normal gene regulation must, therefore, exist. It was to be assumed that, in eukaryotes, translation did not immediately follow transcription; there was a topological imposition on that: transcription took place in the nucleus, whereas the proteins were assembled in the cytoplasm. This compartmentalization had been recognized since Brachet (1947). Perplexing, though, was the plethora of RNAs continuously present in the nucleus; these were denominated heterologous nuclear RNAs, hnRNAs, since they consisted of molecules of extremely variable lengths; unveiling the enigma of their existence would only come about years later (Excursus 21-3).

One of the oddest features distinguishing the gene functions of prokaryotes and eukaryotes, was, unexpectedly, the subdivision of eukaryote genes in regions to be translated (the exons), and those not subjected to the translation process (the introns).

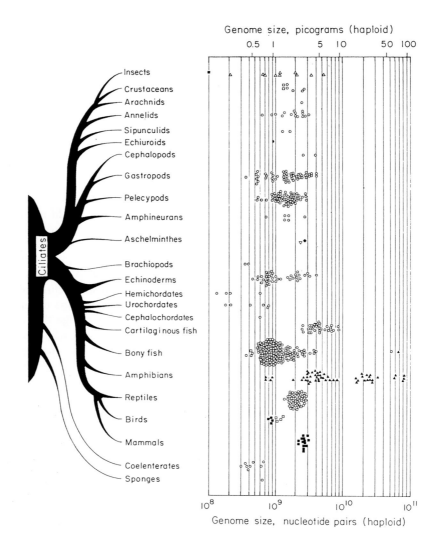

Figure 21.1: The C-value paradox (Britten & Davidson, 1971). It is still not known what causes the huge variations in the genomic DNA contents (DNA complement or C-value) of distinct – but often related – organisms (among amphibia, for instance, the range of variation reaches 100-fold); non-coding, "parasitic" DNA (see Excursus 11-1) "infecting" the various genomes to different extents suggests one plausible interpretation. Certain is that, in general, the DNA content pro eukaryotic haploid genome lies far above that of prokaryotes (the coli genome, as an epitome of prokaryotes, comprises merely $4.6 \times 10^6$ nucleotides pairs). © University of Chicago Press.

The whole gene – exons as well as introns – was transcribed; however, the mRNA became functional only after the introns had been spliced out and further alteration of the primary transcript had taken place (Excursus 21-3).

Another typical feature of eukaryotes relates to its secondary gene action, that is, the expression of a phenotype, after the actual gene product, mostly a protein, has already been forged. Although not unknown in prokaryotes, post-translational modification of proteins is particularly widespread and varied in eukaryotes. In many cases, the fully assembled polypeptides had to undergo posterior alterations – be it phosphorylation (see, for instance, Charbonneau & Tonks, 1992), glucosylation (see, for instance, Parodi, 2000) or proteolytic degradation – in order to fulfil their ultimate function. Besides that, many proteins only become active after having been excreted from the cell, having traversed its membrane (see, for instance, Pryer et al., 1992) – an undertaking depending on intricate regulatory mechanisms. Cell physiology, in peering deep inside such mechanisms, hopes to decipher phenomena that, essentially, are as important as the central dogma: health or disease – be it arteriosclerosis, diabetes, cancer or dementia –, youth or senility (see, for instance, de Boer et al., 2002), life or apoptosis, the programmed cell death (see, for instance, Strasser et al., 2000; Joza et al., 2002; Zhang & Xu, 2002).

[Apoptosis, i.e., physiologically programmed cell death, has in the last decade received increasing attention as a central mechanism in processes as variable as embryonic differentiation, self-destruction of auto-reactive immune cells, and beyond that – quite important for successful financing of research projects – the origin of malignant tumors, whose uncontrolled growth may be viewed as a failure of the apoptotic mechanism (see, for instance, Hakem & Mak, 2001).

Navigating high on the wave of enthusiasm for apoptosis, Weiss (1993) compiled an article in which he described in detail, how, by means of gene technological manipulation, an immortality gene, obtained from nonaging carps, was introduced in mice, turning them eternally youthful (the so-called Dorian Gray mice). The effect was alleged to hold as long as the repression exerted by the genes *p53* and *bcl2* was switched off; otherwise, tissue apoptosis and consequently the individual's death would ensue within 40 hours. The article provoked a certain sensation (see Gullen, 1993; see also letters to the editor in Nature, Nr. 6431): scores of managers of gene technology enterprises, as well as specialists, were not able to immediately recognize what it was all about, although, on the same page the date was: April 1st, 1993.

But, in reality, one ideal subject for the study of apoptosis turned out to be a small (1.5 mm long) worm, *Caenorhabditis elegans*, introduced as a model organism by Sydney Brenner, and intensively studied by John Sulston's group in Cambridge, U.K. In recognition of their work, these two scientists were awarded the 2002 Nobel Prize in Physiology or Medicine, together with Robert Horvitz, MIT, whose analysis of the genes involved in apoptosis showed their importance also for humans.]

**Excursus 21-1**
**THE EUCYTES**

Eucytes, "cells with a genuine nucleus", are, in what concerns their structures and intricate multitude of functions, obviously quite more complex than procytes, the bacteria; and yet, the essential genetic phenomena leading to self-replication are the same. Because in eucytes (Fig. 21.2) almost the totality of DNA is to be found inside the nucleus (less than 1% is located in mitochondria and chloroplasts), whereas protein synthesis takes place in the cytoplasm, there is, consequently, a spatio-temporal discrepancy between transcription and translation processes (see Fig. 21.3).

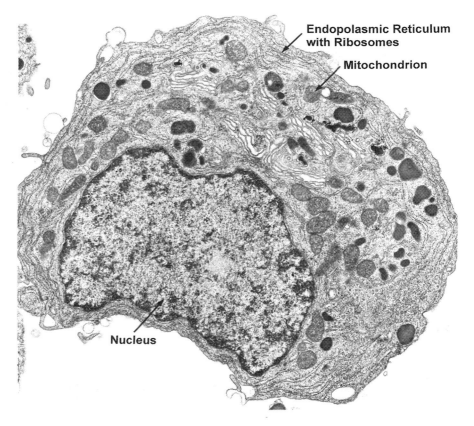

Figure 21.2: Electron micrograph (magnification: 18,000x) of a thin slice of a cell from a rat hypophysis (prolactin cell), representing an eucyte (Salpeter & Farquar, 1981). As reference, the size of a typical bacterium would be, approximately, that of a mitochondrion.

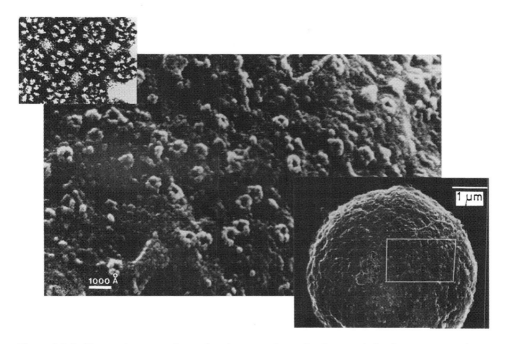

Figure 21.3: The nucleus-cytoplasma barrier: an eukaryotic characteristic. In contrast to those of bacteria, the eukaryotic chromosomes are confined within the nucleus, contained by a double membrane. This nuclear membrane is riddled with the so-called nuclear pores (see, for instance, Rout et al., 2000), complex quaternary structures, traversable by various RNA species (mRNA, tRNA, 7SL RNA), but also by other types of molecules, in their journey to the cytoplasm, and by nuclear proteins (DNA polymerases, RNA polymerases, histones, etc.) migrating in the opposite direction, from the cytoplasm where they are synthesized to the nucleus where they perform their functions (see, for instance, Sweitzer et al., 2000; Komeili & O'Shea, 2001). Ribosomal proteins also travel from the cytoplasm via nuclear pores into the nucleus, where they are bound to the rRNA to shape the ribosomal subunits, which, in their turn, move back to the cytoplasm, crossing again the nuclear membrane through the pores. At the middle and bottom right: a scanning electron microscopic image of an isolated cell nucleus from mouse hepatocytes (Kirshner et al., 1977); top left: a mild treatment with Triton X-100 (a non-ionizing detergent) set free the nuclear pores; negative contrast by uranylacetate allows the observation of their 8 subunits, themselves also intricately structured (Unwin & Milligan, 1982).

Although many bacteria also excrete special proteins into the surrounding environment, the mechanism for protein transport in eucytes is particularly differentiated. Ribosomes churning out export proteins gather on the inside of the so-called endoplasmic reticulum, the lamelliform projections of the cell membrane into the inside of the cell. The peptide chains being assembled on the ribosomes insinuate themselves, still during the synthesis process, through the membrane into the lumen of the endoplasmic reticulum (Fig. 21.2); there, they acquire their characteristic tertiary structures (see, for instance, Pryer et al., 1992). The transport towards the outside of the cell is accomplished by vesicle flow movements provided by the so-called Golgi apparatus.

Genetically fundamental, and typical for almost all eucytes, are the mitochondria, special cell organelles, that – like bacteria – propagate themselves by dividing in two. In mitochondria, similarly to aerobic bacteria, in the presence of oxygen, ATP synthesis takes place. Comparing mitochondria to bacteria is not that far fetched, since mitochondria are direct descendants from originally free-living microorganisms, which in early evolution – perhaps 2 billions years ago – entered a symbiotic relationship with the ancestors of modern eucytes. Back then, eucytes had surely already evolved complex mechanisms for mitotic cell division and perhaps even for meiosis, but they lacked a specialized mechanism to take advantage of air oxygen as an electron acceptor for their metabolism. Oxygen started accumulating in the earth atmosphere in the aftermath of the newly evolved photosynthesis by the ancestors of modern cyanobacteria (these are, misleadingly, also called "blue-green algae"). Some of these cyanobacteria – although somewhat later; maybe 1,5 billion years ago – suffered a fate comparable to that of mitochondria: through symbiosis with eucytes, they became the predecessors of the chloroplasts of green plants. The so-called endosymbiont theory for the origin of mitochondria and chloroplasts had already been formulated half a century ago (see, for instance, Whatley et al., 1979); nevertheless, it only could be convincingly supported by modern molecular biological methodology. A range of similarities between the components of mitochondria and chloroplast protein synthesis and DNA replication mechanisms – ribosomes, tRNAs, and circular genomes – and their counterpart structures of prokaryotes, can only be understood in the light of a common phylogenetic origin (see, for instance, Doolittle, 1980; Kössel et al., 1983; Gray, 1989).

**Excursus 21-2**
**NUCLEOSOMES, CHROMOSOME STRUCTURE**

Dean Hewish and Leigh Burgoyne at Flinders University in South Australia – no one had ever heard neither of the place nor the authors – apparently took their time to process their preparations of hepatocyte nuclei; when they finally did, the DNA had been, to some extent, digested by the enzymes present in the cell nuclei (autolysis). One would expect that under these conditions, the homogenate, upon electrophoresis, would reveal a smear, brought about by the many DNA fragments of diverse, random lengths. Instead, what the authors observed was a clear-cut division of the DNA into small segments of the same length, or else, of precise multiples of this length (Fig. 21.4). This outcome was self-explanatory: "We suggest that the chromatin is composed of iterative basic structures, endowed with endonuclease-digestible regions, repeated in regular intervals" (Hewish & Burgoyne, 1973).

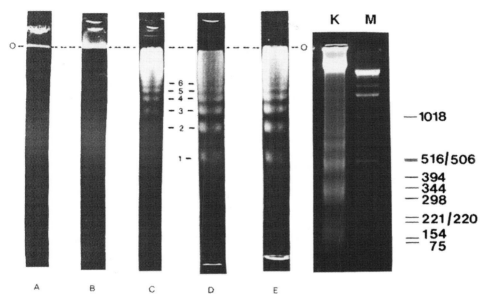

Figure 21.4: The laddering of the DNA from autolytic cell nuclei, after polyacrylamide gel electrophoresis, as a cue for the nucleosome structure of the chromosomal DNA; left: the seminal observation on rat hepatocytes (Hewish & Burgoyne, 1973); right: DNA from apoptotic nuclei from cultured tumor cells (K) (Jürgensmeier, 1993). [The values at right refer to the lengths of the DNA fragments (in base pairs) run as markers (M) on the right lane; see also Fig. 21.9]

This view clashed with the notion accepted for many years that chromatin (Miescher's "nuclein") was a rather homogeneous structure, the DNA being protectively embedded in an outer layer of histone molecules, as if it were an electrical cable enveloped by isolating material (Fig. 21.5).

The astonishingly simple experiment carried out by the Australians triggered an avalanche. All of a sudden, everybody started paying attention to details that had

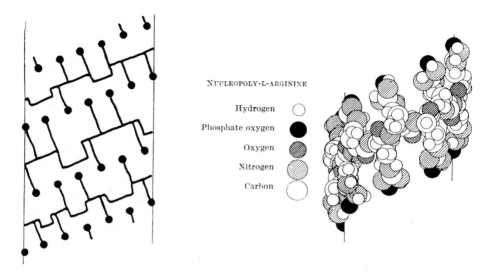

Figure 21.5: Diagram of the association between histones and DNA: it was assumed, till the mid 1970s, that histones wound along the grooves of the DNA double helix. Left: the black dots represent the phosphate residues on the DNA backbone, the dashes stand for the basic amino acids arginine and lysine, components of the histones (Wilkins, 1956). Right: an imaginative spherical model relating to the same topic (Feughelman et al., 1955).

remained unnoticed for years (but, see Yasuzumi, 1955): for example, electron microscopic images clearly displayed the so-called nucleosomes (Fig. 21.6), arranged at regular intervals, lending the DNA its typical appearance (Olins & Olins, 1974). The Hewish-Burgoyne intervals – determining them became an easy task – encompassed around 200 base pairs (see, for instance, Noll, 1974). The nucleosome's protein fraction was shown to be an octameric unit composed of two of each of the histones, H2a, H2b, H3 and H4; around this octamer, fitting tightly, were two windings of the DNA thread. Between adjacent nucleosome complexes the DNA helix continued through a relatively loose stretch, only partially protected by histone H1 – this was the Achilles' heel, susceptible to endonuclease attack (Fig. 21.7).

Many doubts about the structure of chromatin that were still subject to guesswork not long ago (see, for instance, Ris & Kubai, 1970) were settled by a score of fresh information pouring in (Olins & Olins, 1978); the gained insights brought with them a plethora of new questions as well, many still awaiting to be solved: Are nucleosomes distributed randomly on the DNA or, perhaps, on preferential sequences? How do the nucleosomes get assembled? To what extent are they involved with the transcription mechanism (see, for instance, Kornberg & Lorch, 1992;

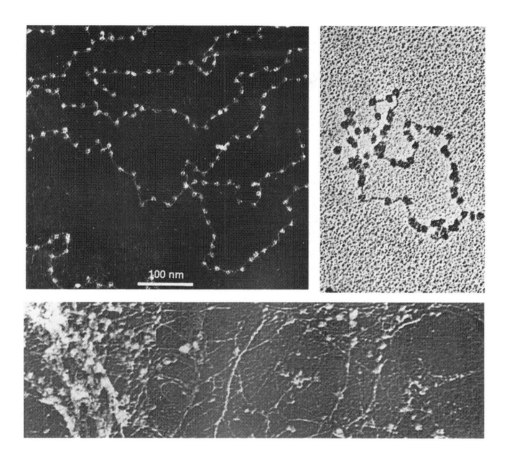

Figure 21.6: The pearls-on-a-necklace structure of chromatin. Top left: hen erythrocyte nuclei dispersed in low-salt buffer solution, revealing the structure of the native chromatin (Olins & Olins, 1978). Top right: the circular, double-stranded DNA from the SV40 virus was incubated with histones, which led to an *in vitro* restoration of the nucleosome structure (Germond et al., 1975). Bottom: nucleosomes were discernible under the electron microscope (Yasuzumi, 1955) two decades before any attention was directed to them.

Näär et al., 2001)? What kind of superstructure will the nucleosome thread form within the chromosome? And what mechanisms are behind it?

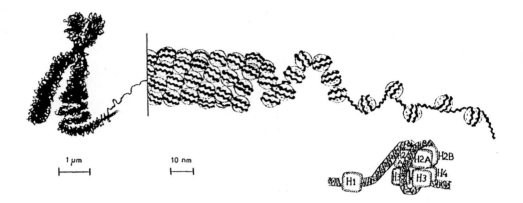

Figure 21.7: Scheme of metaphase chromosome architecture. Each chromatid is composed of a continuous double-stranded DNA molecule which, enwrapping the histone octameres, forms a nucleosome chain. This chain spirales to shape a 30 nm fibril. By forming loops, the latter arranges itself on a proteinaceous central axis which, for its part, also is spiralized during metaphase.

On the basis of these new insights, the assumption for the chromosome's basic structure was that of a continuous DNA molecule many centimeters long (Fig. 21.8), which, arranged in a row of nucleosomes, spiralized into a so-called solenoid; such solenoids, in their turn, folded into loops which attached themselves to a proteinaceous axial core (Fig. 21.7). Although an explanation for the specific banding patterns and the subtle structural and functional variations of the diverse chromosome regions could not be directly derived from this general chromosomal architecture, concrete questions could be posed and worked on. For example, the details of chromosome replication, not only within the streches of the DNA molecule (Kelly & Brown, 2000), but pointedly at its ends, the so-called telomeres; since, in contrast to the circular bacterial DNA molecule, the chromosomal DNA was linear, this situation generated the question: how could the 5' end of the single strand be possibly replicated without a primer? ...and the answer: intricate hairpin structures and corresponding – equally intricate – enzyme systems circumvented the need for a conventional primer (see, for instance, Biessmann & Mason, 1992; McEachern et al., 2000).

The mechanisms involved in cell division, commanding chromosome condensation, could then be successfully analyzed: individual proteins fulfilled several key functions as, say, preventing chromosome condensation before their replication had been accomplished; but chromosome structure and its functional modifications are solely a partial aspect of the complex cell cycle, whose modulation may, alternatively,

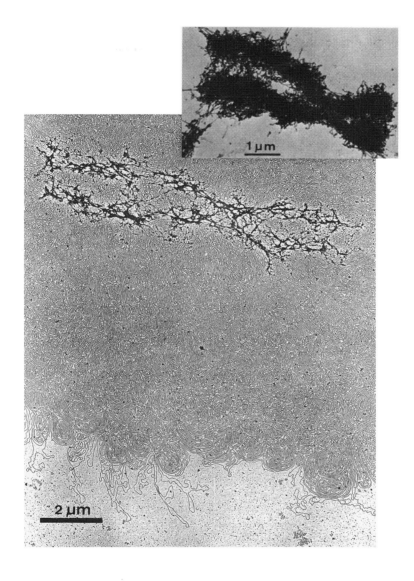

Figure 21.8: Top: electron micrograph of a metaphase chromosome (Mouriquand et al., 1972). Noticeably, threads, roughly 30 nm in diameter, project out of the bulk of the chromosome mass, a seeming jumble. Bottom: a chromosome void of histones, owing to a treatment with 2M NaCl; this procedure brought about the extrusion of a many-centimeters-long thread of the DNA double helix, leaving behind a skeleton of non-histone proteins hinting at the original shape of the chromosome (Paulson & Laemmli, 1977, © Cell Press).

lead to mitosis, to reductional division (meiosis) or to irreversible differentiation blocking further mitoses: a whole range of questions to be addressed by modern biology (see, for instance, Nigg, 2001). Although there is no end in sight for this endeavor, in the last few years the key molecular steps in the cell cycle, and the genes involved, have been characterized: in yeast, mainly by the group of Leland Hartwell, in Seattle, and by Paul Nurse, in Edinburgh, and in sea urchin eggs, by Tim Hunt, working at the Marine Biological Laboratory in Woods Hole, Massachusetts. For this, the three scientists received the Nobel Prize in 2001.

**Excursus 21-3**
**SPLIT GENES: EXONS AND INTRONS**

Isolating the mRNA of genes which were intensively expressed in certain cell types should theoretically not be too difficult (see Chapter 20). For example, reticulocytes, the immature precursors of red blood cells, accumulated, almost exclusively, mRNAs for both hemoglobin chains (the $\alpha$ and $\beta$ chain); in one further specialized cell type, that of the hen's oviducts, around 40% of total mRNA consisted of that for ovalbumin, the main component of egg-white; this mRNA could be obtained in almost pure form by, for instance, precipitating ovalbumin-synthesizing polysomes with specific anti-ovalbumin serum from rabbits. The isolated mRNAs could be used to produce cDNA copies by the action of the reverse transcriptase (see Chapter 20), and the cDNAs, for their part, could be used to produce clones of their corresponding genes.

A promising project seemed to be the location and isolation of a chromosomal DNA fragment with the help of the corresponding mRNA. In addition to the structural gene, also the flanking sequences – possibly significant for deciphering the mechanisms of regulation of gene expression in eukaryotes – would become available in this way. There were a multitude of unresolved items waiting for interpretation, as for instance: Why was the expression of the mammalian $\beta$-globin gene intensified at birth, whereas that of its neighbor, the gene for $\gamma$-globin, was concomitantly curtailed? ...and $\beta$-globin was churned out intensely, in contrast to its almost identical $\delta$-globin; why was that? The expression of the $\beta$-globin gene failed to occur in the victims of certain hereditary diseases (some forms of thalassemia); what motivated this absence? The synthesis of globins was solely carried out in reticulocytes, not in any other cell type, say hepatocytes; what controlling mechanism lay behind that? Why did avian oviduct cells respond to the stimulus of estrogen hormone by synthesizing ovalbumin mRNA whereas other cell types did not? All these were telling points representing model systems for research on various aspects of cell differentiation.

Philip Leder's group, NIH, took up the challenge of cloning a region of mouse DNA comprising the $\beta$-globin gene and its adjacent sequences (Tilghman et al.,1977,

1978). At about the same time, at the university of Strasbourg, Chambon and collaborators succeeded in cloning a segment of hen DNA englobing a large chunk of the ovalbumin gene (Breathnach et al., 1977). For their undertakings, the groups treated their respective DNAs – Leder's the mouse DNA, Chambons' the hen DNA – with restriction endonucleases. This procedure cleaved the DNA in perhaps roughly one million restriction fragments, subsequently separated by electrophoresis, according to their lengths (see Fig. 21.9). The different length classes were individually analyzed; this facilitated the screening of the wanted sequences. (Anyone for checking one million clones?) Isolating and characterizing individual fractions taken from the electrophoresis gel (or, as in the case of Leder, from a chromatographic column), was a task to be done with an impractical, fussy instrument, as the Americans undauntedly did, or else, like Chambons' group, by use of a nitrocellulose sheet to which the DNA fragments, previously alcali-denatured, were transferred from the gel. To achieve that transfer, the nitrocellulose sheet was simply laid on the gel, and covered with a pile of blotting paper; together with the fluid from the gel, the DNA migrated out of the gel but was then electrostatically trapped by the nitrocellulose. [Edwin Southern (1975), at the University of Edinburgh, had elaborated this method two years earlier.] The following step consisted in locating the sought gene sequence. This was done by bathing the nitrocellulose sheet in a buffer solution containing radioactive cDNA in suspension; this so-called cDNA probe specifically bound (hybridized, see Excursus 6-3) to its complementary single-stranded base sequence on the nitrocellulose sheet. An autoradiogram from the latter revealed – through a blackened band on the X-ray film, derived from the DNA radioactivity – the precise location of the desired chromosome fragment. The so pinpointed fragments could then be isolated, cloned and further characterized.

Leder's laboratory did not wait to follow suit. Both groups observed that: first, the chromosomal DNA displayed base sequences which were absent in the cDNA, thence, also inexistent in the mRNA; second, electron micrographs of DNA-mRNA hybrid molecules revealed unexpected loops, conceivably due to DNA regions without correspondence on the mRNA, thus excluded from the DNA-RNA double helix, and visibly projecting out (Fig. 21.10). Apparently, in the DNA of the chromosomal genes, the sequences to be translated, therefore those found in the mRNA, were intercalated by non-translatable ones. It looked as though the coding sequences of the genes were interrupted by non-translated regions. Jeffreys & Flavell (1977), working in Amsterdam, had come to the same conclusion almost simultaneously. Their approach was – without having to isolate the genomic DNA from diverse rabbit tissues – to puzzle together a restriction map of the β-globin gene from the respective restriction fragments (see Chapter 19) gained from a total DNA digest (the restriction fragments aimed at were screened by hybridizing with a radioactive cDNA probe specific for the β-globin gene) and compare the cDNA map with the genomic map; they noticed a surplus of around 600 bases in favor of the genomic map.

And yet, the strict colinearity between the base sequences of a gene and the sequence of amino acids in the proteins – a notion derived from bacteria and phage genetics – had virtually been taken for granted; ... the fact that these seminal observations were not straightaway relegated as artifacts deserves, under these circumstances, special recognition.

At first, the reaction of the scientific community was marked by caution towards the tantalizing idea of interrupted genes. But further instances of split genes, described by a variety of groups, confirmed the generally valid, if still hard to

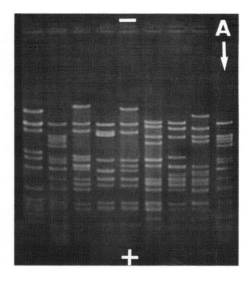

Figure 21.9: Separation of DNA restriction fragments by gel electrophoresis. A thin layer of agarose, with slots for placing the individual probes at its top, is immersed in buffer solution. DNA probes, previously digested by a restriction endonuclease, are separately pipetted into each of the slots; an electric current is applied (+ pole at the bottom, 80-200 V, 1-5 hours; A: start of the run). The larger the DNA fragments, the slower they will progress through the net formed by the agarose molecules. In this example, 9 DNA probes from different but related phages of the T7 group were digested by the restriction enzyme *Hpa*I (recognition sequence: 5'GTTAAC3'). All of the phage genomes comprise around 40,000 base pairs (40 kb). In principle, the same method can be applied to DNAs of higher organisms; according to the DNA source, 1000 to 100,000 times more individual fragments are then originated – discernible by electrophoresis only if special precautions are taken. [Suitable conditions allow, also in the case of higher organisms, the analysis of individual restriction fragment patterns, a process that may have a crucial role in forensics or paternity analysis, for instance.]

believe, observation that within eukaryotic genes, the regions coding for amino acid sequences were intercalated by other, non-coding DNA base sequences. In other words, an eukaryotic structural gene was, in general, no continuous functional DNA segment, but a composite of many parts, interspersed by segments of non-translatable sequences. In the example of the ovalbumin gene [simultaneously analyzed by two independent research groups: Dugaiczyk et al. (1979) and Royal et al. (1979)] there were 8 segments coding for amino acid sequences. Together, they made up 1860 base pairs coding for the 620 amino acids of the protein's primary structure; seven non-coding stretches, comprising ca. 6,000 base pairs, intersected the coding sequences; the complete gene was, therefore, longer than judged by the length of its peptide product. Exons was the name accorded the translatable (expressed) regions, whereas the intervening non-translatable sequences were dubbed introns (Fig. 21.11); Walter Gilbert (1978), who, hitherto, had nothing to do with this research, had suggested these terms, and they took hold.

Alas, the accolade of the Nobel Prize for physiology of 1993, recognizing the importance of the discovery of interrupted genes, went to Richard Roberts and Phillip Sharp. Their research teams had, independently (Roberts' at Cold Spring Harbor and Sharp's at MIT), shown that the mRNA of adenoviruses (agents of upper airways infections) was composed of copies of DNA sequences disposed separately on the virus genome (Broker et al., 1977; Berger et al., 1977). Had their odd observation proved to be a peculiarity of adenoviruses, it would have remained confined to the insider circle of virus specialists; however, as described above, it turned out to be a general principle extending to all higher organisms.

How could this discontinuous genetic information possibly be transmitted to an mRNA conveying a continuous message to be translated into proteins on the ribosomes? Two potential mechanisms could be envisaged: 1) solely the exons were transcribed and the resulting segments were posteriorly pasted together (say, by ligase-like enzymes); 2) the whole gene, namely, exons as well as introns, was transcribed, but the intron sequences were excised from the transcription product before it reached the cytoplasm as mRNA. The last scenario was validated as the correct one: the hitherto mysterious long transcripts, the hnRNA, found in abundant amounts inside the nucleus, were shown to englobe base sequences also existent in cytoplasmatic mRNAs. The processes culminating with intron excision were evidenced as variable, some quite complex: whereas the majority of transcripts were subjected to intricate biochemical reactions catalyzed by complex structures composed of proteins and small specific RNAs, the so-called spliceosomes (see, for instance, McKeown, 1992), some mRNAs underwent a spontaneous intron ejection. The latter indicated that special RNA molecules were also bearers of enzymatic action. For these RNAs, a fitting new expression was coined: ribozymes (see, for instance, Cech et al., 1992; Doherty & Doudna, 2000). Today, ribozymes have gained in significance with relation to a variety of topics, including some wildly speculative ones about the origin of life (see, for instance, Pace & Marsh, 1985).

Interpreting the meaning of eukaryotic interrupted genes remains elusive, its factual existence still puzzling (see, for instance, Doolittle & Soltzfus, 1993). They must, apparently, confer a selective advantage on its bearers lest they would not have evolved; possibly, the less elaborate prokaryotic genome structure would have prevailed if the interrupted genes were not intrinsically beneficial. Alas, making out a decisive profit to be gained from such artfully split gene structures is not an easy matter. One line of thought suggested that, as rare events in the course of evolution, individual exons, through recombination, could be joined to others than their original partners, so creating novel, potentially more advantageous, proteins: it is the module principle applied to generate new genes from previously existent genes in the course of evolution (see, Gilbert, 1978; Dolittle & Stoltzfus, 1993).

Exons belonging to the same gene may also be selectively spliced in diverse manners, so giving rise to various proteins (Fig. 21.12). Taking in account this latter mechanism, a novel version for the "one gene – one enzyme" could be fashioned: "one gene – but many gene products". The gene had, still once more (see Kay, 2000), to be defined anew – but, is it worth the effort? Meanwhile, the concept of fuzzy logic came to imbue our life styles – also our modern computer software; we have learnt to accept unsharp formulations and definitions: fuzzy logic is in (see, for instance, Sangalli, 1992; Kosko & Isaka, 1993).

Figure 21.10: Top: electron micrograph of a hybrid complex consisting of DNA from the cloned mouse β-globin gene and its complementary mRNA. The RNA displays more affinity to its complementary DNA than its corresponding DNA; that is why RNA-DNA heterologous double strands are preferentially assembled, as compared to the equivalent DNA double strands. Owing to this, in a mixture of RNA and partially denatured DNA, the RNA may come to thoroughly replace its corresponding DNA in the double helix structures. The image shows that the mRNA, in one region of the hybrid molecule, failed to find its complementary sequence. The scheme on the right expounds the pushed-out DNA strand as a thin line, the double-stranded DNA as a thicker thread and the DNA-RNA hybrid regions as a thin waved line (photo kindly provided by P. Leder, see also Tilghman et al., 1978; Leder et al., 1977). Bottom: the electron micrograph and, at its side, the interpretative scheme of a hybrid complex formed by renaturation of a DNA single-strand from the hen ovalbumin gene with its counterpart mRNA (punctuated line). The subdivision of the gene in 8 exons and 7 introns is clearly indicated by the 7 loops, A to G (the protruding loops of DNA are made up from sequences lacking complementarity on the mRNA). (The whole gene – exons and introns – is composed of 7600 nucleotide pairs; the mRNA encompasses 1860 nucleotides.) (Dugaiczyk et al., 1979).

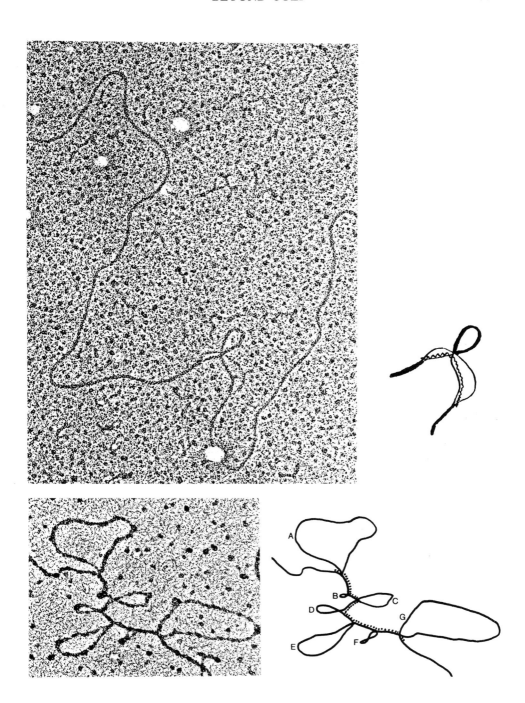

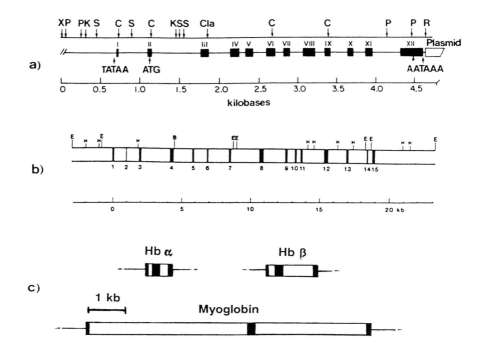

Figure 21.11: *a*) The cloned hen glyceraldehydephosphate-dehydrogenase gene with 12 exons (I-XII). The sequence TATAA signals the transcription start, ATG is the initiator codon (AUG on the mRNA), AATAAA is the signal sequence for the posterior tagging with the poly-A-tail, which helps to stabilize the mRNA. The letters on the upper line indicate the cutting sites of various restriction enzymes (for instance, K = *Kpn*I, see Fig. 19.4) (Stone et al., 1985). *b*) The human gene for α-fetoprotein (a protein, which in the foetus takes up the function of serum albumin). The gene encompasses 15 exons; the amino acids encoded by exons 3, 7 and 11 sport homologies, respectively, to those of exons 4, 8 and 12, but also 5, 9 and 13, indicating that the modern gene evolved from repeated duplications of ancestral sequences. The letters represent restriction sites (for instance, E = *Eco* R1) (Sakai et al., 1985). A similar structure is displayed by the gene for serum albumin. *c*) Correspondence between the gene structures of hemoglobin α- and β-chains and myoglobin (each with 3 exons and 2 introns). All three proteins show clear amino acid sequence homologies, pointing out their common descent from an ancestor's ancient globine (around 500 million years ago, as deducted from comparisons of amino acid sequences and paleonthological evidence). Apparently, a gene duplication triggered the subsequent divergent evolution of hemoglobin and myoglobin. And somewhat later, around 450 million years ago, the primordial hemoglobin gene, for its part, underwent a further duplication, giving rise to the precursors of the α- and β-hemoglobins, which then followed separate evolutionary paths to their respective modern forms. It is significant that the exon limits remained exactly conserved, despite the introns' differences in lengths and base sequences (Blanchetot et al., 1983).

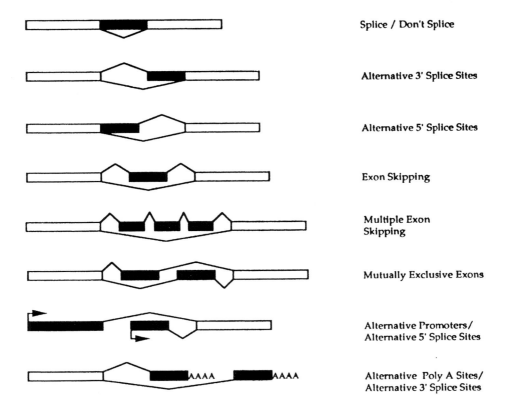

Splice / Don't Splice

Alternative 3' Splice Sites

Alternative 5' Splice Sites

Exon Skipping

Multiple Exon
Skipping

Mutually Exclusive Exons

Alternative Promoters/
Alternative 5' Splice Sites

Alternative Poly A Sites/
Alternative 3' Splice Sites

Figure 21.12: Epitomized alternative splicing (McKeown, 1992; permission from Annu. Rev. Cell Biol. Vol. 8, © 1992 by Annual Reviews Inc.). Diverse mRNA species, translatable into proteins with various functions, can be derived from one sole gene. Innumerable examples of alternative splicing have become known; it actually seems to be rather the rule than the exception.

**Excursus 21-4**
**WHO IS BOB TJIAN?**

TATA-boxes have belonged for many years now to the costumary scientific jargon; these are short sequences located within promoters, some base pairs ahead of the transcription start for RNA polymerase (see, for instance, Mathis & Chambon, 1981). The TATA-boxes occur in prokaryotes as well as in eukaryotes. In the course of evolution, these special sequences – TATAAT, TATTA, or else TATATAA, etc. –

were more or less conserved inside the some-dozens-of-pairs-long promoters. This degree of conservation hints at TATA-boxes performing an essential function in the RNA polymerase binding to the promoter. Defining this function – as just one case in point – is certainly one of the many alluring aspects of modern molecular biology. And here, the question: "Who is Robert (Bob) Tjian at the Howard Hughes Medical Institute & University of California in San Francisco?" stands again as one representative example; he represents, typically, the young generation of dynamic, hard working contemporary molecular biologists, who were still attending school when the promoter, as today it is conceived, was first described (see, for instance, Pribnow, 1975) and who, in contrast to many of the more easygoing pioneers, represent a group of highly stressed professionals. Tjian is not exactly very famous, but he has called attention upon himself in a series of articles published in Nature in which he and his constantly changing group of co-workers dissect the subtle details of TATA-box functioning. For some time, it has been inferred, that in eukaryotes RNA polymerase, by itself, is not capable of initiating the transcriptional events (see, for instance, Zawel & Reinberg, 1993). For that, among many other prerequisites, a so-called TATA-binding protein (TBP) – as the name makes clear – has to bind first to the TATA-region of DNA. Following this initial step, further proteins – the so-called transcription factors – garner on the TBP. Crowley, Hoey, Lin, Jan, Jan & Tjian (1993), characterized a TBP-like protein – the "TBP-related factor", TRF – which, unlike the broadly effective TBP, is only active in certain cells of the *Drosophila* embryo. This means that the TRF has a key function in the selective expression of certain genes of special tissue cells at precise moments of the embryonic development. (But now, how is the synthesis of TRF in these cells brought about?) Yet, TRF occupies merely one rung on the ladder of auxiliary factors collaborating with TBP in the task of rendering some promoters more attractive to the RNA polymerase; Weinzierl, Dynlacht & Tjian (1993) suggest some other interactions which might provide a glimpse of work in progress (Fig. 21.13 ).

And now the crucial question: is this pace going to be sustained forever (see, for instance, Brenner, 1983; Näär, Lemon & Tjian, 2001)? More and more, biochemistry and cell biology – traditional genetics perhaps less – will delve into the depths of the multitude of aspects of the cell's life. Some will feel devastated by the lack of grandiose perspectives without the promise of further revolutionary insights, but others will get inspired just by the infinite details – possibly very important details, as far as applied science is concerned – offered by prospective research themes still to be approached.

Yet, after all riddles which can be scientifically tackled have been deciphered, will the universe, especially the mystery of life in it, be better understood than it is today? Or will we come to recognize that the huge amount of scientific achievements, in spite of all the overwhelming technical advances involved, have only revealed epistemological trivialities, and that, in the end, the world remains as it always has been: inherently inscrutable?

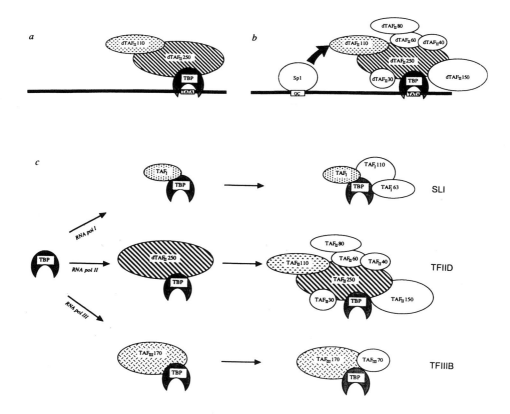

Figure 21.13: Scheme of one possibility of how the TATA-box-binding protein (TBP), by means of additional proteins, the so-called TBP-associated factors (TAF-proteins), creates a specificity enabling the different RNA polymerases of eukaryotic cells to recognize diverse promoters (Weinzierl, Dynlacht & Tjian, 1993).

*a*) *The* dTAF$_{II}$ 250 (d stands for *Drosophila*) is capable of binding on one side to the TBP and, independently, to an additional TAF (dTAF$_{II}$110). *b*) The latter only becomes active upon interaction with Sp1 (a so-called zink-finger protein); for its part, Sp1 must bind to a special base sequence (the so-called GC box), located upstream (namely, in the transcription's opposite direction). Many additional TAF-proteins may then join in. *c*) The complexes SLI, TFIID and TFIIIB (right) formed in this or in a similar way are pivotal for the process of transcription by the RNA polymerases I, II or III.

Taking into account that the binding of specific TAF-proteins is conditioned by other hitherto unknown factors, this scheme actually represents merely the beginning of an increasingly intricate chain of possible interactions between specific proteins and regulatory elements on the DNA (certain crucial base sequences supportive of the actual promoter). The type of interaction will determine the pattern of gene expression in the diverse tissues of higher organisms. Will we be able to scrutinize the whole complex?...and, if yes, will this knowledge bring us closer to "comprehending"?

# CHAPTER 22

# SEQUENCES

In the 1960s, the conviction prevailed that sequencing the bases of polynucleotides was fraught with infinitely more difficulties than sequencing amino acids in proteins. It probably would never be possible to decipher a long row of nucleotides, be they the components of RNA or DNA, let alone the whole sequence of even the simplest of the genomes; that was the undisputed opinion, grounded in what seemed to be a legitimate argument: in contrast to the 20 protein building blocks, there were only four bases as building blocks for the nucleic acids; this implicated less variable neighborhoods and, consequently, a more limited possibility for determining the succession of bases in small fragments derived from larger polymers. (There are $20^3 = 8000$ conceivable tripeptides, but only $4^3 = 64$ distinct trinucleotides.) Moreover, no specific nucleases able to degrade polynucleotides to defined smaller pieces were available; in contrast, for the polypeptides, there were several specific proteases like, say, chymotrypsin, which cleaved behind phenylalanine and tyrosine, or trypsin which cut behind arginine or lysine, their combined use giving rise to precisely defined, overlapping oligopeptides – an essential tool for protein sequencing.

Among all nucleic acids, the most viable candidates for a tentative sequencing were the tRNAs; these were – judging from their structure – the most similar to proteins. (Pointedly, Crick opined: tRNAs are Nature's attempt to construct enzymes by means of nucleic acids.)

Robert Holley (1922-1993), working at Cornell University in Ithaca, N.Y., and supported by a small group of co-workers, had isolated total tRNA from yeasts and attempted to separate individual amino acid-specific fractions. His approach was by means of the so-called counter-current procedure, which takes advantage of the fact that the components of a mixture have slightly different solubilities in diverse solvents. The purest fraction was that of alanine tRNA. That result meant that the project of sorting out polynucleotides by means of biochemical methods was feasible. A chemist in possession of a purified product, inexorably takes up the next challenge of analyzing it. In the case of tRNA$^{Ala}$, this meant sequencing its several dozens of bases – 77, as we know today. That tRNA could be cleaved by different

RNAases, yielding two distinct and reproducible groups of oligonucleotides; this bode a good omen. The pancreatic RNAase cleaved behind pyrimidines, namely C and U, whereas the RNAase T1 (also called tacadiastase) cut behind G and, at lower temperatures, even more selectively, behind only certain Gs of the sequence. Thus, Sanger's method for protein sequencing, worked out in the 1950s, could be adapted to tRNA: enzymatic cleavage of a large polymer into many overlapping fragments, sequencing the small number of building blocks in each fragment, and finally deriving the original order of the fragments by puzzling together the overlapping regions (Holley et al., 1965). Rare modified bases, placed at special sites of the tRNA, had served as orientation marks within the monotonous landscape of Gs, Cs, Us and As, so facilitating the procedure. Actually, one of the hardest nuts to crack in the whole project was identifying these very bases: dihydroxyuracil, inosine, methylguanine, ...

Zachau and his group in Cologne, Germany, engaged themselves in a parallel effort to sequence two different yeast tRNA$^{ser}$. One and a half years earlier, and they would have attained the goal first (Zachau et al., 1966).

Thus, it was Holley who was to be bestowed with the 1968 Nobel Prize for having sequenced the first biologically significant polynucleotide. However, as verified later, Holley's sequence was not flawless, whereas Zachau's was correct to perfection. Unfortunately for the latter, these details were revealed too late to matter.

In Cambridge, Fred Sanger had already made concrete contributions for the development of the RNA sequencing technique, especially the use of paper electrophoresis to distinguish and characterize $^{32}$P-labeled oligonucleotides (Sanger, Brownlee & Barrell, 1965). And, after joining the bandwagon of rRNA sequencers, he and his team delivered and published a monumental work exposing the complete sequence of coli 5S rRNA's 120 nucleotides, documented by overlapping sequences of a seeming infinitude of segments obtained enzymatically. It was, at the time, the longest nucleotide sequence yet resolved (Brownlee, Sanger & Barrell. 1968).

Sanger's group started on their next project: the genome of phage R17 (Adams et al., 1969), from whose 3500 nucleotides they managed to sequence 100 (Jeppesen et al., 1972). Apparently, they originally aimed at cracking the genetic code by comparing the genome's sequenced bases with the order of amino acids in the phage coat protein. Alas, in the meantime, the code problem had been solved (see Chapter 10)! Moreover, some weeks ahead of them, a rival research group had disentangled, in almost its entirety, the sequence of amino acids in the coat protein of phage MS2, a close relative of R17 (Fiers et al., 1971). For Sanger's group, it had been a truly frustrating, very laborious and time consuming undertaking.

Sequencing DNA was fraught with still more difficulties. First, at the time, no sequence-specific cutting enzymes were known. Second, the modified bases commonly found in tRNA, which provided additional orientation, were mostly absent from DNA.

Despite all these handicaps, Ray Wu at Cornell University, together with Dale

Kaiser at Stanford, dared to make an attempt. Their effort to define the few bases at the protruding single-strand ends of phage λ DNA caused a small sensation with a long publication in the renowned Journal of Molecular Biology (Wu & Kaiser, 1968). The corresponding procedure had been lengthy: the protruding ends served as templates for DNA polymerase; in each round of successive experiments, only one of the four DNA building blocks was offered and its eventual incorporation monitored. Wu and Kaiser reckoned the so-called "sticky ends" of phage λ to be 20-nucleotides-long; three years on, after having access to endonucleases to cleave the radioactively labeled, newly synthesized short sequences, and analyzing them electrophoretically, Wu realized that the sticky ends had only 12 bases, not 20, and that the sequence, as originally published, was largely wrong (Wu & Taylor, 1971). What is more, the technique was not suitable for general use.

Enter Sanger; he decided to extend his RNA sequencing strategy, effective albeit strenuous, to DNA. The outcome was modest: some dozens of bases of the small phage f1, a single-strand DNA-containing oddity, were unveiled (Sanger et al., 1974).

The use of highly radioactive DNA fragments greatly facilitated this sort of experiments; such fragments were obtained by means of DNA polymerase and highly radioactive dATP. Because the availabililty of such radioactive material was rather limited, the practice of mixing it with non-radioactive dATP was common. However, in order to attain the highest possible specific activity, Sanger chose to dismiss this practice. Luckily, he did not miss observing that, under this condition of restricted availability of dATP, many of the newly synthesized DNA molecules were markedly shorter than expected. This observation fed the hunch that the polymerase reaction had been prematurely interrupted and all further synthesis arrested, as soon as the short supply of dATP had been used up. If this suspicion held true, then the synthesis of all these short segments was aborted at a point when the building block A had to be incorporated, but failed to be present. Perhaps, similarly, smaller pieces of DNA would result under shortage of C, G, or T, too. The so-called "minus" method was born, with extended expectations, supported by the help of the – in the meantime discovered – restriction enzymes (Chapter 19). The trend-setting method, grounded in the separation of polynucleotides of diverse lengths, whose next-to-be-incorporated base was known, was a far more elegant system than that relying on the same principle as that involving protein sequencing. The latter suddenly appeared archaic, like the technology of charriots transposed to the era of the automobile. Pivotal for the modern method was the immense resolution power wielded by electrophoresis: variably sized single strands of DNA could be accurately separated according to their exact number of nucleotide building blocks. Even the difference of one single nucleotide between molecules led to their visible separation. Four parallel synthesis assays – each of them with a different limiting nucleotide building block – sufficed to create an electrophoresis ladder from which the sought sequence could be directly deduced. Besides the "minus" method, the "plus" technique yielded similar results (Excursus 22-1); its strategy was set by eliminating 3 bases from the synthesis reaction, so that only one base remained available to be incorporated before the reaction came to a halt (Sanger & Coulson, 1975). Suddenly, by means of a few gel-electrophoretic runs, hundreds of bases could be sequenced (Sanger et al., 1977) – an undreamt-of breakthrough.

Figure 22.1: Frederick Sanger (born in 1918).

Walter Gilbert at Harvard, after having isolated the *lac* repressor, together with Benno Müller-Hill, zeroed in on the *lac* operator, targeting the sequence of its nucleotides.

Isolating the operator was feasible because of its repressor-binding characteristic; the repressor, namely, acted protectively against the degrading effect of DNAases. Gilbert & Maxam (1973) arrived at sequencing the 25 base pairs shielded by the repressor, after a years-long struggle with the RNA sequencing method of Sanger et al. (1965). (Of course, applying this method required an RNA copy of the operator sequence to be first produced.) Further enticing targets were then the other signal sequences of the *lac* operon, particurlarly the promoter. But then the group around Reznikoff (Dickson et al., 1975) came marching in in front of the Harvard team, having accomplished already what Gilbert's group was planning to do.

Unlike Sanger, Gilbert's curiosity was attracted not merely by the sequences per se, but primarily by the issue of gene expression and gene regulation, specially in what concerned the *lac* operon. However, once the important regulatory sequences were laid bare – owing to the rival's successful venture – a further particular aspect of gene regulation became the focus of his interest: the details of the molecular interaction between regulatory proteins and their target sites on the DNA, especially the adjustment of the repressor to its operator.

As so often in science, progress was due to a chance meeting (see Gilbert, 1981). A certain Andrei Mirzabekov, a visiting scientist in Harvard, described how he got certain specific DNA purines (guanines and adenines) to be methylated by dimethylsulfate. He aimed at revealing the contact points of histones on the DNA; the DNA, being partially shielded by the histones at those sites of contact, was selectively refractory to methylation. Extending the principle to unravel the points of contact between *lac* repressor and operator seemed promising.

A 55 bases long restriction fragment, encompassing the *lac* operator, isolated some time earlier in Gilbert's laboratory, helped in the pursuit of the alluring, though yet to be ripened, concept. Dimethylsulfate methylated the purines, and, as Gilbert then remembered, once methylated, the purines could be easily detached from the DNA by heating. The deoxyribose groups devoid of their bases became susceptible to alkaline hydrolysis, on account of the gaps along the phosphate-sugar backbone yawning as a result of the removed purines. If the methylating reaction was controlledly performed, few purines – or even merely one purine – of the short DNA segment was to be targeted. And, after the alkali treatment, a series of still shorter fragments could be gained. The strand with the critical segment, one of whose purines had been methylated, had to be marked at one end with radioactive phosphate, for later identification by autoradiography. Even the first trial runs yielded excellent results; apparently, the method not only specifically targeted purines, but it also allowed Gilbert's group to distinguish between adenines and guanines, because guanine reacted markedly better with dimethylsulfate than adenine, and besides that, the methylated adenines were more readily excised from the sugar. With these cards in hand, one could move on to optimize the procedure, in order to precisely differentiate between both purines.

By treating similarly each of the two complementary DNA strands, namely by first labeling their ends with $^{32}$P, then subjecting them to dimethylsulfate and finally cleaving them with alkali, it would, in principle, be possible to ascertain their exact base sequence: each of the double strands' purines matched with the corresponding pyrimidine in the complementary strand.

However, not fully satisfied, Alan Maxam, Gilbert's co-worker, decided to pinpoint the pyrimidines' positions (cytosines and thymines) directly; he knew that hydrazine reacted with them, thus cleaving the phosphate-sugar backbone would be possible also in this case. However, both pyrimidines reacted with equal intensity with hydrazine – till Maxam verified that, at higher salt concentrations, hydrazine reacted only with cytosine. Thus, two parallel runs, one with, the other without salt, made the trick of differentiating between them. This advance rounded the concept of sequencing by means of base-specific chemical splitting of the DNA. Optimizing the technique to complete the task followed as an anticlimax (Maxam & Gilbert, 1977).

This newly developed, generally applicable sequencing method, taking advantage of selective chemical modifications of the DNA bases, soon eclipsed the original idea, namely that of characterizing DNA sequences through the pattern of protection of their bases accorded to them by specific proteins bound to that DNA. However, this later tactic attained a fresh perspective through the "footprint" technique, designed to reveal specifically the contact sites for DNA-binding proteins (Excursus 22-2).

Nevertheless, the main scientific interest remained focussed on DNA sequencing itself, by then, supported by well delineated, routinized and broadly employed techniques. Uncounted sequence analyses followed in its wake; soon, many thousands of base sequences deluged the scientific scene.

And still, there was plenty of room for improvement. Sanger's plus-minus method was restricted to solving relatively short sequences, as a certain degree of synchronization of the DNA polymerase reaction on the individual template molecules limited the range of lengths of the newly assembled strands. Thus, without resting on his laurels, Sanger pursued further refinements of his method (Sanger, Nicklen & Coulson, 1977): the DNA polymerase reaction could be thwarted not only by the absence of the next-to-be-used nucleotide, but also – and still more elegantly – by the use of an analogue whose incorporation led to premature curtailment of the ongoing polynucleotide synthesis. Ideal analogues were the 2',3'-dideoxytriphospho-nucleosides; the lack of a hydroxyl group in the sugar's 3' position entailed that no further ester binding to the phosphate of a next neighbor-to-be nucleotide could take place. The technique was extremely easy to perform, with a caveat: 2',3'-dideoxy-nucleotides were not available, they first had to be strenuously synthesized. (Perhaps not that strenuously; Sanger, later, in 1988, hinted that the changes introduced in his laboratory work by the little bit of pure organic chemistry, in effect, brought about a lot of entertainment.) This handicap is confined to history, though: today, innumerable biotech and pharma companies compete for customers for their dideoxytriphosphonucleosides. The sequencing boom got its real head start with the dideoxy strategy, not wanting to belittle the remarkable accomplishments reached by Maxam & Gilbert's tactic. An infinite series of sequences ("megabases") flooded the daily life of molecular biologists – a scourge or a bonus? In the wake of the first horror scenario of being overwhelmed by the sequencing mania, science recovered to the point of accepting this abundance as a benefit.

The 1980 Nobel Prize for chemistry recognized the merits of Frederick Sanger and Walter Gilbert in advancing their DNA sequencing techniques. Sanger and Gilbert shared that year's Nobel accolade with Paul Berg, who was honored for his work, especially in gene technology, on the biochemistry of nucleic acids. (Fred Sanger had been already distinguished with one Nobel Prize in 1958, for his achievements in protein sequencing; see Chapter 1.)

Truly, the potential tally of sequences to be resolved is certainly larger than the number of sand grains in the desert. This translates to an infinite expansion of molecular biology's working field, touching countless generations of researchers yet to come. The hitherto usual means of publication, the specialized scientific journals, were rendered fully inadequate for dealing with the avalanche of newly revealed sequences; therefore, a novel branch of research was created: Bioinformatics, in order to elaborate a world-wide system for computer-based storage and analysis of billions of base sequences. Who will have access to it? That is tricky; the issue is still controversial, because of the efforts being made to assure the patent rights over worked-out sequences – even if their function and position in the genome are still unknown. Debating the virtues of such patenting practices in editorials of scientific journals has become a fashionable endeavor (see Excursus 23-1).

**Excursus 22-1**
**SANGER'S DNA SEQUENCING TECHNIQUES**

**1. The plus-minus method (Sanger & Coulson, 1975)**
A sample of single-stranded DNA, whose sequence was to be determined, was taken as template for the *in vitro* synthesis promoted by DNA polymerase I. Pivotal for the method was the use of a precisely defined and homogeneous primer (as, say, a synthetically assembled oligodeoxynucleotide – an idea contributed by Hans Kössel at the University of Freiburg in Germany – or a strand from a short restriction fragment hybridizable with the template). The starting situation would be as depicted:

template

specific primer

First, a normal polymerization was allowed to go on briefly in the presence of all 4 triphosphonucleosides – at least one of them radioactively labeled with $^{32}$P-phosphate. In this initial step, radioactive strands of various length were produced, complementary to the template (the different lengths were due to the fact that the polymerization reactions running along the many template molecules were not fully synchronous):

radioactive product

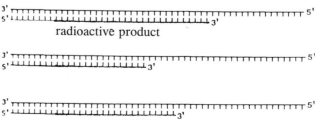

The total DNA (that is, the whole complex of template, primer and newly synthesized DNA) was subsequently chromatographically separated (in agarose columns) from the added nucleotide building blocks, and further treated differentially in four parallel assays: in each, only three of the four building blocks were added (in each one, either dATP, or dCTP, or dGTP, or dTTP, was missing). In each of the four assays, a newly supplied DNA polymerase exerted its action up to the point where the absent nucleotide should have been incorporated. The strand pairs (template plus the newly synthesized radioactive strand) were then separated by means of a short heat-denaturing procedure. The products of the four so-called minus experiments were then resolved electrophoretically, each running in a distinct lane of a gel slab. This gel slab was then placed on a X-ray film and exposed in the dark (autoradiographed) for a certain period of time (hours till days). After developing, the autoradiogram revealed the lengths of the molecules synthesized in each of the four parallel assays by the position of darkened bands (see Fig. 22.2). For each of these molecules, it could be ascertained which of the four bases would have been the next to be incorporated: exactly that one missing in the respective experiment. So it was possible to "read", directly on the gel, a certain partial sequence of the strand complementary to the template used.

In order to address the arising uncertainties, the minus method was complemented by the parallel use of the plus method. The procedures were similar. The difference lay in the nucleotides offered after the radioactive labeling in the presence of all four building blocks: instead of three nucleotides (the fourth missing), only one was added before continuing the reaction with the T4 DNA polymerase. This specific polymerase is apt, in the absence of triphosphodeoxynucleosides, to degrade the DNA product again, starting from the 3' end. This reaction is halted as soon as the point is reached where the next nucleotide is the one available to the reaction. In the plus method, thus, all radioactive molecules within each of the four parallel experiments ended with the same nucleotide, namely with the last and only one to be offered.

## 2. The chain-terminating method (Sanger, Nicklen & Coulson, 1977)
Similarly to the plus-minus method – by now virtually abandoned –, in this approach a labeled primer is also lengthened by means of the DNA polymerase, along the single strand template to be sequenced. The four parallel assays are provided with all four triphosphodeoxynucleosides; however, to each assay a small amount of one of the triphospho-2'-3'-dideoxynucleosides is also added (in a concentration ca. 1/100 of the concentration of the respective regular building block). Chain growth is thwarted at the point where the analogue is incorporated. When, say, 2'-3'-dideoxy-GTP is supplemented to the reaction, a population of polynucleotides accumulates, whose last base is guanine. As in the plus-minus method, the sequence can be directly read after electrophoresis and autoradiography (Fig. 22.2 and 22.3).

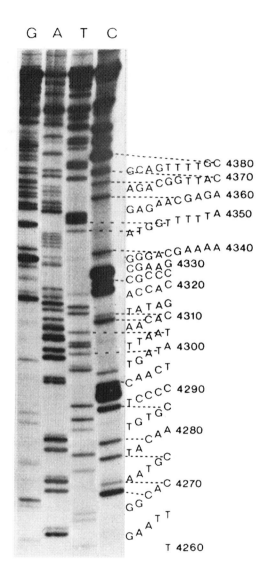

Figure 22.2: The original work's autoradiograph of a sequence gel, dealing with the chain termination method (Sanger, Nicklen & Coulson, 1977). The shortest segments are at the bottom, on account of their moving the fastest during the electrophoresis in a polyacrylamide gel (anode at the bottom). The sequence is to be read from bottom to top in the direction 5' → 3'; the last nucleotide incorporated (namely the 2'-3'-dideoxynucleotide) is indicated above the respective lane.

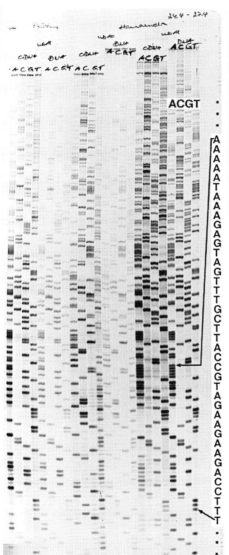

Figure 22.3: Sequence gels as routinely read today, in ordinary, nonspecialized biological laboratories. (Such a "by-hand-sequencer" would not be able to work for more than two years on such a job – opined W. Gilbert.)

The chain-termination method has, in the meantime, been adapted for large throughput automatic sequencing. Instead of the four different assays, one for each base, the polymerase reactions with the 4 dideoxynucleotides can be run in one assay mixture and subject to electrophoresis in one single capillary tube; as a precondition for this procedure, each of the four dideoxynucleotides has to be labeled with a different fluorescent dye and the minute bands are then identified by laser-photometry (Fig. 22.4)

Figure 22.4: Optimizing the automation of the sequencing technique is quite advanced: bases today are sequenced at rates of 100 or even 1000 times speedier than only a few years ago. Many laboratories already employ routinely automatic sequencing devices, the DNA synthesized for the process being labeled with nucleotide derivatives that render them fluorescent in different wave lengths: this allows the laddering on the sequence gels to be photometrically monitored; the outcome is then represented graphically. To tell apart the four curves on the graphic representation, four distinct colors, or as depicted here, different dashline patterns, are used. The sequence displayed here was obtained by Gabor Igloi at the University of Freiburg in Germany by using an A. L. F. DNA sequencer (Pharmacia). In high throughput centers, as those involved in the Human Genome Project, fully automated sequencing is carried out by use of capillary electrophoresis.

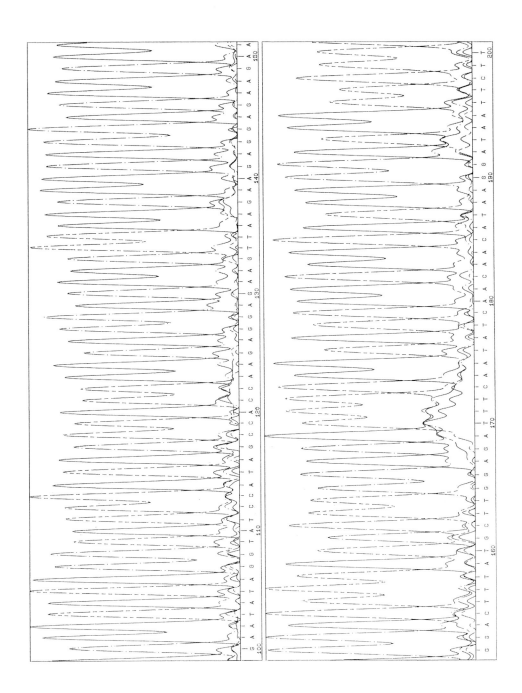

**Excursus 22-2**
**"FOOTPRINTING": DEMONSTRATION OF CONTACT SITES BETWEEN**
**DNA-BINDING PROTEINS AND THEIR SPECIFIC DNA SEQUENCES**

The sequencing method developed by Maxam & Gilbert (1977) presented a way of characterizing specific protein-DNA interactions, known to occur, as for instance, between operator and repressor or between promoter and RNA polymerase: the lodging of the DNA-binding proteins hindered the access of DNA-damaging agents. A DNA fragment to be sequenced, if protected by a protein, was refractory to cuts along the covered region. This could be recognized on the autoradiogram from an electrophoresis gel as a gap void of bands. If the nucleotide sequence of the fragment in question was already known, then the base specificity of the break points could be overlooked and unspecific endonucleases, like DNAase I, which cut between any two nucleotides, could be used instead. Gilbert selected, as a model system, the *E. coli* RNA polymerase and its cognate promoter. A detailed analysis of the contact sites between promoter and polymerase was finally reached (Siebenlist & Gilbert, 1980; Siebenlist, Simpson & Gilbert, 1980). But the term "footprinting", today a jargon for this technique – proposed as a jocose alternative to "fingerprinting" (see Excursus 8-1) – was avoided by Gilbert's group: the expression had been coined by two unknown young colleagues from the University of Geneva, who, in the meantime, guaranteed the priority for the description of the novel DNAase footprinting method, through the successful visualisation of the contact sites between the lactose operator and its corresponding repressor [Galas & Schmitz (1978), see Fig. 22.5].

Figure 22.5: DNAase I "footprint" after the binding of the *lac* repressor to its operator (Galas & Schmitz, 1978). A radioactively labeled restriction fragment encompassing the *lac* operator was incubated with the *lac* repressor and subsequently subjected to a precisely dosed DNAase I digestion (cuts endonucleolytically 3'-sugar-phosphate bonds); from the lack of DNA fragments of certain lengths, reflecting the shielded regions, the DNA-repressor contact sites could be inferred. -R: banding pattern after DNAase I treatment without repressor (control run); +R -I: banding pattern with repressor, without inducer; +R +I: banding pattern with repressor and inducer, though, apparently too little of it (the same authors, had, in fact, also demonstrated a clear inhibitory effect of the inducer on the repressor's binding capability); both last lanes refer to a sequencing experiment with the same fragment (performed by a simplified variant of the technique of Maxam & Gilbert, 1977). The method here described, presented in a pioneer experiment, has meanwhile been refined repeatedly and further optimized (see, for instance, Tullins, 1989).

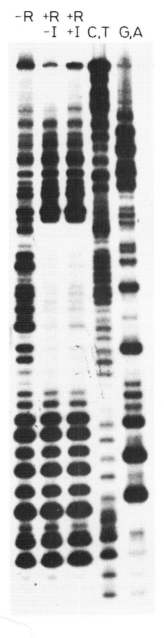

# CHAPTER 23

# HUGO & ELSI

The newly developed sequencing methodology cleared the way for tantalizing prospects. The *E. coli* chromosome? Sure, an enticing goal, but... Charles Weissmann from Zurich suggested at a meeting in Heidelberg, Germany, at the beginning of the 1980s that the veritable challenge for modern science would be the sequencing of the 1000 times longer human genome. However, resolving its 3 billion bases by the standards of the time seemed to be a nearly impossible dream. Nearly!

It was probably Robert Sinsheimer (University of California, Santa Cruz) who, in 1985, made the first serious proposal to sequence the human genome by inviting to Santa Cruz a group of experienced sequencers, like Walter Gilbert and John Sulston, to discuss the matter. Although the idea of sequencing the whole human genome was endorsed with enthusiasm by Sinsheimer's guests, they also felt that that ideal was just what it was: an utopian ideal, impossible to be attained. With their roots in traditional biological research, they had in mind work being done by independent individuals in small university laboratories, as a kind of cottage industry, the way it always had been.

This was not the way of thinking at the Department of Energy (DOE). This large institution, originally responsible for all aspects of atomic energy, traced its involvement with biology to its task of monitoring the effects of radiation on the survivors of the atomic explosions over Hiroshima and Nagasaki. Since radiation damage affected primarily the human genome, and since the DOE had the tradition of "thinking big", it was only natural that some of its officials, like Charles DeLisi, felt that sequencing that very genome was a project tailor-made for the DOE – especially since, in the relative quiet of the armaments control era, there was a lot of unused resources (see Roberts, 1988; Watson, 1990). Considering the Manhattan project or the drive for the landing on the moon, such an undertaking, if honed by committed efforts, could very well be accomplished. However, most scientists of the time did not take seriously such arguments, or mocked them as preposterous and megalomaniac dreams ("Is megasequencing madness?", see Newmark, 1985).

The mere intention, proposed by the DOE, was enough, however, for the sequencing idea to gain momentum in the traditional molecular biologists' community.

The idea that it was impossible was slowly replaced by the thought that it might be possible, but that it was not worthwhile (David Botstein: The problem is not big science so much as bad science. Luria: If something is not worth doing at all, it is not worth doing well.) Akin to Eugène Ionesco's "Rhinocéros", though, more and more biologists converted to "rhinoceroses" till the point was reached where scarcely a critical voice was still being raised. How did this transformation come about?

Slowly it had become clear: Science was not what it used to be 20 years ago, a claim which became represented by the idea that big science was "in" after all, individual thinking and individual research were both "out". The sequencing of the human genome stood as a symbolic, overwhelming ploy (see, for instance, Watson, 1990; Cantor, 1990) to put aside scientific objectivity, cost-benefit assessment and evaluation of alternative options. Several billion dollars were to be earmarked for financing a game which dealt as much with power, national prestige and all sorts of ambitions as with a real scientific challenge. The motor behind it all: alliances between parties – scientific and political – once more concentrated around mutually sustained interests, complemented and reinforced by each other. Once standing, these alliances needed a media-effective and straightforward symbolic goal, ideally represented by the human genome's base sequencing.

When the DOE gained support from the American Congress, starting a Human Genome Initiative in 1986, the National Institutes of Health (NIH) could no longer sit on the fence and in 1988 the National Research Council endorsed a Human Genome Sequencing project by the NIH (which later gave the project a home in one separate new institute, the National Human Genome Research Institute). At the same time, voices rose in favor of expanding the human genome project to a global enterprise, including not only Japan, England, France [which became, overnight, a pioneer in the subject on account of the robotics company Généthon (see, for instance, Anderson, 1992; Guainville, 1992)] and Germany, but also, in a lesser scale, any country willing to join in as a partner. All efforts were to be coordinated in an international enterprise, the Human Genome Organization (HUGO). Ambitious prospects – therapy for hereditary diseases, gauging the number of genes and their genomic locations, revealing the amount of useless DNA being carried around, solving the intrinsic mystery of being human, etc. – would be used to justify the high costs of such an undertaking.

Full scale megasequencing of the human genome was to be postponed, though, until maps of the human chromosomes were available and until technical progress had made very fast sequencing a reality. Such progress was to be tested with model organisms like *E. coli*, yeast and the nematode worm *Caenorhabditis elegans*. The mouse genome would be especially valuable for medical and pharmacological research.

The NIH genome project then gained congressional funding as well (Jim Watson, Walter Gilbert, David Botstein, Norton Zinder, Paul Berg, and all those wielding some sort of influence had mounted a skilful lobby), dispelling many fears that financial resources for the project would be raised at the expense of traditional research. The project got a prestigious boost when Jim Watson was named head of the new NIH office (Excursus 23-1).

Progress in mapping was noticeable; Sulston's group in Cambridge, U.K., had already a nearly complete map of *C. elegans* (Coulson et al., 1991); more and more genetic markers (mainly as restriction sites) were being placed on human chromosomes. Soon, large scale sequencing would become feasible...

This general consensus was only disturbed in 1991 when Craig Venter, running a large sequencing lab at the NIH's Institute for Neurological Disorders and Stroke, came up with an obvious idea: Was it really necessary to sequence all of the human DNA? The sequences of utmost importance were those coding for gene products, for proteins. And these sequences corresponded to only about 1% of the genome. And those sequences could be selectively found by isolating the mRNAs from tissues and they could be analyzed as cDNA copies. One use of these cDNAs was to create expressed sequence tags (ESTs) on the chromosomal maps by hybridizing labeled cDNAs with chromosomal DNA. Venter did not hesitate to immediately take advantage of this technique and announced that his new approach would allow him to find up to 90% of the genes, the really important sequences, within a few years, at a cost which was a real bargain compared to that of the genome project – only to earn the sour-grapes comment that this was a cream-skimming approach. Venter was so

Figure 23.1: Yesterday a theme for cartoons [see Trends in Genetics, 4,31 (1988)] – today a reality? In effect, bio-enterprises already started offering to fans of rock stars bracelets containing their idols' short DNA sequences (see The Observer, 26th of Dec., 1993).

convinced of the usefulness of his idea that he persuaded the NIH director, Bernadine Healy, to favor filing patent applications on the partial gene sequences he was identifying at an incredible rate – he had already several thousands of brain-specific sequences, even if only partial sequences (it was difficult to isolate complete mRNAs). Watson (he had never been close friends with Venter) exploded: patenting such sequences was "sheer lunacy". Healy held another opinion – the following dispute resulted in Watson quitting the NIH (see Excursus 23-1).

Who also quit NIH was Craig Venter, although for quite different reasons. In 1991 he was offered nearly 100 million in venture capital (from whom?) in order to pursue his gene fishing strategy. Thus, a private enterprise, The Institute for Genome Research (TIGR) was founded. This was the point of departure for Venter's unconventional private genomics activities which would cause considerable turbulences. TIGR soon came up with a stunning success: The shotgun sequencing of *Haemophilus influenzae* with a 1.8 megabases genome (Fleischmann et al., 1995).

And Venter was boldly prepared to apply the same procedure to the human genome, more than 1500 times bigger: simply tear the whole DNA to pieces, sequence innumerable random fragments, feed the sequences into a computer programmed to look for overlaps, and join partially overlapping pieces together to form increasingly larger contigs.

Meanwhile, Watson's successor at the NIH genome project, Francis Collins, a physician from the University of Michigan – who had formidable credentials, having already identified several disease genes, including that for the most common human Mendelian disease, cystic fibrosis –, was still committed to the methodical approach of sequencing DNA pieces after their having been laboriously located on chromosome maps. The shotgun sequencing of *Haemophilus* by TIGR caused a minor earthquake at the NIH (which insistently had been claiming that shotgun sequencing would not work). Nevertheless, the methodical NIH approach, endorsed by HUGO, the international sequencing program, especially by the Sanger Centre at Cambridge, U.K., which was led by John Sulston, had also a splendid accomplishment to counterbalance TIGR's success: The sequencing of the yeast genome (Goffeau et al., 1996).

But the bold challenge by a private competitor for the glory of sequencing the human genome could not be ignored. NIH had to speed up sequencing, which was lagging behind schedule (only a few percent of the total genome had actually been sequenced by 1996). The arbitrary standard of 99.99% base sequence accuracy was lowered by a factor of 10, thus, 99,9% would do (which meant that there would be errors in practically every gene). Besides that, the search for technical improvements was intensified and costs reduced dramatically.

To no avail: Venter, committed to the brute force shotgun technique, landed a new coup: in May 1998 he announced that, with his method, he would sequence the entire human genome in just 3 years for a bargain, a mere 300 million dollars and several years ahead of schedule, in 2001. The participants in the publicly funded project were totally upset by this obviously impossible claim. But Venter explained: He had negotiated with Perkin-Elmer Corp. to set up a new private enterprise, Celera Genomics, to be located in Rockville, Maryland, whose aim it was to offer information on human genome sequences, for a price, to subscribers only (for instance, pharmaceuticals companies). Perkin-Elmer, a manufacturer of the most advanced sequencing machines, was to deliver 300 of those machines at a price tag of 300,000 dollars each and the sequences spit out round the clock by these machines, at an unprecedented speed of millions of bases a day, would be puzzled together by the fastest computers using the most sophisticated software.

Now the public program had to be speeded up again if Francis Collins, John Sulston, and all the others responsible for it wanted to avoid a humiliating defeat. In September 1998, they announced that their goal would be attained by 2003 – two years ahead of schedule. The point was: There still remained many doubts as to whether Venter's approach would work, although by February 2000 Celera had tested the shotgun strategy on *Drosophila,* whose 180 megabase genome it had sequenced (Adams et al., 2000), although with countless serious errors, as it was shown later (Karlin et al., 2001). The quality of the data was not assured and there would be innumerable gaps impossible to bridge by shotgun sequencing. The public consortium thus saw a chance in, at least, coming to a draw with Celera by again lowering their standards: 90% of the genome would do for producing a "rough draft", and this by the spring of 2001 – Celera certainly would not do any better. One important goal would be attained by doing this, namely to prevent Celera from filing patent applications for large portions of the human genes.

This issue of data access, the dispute over quality standards, and personal acrimonies poisoned the merciless race, and increasingly became an embarrassment to the scientific community, which cautiously started to urge the contenders to find a way to put an end to the feud and cooperate, at least to a certain point. Mainly through the efforts of Eric Lander (Whitehead/MIT Genome Center) and Ari Patrinos (DOE), a truce was signed under which both groups would announce their "rough drafts" simultaneously (Fig. 23.2). In a rather spectacular PR initiative, culminating with President Clinton's congratulating both rivals for their unique accomplishment, the race, which after all had benefited from this grotesque competition, came to appear honorable to both Celera and the publicly funded consortium. But a last minute squabble over data access led the two parties to publish their work in different journals, Celera in Science (Venter et al., 2001), and the publicly founded consortium in Nature (International Human Genome Sequencing Consortium, 2001).

Why "rough drafts" should be acclaimed with such pompous hype is not understandable to everybody (the "polished versions" are not expected before 2003), but nevertheless, it is certainly worthwhile to give a short overview of what conclusions and surprises they offered.

The genomes presented by Celera and the Human Genome Project, in February 2001, could as well have been from two different organisms: Celera had found 26,383 genes, while the public consortium had 31,778; these numbers were actually one of the biggest surprises deriving from the projects, since, with an estimated total between 30,000 and 40,000 genes, the human genome apparently has only about half as many genes as previous (quite arbitrary) estimates suggested. For comparison, it is noteworthy that the genome of the tiny flowering plant, *Arabidopsis thaliana*, was found to contain 25,498 genes, that of *C. elegans* 19,099 genes, and that of *Drosophila* 13,601 genes.

Besides a few thousand genes for housekeeping enzymes and basic cell components (as cytoskeletal proteins), there are genes for nearly 2000 putative transcription factors, far more than 1500 receptor proteins, nearly 1000 kinase genes, hundreds of genes for ion channels, protooncogenes, immunoglobulins, extracellular matrix proteins, etc. But more than 40% of the putative proteins could not be assigned to a functionally defined protein family.

Figure 23.2: Ari Patrinos (center), enjoying the results of his behind-the-scenes efforts: he had architectured a (fragile) truce between Venter (left) and Collins, who at a White House ceremony with President Clinton, in June 2000, agreed to publish their results at the same time.

The non-coding sequences – more than 98% of the total – are quite a problem of their own, encompassing transcribed but not translated sequences, as introns, genes for rRNAs, tRNAs, and several signal RNAs, as well as a majority of untranscribed DNA which is, to a large extent (but not exclusively: consider pseudogenes, promoters, transcription factor binding regions, etc.) composed of repetitive sequences and retroposons whose possible viral origin lies many millions or hundreds of millions of years back in the evolutionary past of our species. Special mention might be made of the so-called Alu sequences, pieces of about 300 base pairs apparently derived by retroposition from cellular sL RNA and which constitute about 10% of all human DNA. It cannot be said that all these insights, as interesting as they are, came as stunning surprises (Alu sequences, for example, have been studied in detail for more than 20 years).

The main task is now, of course, still ahead and refers to what has become known as "annotating" the raw sequence data; i.e., genes, with their exons and introns, have to be identified in detail, noncoding sequences have to be analysed with regard to their possible functions and origins, comparisons with the genomes of other organisms have to be carried out... there is no end in sight of what to do with the human sequence.

A series of projects, for instance, involves the genetic analysis of whole populations. Thus, the Human Diversity Project, initiated by Stanford's Luigi Cavalli-Sforza, aims at genetically characterizing indigenous populations and tracing the historical migration paths of human tribes (see Butler, 1995; Ingman, 2000). On the other hand, Kari Stefansson's enterprise deCode, associated with the pharmaceuticals company Hoffmann-La Roche, excels in analyzing the genetic make-up of the ethnically homogeneous population of Iceland, hoping to glean medically important insights on the inheritance of a series of the most varied diseases (see Enserink, 2000a).

Of course, even within one ethnic group, no individual is genetically identical to any other individual (except in the case of identical twins). To a large extent, such individual differences are due to the so-called single nucleotide polymorphisms (SNPs), i.e., single base differences, distributed over the whole genome, which are found by comparing homologous DNA sequences from various individuals of a given population. By now, millions of such SNPs have already been identified, and – one more sign of our times – there is no lack of efforts to make money out of that information, since it is likely that particular SNPs might predispose their bearers to certain diseases (see Adam, 2001; Robertson, 2001).

[One question worth mentioning, in this regard, refers to the number of people whose DNA was to be examined in the course of sequencing the human genome. Of course, no single individual was to be chosen for that purpose (although Jim Watson had been suggested – Fig. 23.1). Craig Venter said, his data were based on the DNA of five individuals (one of whom is himself, or so says the rumor that he never disclaimed). No further details are known about these individuals. (Are minority groups represented?) This is an astonishing lack of information; no work on mouse genetics would be accepted for publication without an exact reference to the strain used.]

The accumulation of data regarding genetic details at the genomic level is likely to proceed explosively for quite some time. What nowadays is known as the university-industrial complex will be one of the main driving forces in this process. Soon the number of sequenced bacterial genomes will be counted in hundreds, not dozens, as of today (see Nelson et al., 2000), among them most human pathogens. The first plant pathogenic bacterium, *Xylella fastidiosa,* has been fully sequenced by a Brazilian consortium (Simpson et al., 2000). And, regarding eukaryotes, the trend is bound to be similar: After the publication of detailed drafts of the genomes of the yeasts *Saccharomyces cerevisiae* (Goffeau et al., 1997) and *Schizosaccharomyces pombe* (Wood et al., 2002), the nematode worm *Caenorhabditis elegans* (The *C. elegans* Sequencing Consortium, 1998), and the fruit fly *Drosophila melanogaster* (Adams et al., 2000), unicellular pathogens such as the causative agents of malaria, sleeping disease and Chagas disease will follow. After the genome of *A. thaliana* had been sequenced (The Arabidopsis Genome Initiative, 2000), an important step was the completion of the genome sequences of two varieties of rice, one by a public Chinese consortium (Yu et al., 2002) and one by a private company, the Swiss-based Syngenta (Goff et al., 2002). The fact that the sequence data obtained by this private group were not made fully available to the scientific community has again stirred the bitter controversy regarding the freedom of science. ["This goes to the heart of what science is all about, the exchange of ideas, data..." commented the director of the Cold Spring Harbor Laboratory, Bruce Stillman (quoted by Butler, 2000).] Among the lower vertebrates, the zebrafish and pufferfish genomes are scheduled to be sequenced soon. A draft assembly of the mouse genome has been made public by the Mouse Genome Sequencing consortium (see Marshall, 2002). And the clamor to sequence the chimpanzee genome is intensifying. The idea is to pinpoint the genetic differences between humans and their closest animal relations at the molecular level. Humans and chimpanzees differ in only about 2% of their DNA bases (see, for instance, Gibbons, 1990), but what are the crucial genetic differences underlying the chimpanzees' intellectual shortcomings? [Let us sequence the chimpanzee (see Cyranoski, 2001), in order to know why he can't speak; let us sequence the parakeet (see Vogel, 2001), in order to know why man can't fly!]

Summarizing: most of us will live to see many thousands of genomes, small and large, having been fully sequenced. But, even among the most enthusiastic genomic visionaries, none is so naïve as to believe that our insights into the mysteries of, and our power over, nature are going to increase in a way proportional to the number of mega- and tera-base pairs deposited in the worldwide data files.

Nevertheless, the best surely still lies ahead: Genomics will lead to proteomics, i.e., the study of the total set of proteins of one organism, their structure and function (see Fields, 2001).

Upstart private companies are quite hopeful about selling proteomic information to pharmaceuticals companies planning to use this information for drug design. High-throughput strategies might – or might not – revolutionize protein structure analysis (see Fletscher, 2000; Harris, 2000). This approach constitutes quite a gamble – on the one hand, thousands of proteins will certainly be unsuitable targets for new drugs; on the other hand, one single protein, playing a key role in heart disease, cancer or a neurodegenerative condition like Parkinsons's disease or Alzheimer's syndrom, might be the source of a multibillion dollar bonanza. As one expert remarked at the Second International Structural Genomics Meeting at Airlie House, Viriginia, on April 2001 (see Marshall, 2001): "It reminds me of the lottery, very few people will win, but everybody dreams they will."

But, what if, already in the foetus, the predisposition for a future disease, disability or disadvantage – diabetes, vascular abnormalities, cancer or, who knows, even baldness at 40 – will emerge in a genetic profile? Is it conceivable that potential sufferers would be discriminated against by work contracts or insurance policies or even simply downright aborted to avoid such issues coming up in the first place? An eugenic abortion wave could very well come in the wake of genetic profiling, dictated by society's expectations (Watson: How much does one Alzheimer patient cost?) or by government planning. To brave this new juncture and the evolving standards of moral conduct, a comission for "Ethical, Legal and Social Implications", ELSI, was called into existence to which 3% of HUGO's research budget was initially allocated, and later still more. Not only one, no, but innumerable ethics comissions have in the meantime convened and held international congresses (Aldhous, 1991; Jeffords & Daschle, 2001; Robertson, 2001). Let us face the future confidently...

**Excursus 23-1**
**SOME GOSSIP: WATSON'S DOWNFALL (OR A MODERN FAIRY TALE)**

In 1987, Jim Watson announced that he could not think of a job which he would enjoy less, only to be anointed as the boss of the National Institutes of Health's (NIH) "Human Genome Initiative" one year later. Its ultimate goal was laying bare the total human genome, for which the project drew from a budget worth a small kingdom. In the spring of 1992 he threw the towel. Bernadine Healy, the wicked Empress, was

Figure 23.3: Bernadine Healy's nomination as the National Institutes of Health's director, overlooking 15,000 co-workers, was a politically motivated action of the Bush administration. A score of candidates (all male) had already declined the job – specially over the issue of research on fetal tissues, whose support by federal funds was forbidden under the Reagan-Bush doctrine –, when the cardiologist Healy decided to go along with this policy, even if not really enthused by it. Although her administration was accompanied by adverse critical pronouncements, her positive achievements are undeniable, especially such pioneering programs as the "Woman's Health Initiative" (600 million dollars) directed to combat diseases

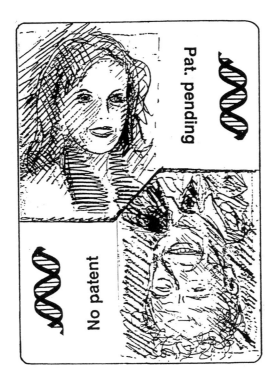

as breast cancer and osteoporosis, or the "Human Genome Project" (over 1 billion dollars), but also the emphasis on an application-oriented research perspective for the NIH. Healy defended the patenting of cDNA sequences even without known functions because – according to her reasoning – only this way would such sequences be published in the first place without facing the risk of being appropriated by private enterprises. Jim Watson, her opponent in this controversy, stood for a patent-free publication of such sequences by the NIH; in this he was backed up by most scientists. The dispute led to Watson's resignation as head of the genome project, a job he had been holding simultaneously with his position as director of the Cold Spring Harbor Laboratory. As Watson's successor, Healy hired the physician Francis Collins, the co-discover of various human defect genes, such as the gene for cystic fibrosis – one of the most common human hereditary diseases. Upon his inauguration, President Clinton revoked the Fetal-Tissue-Decree. Yet, Healy held her position for one further year; Harold Varmus (Nobel Prize of 1989 for his work on the retrovirus replication cycle) came to succeed her, bringing along his marked preference for basic research. Ironically, also in the "science métier", yesterday's heresy is today's dogma: in the meantime Watson – he, who so vehemently opposed Healy on this issue, as many other colleagues, seems quite comfortable with private companies actively taking up the profit-shedding business of sequencing (Anderson, 1993).

solely to be blamed; she, the highest ruler of the NIH Empire, was formally the boss of prince Braveheart in the quest of biology's Holy Grail (Gilbert, 1986): the human base sequence. Watson had affirmed to anyone willing to listen, that he considered the patenting of base sequences to be "idiotic", especially when no one knew their meaning. Healy thought otherwise; she was convinced that someone somehow could take advantage of published sequences – for instance, by using gene-technology to achieve a cure for Alzheimer's (the critical sequences were brain specific). If protected by a patent, the NIH would benefit from the royalties due. [The U.S. Patent and Trademark Office had refused to grant patent rights on sequences; the NIH wished to appeal the sentence, but finally complied with the verdict; the question is still open, depending on other court rulings; see also Enserink (2000b).] Watson, on the other hand, deemed that the eventual profit to be made by Healy's Empire did not compensate for the damage inflicted on the biotech industry and the harm suffered by international scientific relations. And then, instigated by the ambitious Craig Venter, isolater and sequencer of brain specific cDNAs, Healy ordered Watson to the Throne Chamber to let him know that, if dissatisfied with her opinion, he should rather tell her personally before going to the press. And then, Venter was nominated Healy's Court Counselor for the genome project, a move meant to show who had the upper hand in the Machiavellian power play. The gauntlet had been thrown. Rules of court-etiquette, though, asked for a nobler excuse to explain Watson's falling into disgrace. At this point, a certain Mr. Bourke, a speculation-loving tycoon, stepped in and became, unknowingly, the crucial pawn in the opaque conspiracy which followed: namely, he had sent Healy a furious letter accusing Watson of insulting his person. The alleged motive was the fact that two of Watson's star sequencers were seduced away by Bourke for his private and adventurous sequencing company, a deed driven by nothing but the greedy intention of financial profit – as though such an attitude was not perfectly legitimate in Uncle Sam's free-enterprise realm. The insolent letter also raised piercing rhetorical questions as: "does not Watson's participation in various gentech businesses evoke a conflict of interests incompatible with his position?"...and Watson strategically got to see this letter – at the bureau of NIH Ethics-Court-Counselor, Jack Kress. In the opinion of the latter, Watson's gentech shares did however not pose any obstacle whatsoever for the altruistic and faithful fulfillment of his duties. This was, without any doubt, a final and direct provocation. Why should Healy, already weeks before, have passed Bourke's mean letter, treacherously, to Kress? There was no other way out: Officially, Watson declared: "Having reached my goal of initiating the project, my work has been successfully accomplished". And Healy returned: "Dr. Watson is a historic figure in the annals of molecular biology and his leading force served well the NIH" (see Roberts, 1992).

# CHAPTER 24


# OUTLOOK ON THE TWENTY-FIRST CENTURY


At the beginning of the 20th century, when Mendel's observations were rediscovered and confirmed, no one could have imagined what progress genetics would make by the end of that century. The same holds true for the time when half the century had passed, as documented by the symposium volume, published in 1951, with the – in retrospect – unintentionally humorous title "Genetics in the 20th Century", held in honor of the anniversary of the rediscovery of Mendel's laws and where, as stated in Chapter 7, Avery's discovery of DNA as the transforming principle was ignored by all of the several dozen luminaries of science who contributed to that volume (Dunn, 1951).

Remembering this, one should be wary of advancing any predictions of what the coming decades will witness, especially in view of the developments of molecular biology which were unforeseeable even in the 1970s, i.e., at a time when all the fundamental insights into the principles of biological self-replication had already been made.

But, nevertheless, let us try to elaborate some thoughts on the future – if only to give the next generations something to muse about.

Taking for granted that functional genomics and proteomics will be the driving force in molecular biology, let us consider three aspects of its possible impact.

- What additional basic scientific knowledge will come from extensive genomic analysis?
- What medical and social benefits are to be expected?
- How is the view of ourselves, how is our philosophical world view going to be changed?

Functional genomics and proteomics seem to promise everlasting studies on millions of macromolecular interactions of biological importance. Beginning with the action of the so-called housekeeping genes – those genes responsible for keeping the essential cell metabolism running – there is no limit to the ever increasing complexity of these interactions, as one can imagine by contemplating, for instance, the subtleties of embryonic development. How, exactly, do all these tens of thousands of gene products act and interact to produce particular individuals with all their unmistakable features? How many thousands of tissue-specific and developmental stage-specific transcription factors, receptors, channel proteins, adhesion proteins, etc., are going to be scrutinized?

But still, the fundamental insight of specific macromolecular interactions as the mainspring of life processes will remain an accomplishment of the last century and no amount of additional examples in ever more sophisticated detail is going to change that principle. No basic new principles are expected to emerge in the future, and, actually, they are not necessary to explain biological processes. Nobody is dreaming about any new laws of nature, as was done in the 1940s and early 1950s. All these thousandfold new gene actions will surely be conceptually accommodated within the well established principle of specific interactions among information-carrying macromolecules. It is that principle, established in the 1960s, that was the real revolutionary – some might forgive me the term – paradigm shift. Of course, as inveterate adherents to Hegel's philosophy claim, a lot of quantitative changes turn out to be qualitative changes, and it is in this sense that genomics and proteomics will come to represent some sort of revolution in our basic knowledge.

The huge numbers of genetic and proteinic functions to be explored in the wake of this new revolution refer to a vast, albeit limited arsenal of biological phenomena. And probably long before this full arsenal is exhausted, a phase of diminishing returns will commence to erode the interest and enthusiasm of many of the more innovative and inquisitive scientists.

The same thought is applicable to one additional aspect of genomics: its exquisite usefulness for elucidating evolutionary pathways at all levels, from evolutionary events that occurred billions of years ago, to those occurring in front of our eyes, as for instance, pinning down the paths of recent migrations of human populations. Similarities and differences among thousands of genes are going to increasingly clarify their evolutionary origins, revealing ancient gene duplications with ensuing divergent evolution (giving rise to so-called paralogous genes) as well as exon shuffling as the cause of additional variation. It will become clearer how new genes have their origin in other genes and how many of these old genes have homologous (orthologous) counterparts in a large variety of organisms, such as worms, flies, and even bacteria. The fundamental unity of all life on earth, as documented genetically since the 1960s, when it was shown that the genetic code is universal, will be corroborated in unexpected detail by the identification of sequence motives which may be considered as molecular fossils shedding light on the genetic make-up of the very first cells on our planet.

Regarding the medical and social benefits of genomics, very high hopes have been raised, from the start, perhaps by conviction of the main proponents, but also certainly out of tactical motives, since it was only possible to raise funds – from government or private sources alike – by calling attention to wide-ranging benefits to be expected from genomics in areas like medicine, agriculture, etc.

Will there be any huge returns derived from drug design? How many rewarding target genes will be identified? Up to now, one has to observe a striking lack of convincing examples for which the new approach would seem feasible. Even more vague are the much heralded claims of custom-made drug prescriptions based on individual genetic profiles of patients, or even of healthy persons with an increased genetic susceptibility to develop specific ailments in later years. Of course, substantial benefits in the quality of life will continue to be observed in the case of many hereditary diseases, possibly based on gene therapy. But one has to keep in mind that for the most common monofactorial genetic diseases, such as cystic fibrosis, spinal muscular athrophy, or Duchenne's muscular dystrophy, prospects of a cure based on genomic research are rather dim. In the case of one of the most frequent severe genetic defects, trisomy 21, genomics will probably remain irrelevant for a very long time. The same holds true for the all-important multifactorial diseases, such as mental disorders, circulatory conditions and cancer.

Besides, whatever scientific breakthroughs might happen (including totally unrealistic magic-bullet cures for cancer and heart disease), the average life expectancy in industrialized countries is not going to rise above about 85 years, i.e. only a few years above the present life expectancy of about 78 years (Olshansky et al, 2001; Strohman, 2001). This is in sharp contrast to the doubling of life expectancy which took place in the course of the 20th century, which saw a drastic reduction of infectious diseases. Thus, with regard to medical progress, many of us will live to see that an insurmountable limit has been reached, as far as life expectancy is concerned, even if late onset demographically important diseases as Alzheimer's and a series of cardiovascular conditions eventually benefit from genomics-based developments. This is so, because there is no way of avoiding our biologically programmed aging; and even if more than just a few years should be added to life expectancy in the present century (see Oeppen & Vaupel, 2002), these years, added to old age, will not be the best of our lives. A genomics-based extension of youth will remain an utopian dream.

Finally, regarding our world view:

Charles Darwin's theory of evolution is arguably the most influential landmark in the history of biology. His 1859 "Origin of Species" dealt a heavy blow to the prevailing view of "design", meaning that living organisms had their origin in a special act of creation by a Supreme Being, whose infinite wisdom had conceived and put into existence all living organisms in a way most suited to perform their roles, each in its own place within the great chain of being. Darwin's theory of natural selection suddenly obviated direct interventions of God in order to explain the existence of a multitude of diverse organisms, all seemingly tailor-made to perform in their particular ways of life. The demise of God as an intelligent architect of living things profoundly shook the self-image of 19th-century intellectuals: Acts of God, as explanations for natural phenomena which escape our understanding, became unacceptable in biological sciences after Darwin. But, although many modern theologians frown upon the idea of a so-called gap-filling God, even our present culture has not entirely come to grips with the full consequences of evolutionary thought. What 19th-century critics of Darwin's work had to say – Darwinism was causing the breakdown of morals, the ruin of society, etc. – is still accepted by many today. The malaise caused by evolutionary ideas is so pronounced that very many intellectuals, especially in the U.S.A. and in Moslem countries, are led to simply deny the fact of evolution. What will change now?

Darwin had based his conclusions mainly on comparative anatomy, including fossils. Later, his conclusions were corroborated by comparative physiology, embryology and immunology. But with the advent of molecular biology and genomics, an additional, entirely new avenue of information – sequence information – became available in support of evolutionary ideas and as a means of clarifying, in a totally new dimension of insight and of detail, the paths of the ancestry of organisms. Nevertheless, this huge additional evidence based on molecular evolution has had no perceptible impact on the quality of the controversy which is shaped more by emotional factors than by rational arguments.

Emotional views are also likely to continue to shape the image we have of ourselves after the impact of genomics. That the knowledge of the human base sequence will finally tell us who we are is a statement so preposterous as to be beyond any discussion, no matter how qualified the scientists who made such statements might be. The controversy regarding genetic determinism (is everything in our genes?) will probably go on in the future, very much along the same lines as in the past (Francis Collins: "Free will will not go out of style once the sequence is done").

One should perhaps not forget that observable phenomena should be described in terms and along lines of thought adequate to their level of complexity: Elementary particle physics is inadequate to describe biochemical phenomena as well as biochemistry is inadequate to describe human society and human thought. Science, as a whole, might actually be incapable of really defining human beings, whose adequate level of description might rather be found in drama and lyrics (Rainer Hertel, personal commun.).

Should we paraphrase Gunther Stent (1968): "That was the genomics/proteomics that was"?

# GLOSSARY

(Terms already explained in the text are cited in the index.)

**Å**: The Swedish physicist Anders Ångström (1814-1874) studied the spectral lines of the elements – for example, he demonstrated the presence of hydrogen in the sun – and measured their wave lengths in units of $10^{-10}$ m (0,1 nm). In 1905, this unit was officially called Å (Ångstrom). Today, this unit is held to be "illegal" because it is not considered by the SI (Système International d'Unités). Despite that, it sturdily persists for conveniently describing atomic and molecular dimensions (for instance, the van der Waals radius of a hydrogen atom: 1,2 Å).

**Adsorption**: The first tight contact between a virus and its host, based on fitting surfaces in both, the virus and the host. The term is used more broadly to describe all non-covalent attachments of macromolecules or other small particles to other particles or surfaces; not to be mistaken for absorption, which describes the uptake of one substance into the interior of a structure (one also speaks of absorption of light or other radiations by matter totally or partially impenetrable to these radiations).

**Agar**: Previously called agar-agar, is a polysaccharide obtained from red algae in the form of a mixture of hemicellulose-like substances. Prepared as a granulate, it will melt by heating at 80 to 100 °C in a culture medium, which solidifies to a gelly-like consistency when cooled under 40 °C. In 1876, Robert Koch introduced solid culture media to bacteriology – then prepared with gelatine – thus allowing, for the first time, the isolation of individual bacterial colonies. Alas, in an era of budding nationalism, Pasteur considered Koch's novelty as a humiliation for France. Replacing agar by such artificial substances as polyvinyl alcohol has been repeatedly attempted, though hitherto without success.

**Allele**: A certain conformation of a gene. One speaks, for example, of a wild type allele and its corresponding mutant allele. The term is an abbreviation of the original expression "allelomorph gene". All allelomorph genes (or today: alleles of a gene) occupy the same locus on the genetic map. The words "gene" and "allele" are often used as synonyms. "Gene" is actually an abstraction; it stands for all its allelic forms and will appear under the form of one of them. Because no general symbol exists to describe a gene as a collection of all its possible alleles, mostly one of its mutant alleles is taken to represent it – which is only too logical, taking into account that, for the most part, the existence of a gene is first disclosed when a mutation occurs in it, as for example, the *trp*A gene; this practice may lead to misunderstandings (see also under Marker).

**Alzheimer's disease**: After the German neurologist Alois Alzheimer (1864-1915). A relatively common form of senile dementia with progressive memory loss and, at the final stage, complete personality deterioration; there is a genetic predisposition for the appearance of some forms of this disease.

**Analogue (or analog)**: A substance similar to another one in its chemical structure, but distinct from the latter in a pivotal point.

**Ascus, asci**: Group of four or eight ascospores (haploid cells) in fungi, resulting from meiotic division (eventually followed by a mitotic division), originated from a diploid cell (zygote); budding of spores leads to the formation of new haploid mycelia (filamentous fungal colonies).

**Autoradiography**: Autorepresentation of the distribution pattern of radioactivity in any type of preparation (as electrophoresis gels, tissue slices, chromosomes, etc.). On account of the radioactivity, one will observe blackening of a photographic film (X-ray film) on which the preparation in question had been placed, and then exposed – always in the dark, obviously – for some time (minutes to months; most commonly hours or days).

**Auxotrophy, auxotroph**: Characteristic of a microorganism whose growth depends on the addition of a specific organic substance (amino acid, vitamin or nucleoside) to the medium, besides mineral salts and a source of carbon and energy (generally a sugar). Organisms which do not need any such supplements to minimal medium are referred to as prototrophs. The terminology of auxotrophic  mutants is not standardized; mostly a distinction between phenotype and genotype is made. In bacterial genetics, a three-letters symbol is used when the phenotype is meant, starting with a capital letter; superscript "minus" signal indicates the deficiency, whereas the corresponding "plus" sign refers to the wild type (prototroph). The symbols for genetic markers are used (see under Marker) when the genotype is meant. In bacterial genetics, often no clear distinction between phenotype and genotype is made; in the case of haploid microorganisms, this usually does not matter much. (The pairs of terms auxotroph – prototroph, and autotroph – heterotroph, are not to be mixed up. Autotrophic organisms use $CO^2$ as carbon source, whereas heterotrophs depend on organic compounds as, for instance, sugars. Both auxotrophs and prototrophs are heterotrophs.)

**cDNA**: With the help of a retrovirus (avian myeloblastosis virus) reverse transcriptase, and with oligo-dT as primer, which paired with the 3'-end poly-A tail of a rabbit globin mRNA, Ross et al. (1972) obtained a single-strand DNA, complementary to the complete globin messenger. The synthesis of such cDNA (complementary DNA) was, in the same year, further described by a series of other authors; Rongeon et al. (1975) employed DNA polymerase to synthesize a double strand (ds-cDNA) from the β-globin cDNA, which was then cloned in *E. coli*. Complementary DNA, or cDNA, was later termed complementary copy DNA and today – simplicity over priority – it is just called copy DNA. Neither expression quite fits the bill, since, in effect, any DNA is a complementary copy of its template; what is essential here is that cDNAs are copies of mRNA molecules.

**Chaperone**: Proteins which guide a polypeptide's folding into a defined tertiary structure. This process may run parallel to the synthesis itself, directly on the ribosomes, or occur while the polypeptide is being transported through a membrane.

**Chromatography**: A method of separating various substances in a solution, taking advantage of their differential flow characteristics when the solution is passed through a solid phase. This solid phase can be absorbent paper (chromatographic paper) or else such materials as cellulose, calcium phosphate or synthetic resins, stuffed in a hollow column through which the solution to be fractionated slowly flows (column chromatography). The term chromatography is reminiscent of the pioneering years, the 1940s, when, often, mixtures of pigments were resolved into their single components.

**Conjugation**: Contact between two cells, leading to unidirectional (as in bacteria) or mutual (as in *Paramecium*) transfer of genetic material. After the process of conjugation, the two cells, the so-called conjugants, separate.

**Counter-current separation method**: A technique for isolating individual components of a solution, through successive steps, by gradually enriching one of its components. In each step the subtle differences in solubility of these substances in two non-blending liquids are exploited in order to enrich one substance in one of the liquids. A better resolution than in chromatography is attained on account of the repeating, automatized steps.

**Crossing-over, crossover**: Exchange, during the pairing of chromosomes in the prophase of meiosis I, of homologous segments of two chromatides (chromatides: the still not fully separated halves of an already duplicated chromosome, remaining attached to each other solely by the centromere. In a broader sense, any exchange between homologous DNA sequences is also called crossing-over.

**Denaturation**: Dissolution of superstructures, namely the spatial organization of atoms of macromolecules, without the destruction of covalent bonds. Denaturation is mostly brought about by the action of heat or extreme pH values. By denaturation, proteins loose their tertiary structures and concomitantly their function, for instance their enzymatic activity. DNA denaturation means the dissolution of the hydrogen bonds between its bases, leading to separation in single strands.

**Electrophoresis**: Method for separating different substances in a solution according to their electrical charge. The solution to be analyzed is subjected to a continuous electric current, so that the negatively charged molecules or ions migrate to the anode (positive pole) whereas the positively charged ones move to the cathode (negative pole). The solution to be analyzed can be placed on absorbent paper (paper electrophoresis), but, more commonly, it is embedded in a gel-matrix, such as polyacrylamid or agarose (a specially purified agar): gel electrophoresis. The mesh sizes of the gel substance can be chosen so that the differently sized molecules to be analyzed migrate with different speeds through the pores of the gel matrix.

**Endonucleases**: Enzymes able to cut nucleic acid molecules into smaller fragments by hydrolyzing the phosphodiester bonds within a polymeric sequence (namely, an activity other than that of exonucleases that split away nucleotides located at the ends).

**Gradient**: Any continuous variation of a physical unit as a function of distance (such as atmospheric temperature or barometric gradients); one can speak of steep or slight gradients.

**Homology**: In comparative anatomy, one speaks of homologous structures when these can be phylogenetically traced back to a common ancestral form (as humans' arms and birds' wings). In this sense, homology has a qualitative meaning: two structures are either homologous or not homologous. In classical genetics, the same principle holds true, for example, in the case of chromosomes: homologous chromosomes (of either paternal or maternal origin) display the same genes (possibly with different alleles), and pair up during meiosis. Chromosomes from distinct but closely related species, derived from a common ancestral chromosome, are also called homologous. In molecular biology, the expression "homology" is also used to quantitatively describe the grade of phylogenetic relatedness; if two macromolecules have identical base or amino acid sequences, they are considered to be 100% homologous; accordingly, when 80% of the building blocks share the same position, the structures are said to be 80% homologous, and so on. These different definitions of "homology" have led to acrimonious semantic debates.

**Hybridize**: To cross two individuals, distinct in a large number of characteristics, as occurs in a cross of two different species; in molecular biology the term hybridization refers to the annealing of polynucleotide single strands of different types, or origins, to form double strands; one especially speaks of DNA-RNA hybrid helices.

**Hydrogen bond**: In polar compounds like water, the electrons' charges are not evenly distributed over the whole surface of their molecules. Instead, there are distinct electrostatically positive and negative poles. Such positively and negatively charged areas of neighboring polar molecules tend to attract each other. The bonds formed in this way are weak ones, though (activation energy: ca. 20 kJ/Mol, in contrast to ca. 200-400 kJ/Mol for the covalent bonds holding the atoms within a molecule). Single hydrogen bonds are quickly dissolved by the molecules' thermic motion. The half-life of a single H-bond between a water molecule's oxygen atom (slightly negative) and the hydrogen atom (slightly positive) of its neighboring molecule amounts to a mere $10^{-11}$ seconds. Nevertheless, hydrogen bonds between water molecules are the explanation for water being a fluid at room temperature, instead of a gas, like methane, for instance, a compound of similar molecular weight as water, but non-polar. A stable connection between two molecular structures can be guaranteed by the sum of several hydrogen bonds even in the absence of covalent bonds.

**Marker, genetic**: Mostly a mutant allele, ideally with a known map location, whose phenotype allows easy monitoring of its presence. In bacterial genetics, auxotrophic alleles are preferably used as genetic markers; the alleles are symbolized by three corresponding lower case letters, mostly followed by a capital letter which specifies the locus (for instance, *trpA*, *trpB*). A mutation inside a locus is indicated by a number (for instance, *trpA87*). Whereas the respective wild type allele is represented by a superscript "plus" sign, no "minus" sign is used to point out the defective mutant alleles. Similarly, convenient symbols are used to characterize the fermentation of sugars. Many of the texts in this book show the historic notations in order to facilitate a direct comparison with the original literature.

**Monofactorial**: A characteristic dependent solely on one genetic factor (gene) (for instance, the petal's color of Mendel's peas), in contrast to multifactorial or polygenic features (the inheritance of human skin color or many economically relevant characteristics such as, for example, a cow's daily milk production).

**Mycelium**: A mold colony originated from one spore, compounded of filamentous groups of cells growing radially.

**Native**: The original and functional state of macromolecules preserving their intact spatial structure (protein's tertiary structure, DNA's double helix).

**Nobel Prize**: Alfred Nobel (1835-1896), the inventor of dynamite (the "king of dynamite"), was a Swedish chemical engineer and tycoon; explosives and matches were the main commodities he traded in. He bequeathed a considerable fortune, whose interests, since the beginning of the 20th century managed by a foundation, have been distributed in awards accorded to persons with exceptional achievements concerning the well being of humanity in the fields of literature, physics, chemistry, physiology-medicine, and peace. Nobel, who revolutionized the art of warfare, was indeed a man of paradoxical character. Personalities to be honored with the prize are chosen yearly by a Nobel committee of experts; the prizes – often shared by two or three selected people – are personally handed out by the Swedish king in a festive ceremony in Stockholm, and comprise several hundred thousand dollars. The Nobel Prize for peace, awarded by a Nobel committee in Norway, is presented in Oslo. Since 1968, the Royal Bank of Sweden grants a further Nobel Prize, for economy – in memory of Alfred Nobel.

**Nucleolus**: A region of the cell nucleus where the rRNA genes are located; in the nucleolus the synthesis of rRNAs takes place.

**Oligo-**: This prefix means: little or few. In molecular genetics, the prefix usually refers to polymeres with relatively few building blocks (10 to 20), such as oligopeptides or oligodeoxynucleotides. The latter, mostly simply called "oligos", generally are assembled synthetically, to be used as primers or probes with different aims: sequencing, PCR, detection of special nucleotide sequences on gels or nitrocellulose sheets, etc.

**Petri dish**: The most common and most simple of all microbiological laboratory devices, the petri dish or petri plate, used to culture bacteria, is a shallow dish (originally of glass, today mostly of disposable plastic) with a corresponding cover, having a diameter of ca. 10 cm and a depth of 1,5 cm. Its invention, in 1887, by Richard Julius Petri, an assistant to Robert Koch, immortalized his name. Petri dishes are half-filled with hot, molten agar medium which, upon cooling, hardens to an ideal solid support for bacterial growth.

**Pneumoccocus**: The causative agent of pneumonia; the old nomenclature, *Pneumococcus pneumoniae*, is now replaced by *Streptococcus pneumoniae*.

**Poly-A tail**: A row of ca. 200 nucleotides with the base adenine, added, in eukaryotes, to the 3'-ends of mRNA in order to protect these ends from exonucleolytic degradation.

**Polysaccharide**: A polymer of sugar units, such as starch, cellulose, hemicellulose, as well as the capsular material of many bacteria.

**Postdoc**: The usual abbreviation for postdoctoral fellow, referring to young scientists who, after obtaining a Ph.D., spend 2 or 3 years with a new research group in order to acquire wider experience before taking a more permanent position.

**Probe, DNA probe**: Short DNA pieces (encompassing up to about hundred nucleotides or more) labeled radioactively or by some other means (e.g. with fluorescent nucleotide derivates). These DNA segments are deployed in various types of preparations (as nitrocellulose sheets, tissue sections, etc.), in order to detect the location of complementary sequences with which they hybridize.

**Prototroph**: See auxotroph.

**Renaturation**: Reestablishment of the native form of a denatured macromolecule; in the case of polynucleotides, it refers to the reassembling of single strands of complementary, or partially complementary, base sequences into double helix structures.

**Steroid receptors**: Proteins which recognize steroid hormones (sexual hormones, for instance) as effectors, being activated by them to bind to special DNA sequences (the so-called hormon responsive elements). This activation allows certain genes to be transcribed.

**Tautomers**: Easily interchangeable isomeric forms of a compound, so that, in a solution, their proportions will remain in equilibrium.

**Template**: In molecular biology, it refers to a polynucleotide strand whose base sequence serves as a pattern for the synthesis of a complementary polynucleotide strand.

**Ultracentrifugation**: Common centrifuges reach 10,000 to 20,000 rotations per minute (rpm), sufficient to efficiently sediment bacterial cells within a few minutes, but not particles like phages or ribosomes. In contrast, ultracentrifuges rotate with a speed of over 50,000 rpm which, depending on the rotor's radius, corresponds to over 500,000 times the earth's gravity (500,000 x g). The ultracentrifuge's rotor must be placed in a vacuum chamber in order to avoid effects otherwise caused by air friction (overheating, braking action). Although ultracentrifugation can cause macromolecules to settle at the bottom of the centrifuge tube, the method of choice for analyzing macromolecules is to separate them in distinct fractions, according to their density (density gradient centrifugation, see Excursus 6-2) or their sedimentation velocity (zone centrifugation).

**Wild type**: The form of a species typically occurring in nature; a laboratory reference type isolated from nature. Wild type alleles are mostly represented (especially in microbiology) by a superscript "plus" sign.

**Zinc finger proteins**: DNA-binding proteins, whose peptide chains form loops (fingers) stabilized by 4 cysteine residues or 2 cysteine and 2 histidine residues, and fixed by a zinc ion. These zinc fingers fit inside the large groove of the DNA double helix, allowing the binding of further proteins such as transcription factors. Some zinc finger proteins may have relatively specific binding sites (for instance, the so-called hormone responsive elements for steroid receptors) or else they may bind to sequences with similar bases [such as GC-rich DNA, to which, for instance, the Sp-1 protein can then bind (see Excursus 21-4)].

**Zygote**: Diploid cell formed by the unification of two haploid gametes. In the case of haploid organisms, this zygote undergoes a meiotic division, giving rise to 4 haploid meiosis products (spores). In haplo-diploid and diploid organisms, the zygote goes through cycles of mitotic division before some of the diploid cells divide meiotically (gamete formation).

# LITERATURE

Adam, D. (2001): Genetics group targets disease markers in the human sequence. Nature 412, 105.

Adams, J. M., Jeppesen, P. G. N., Sanger, F. & Barrell, B. G. (1969): Nucleotide sequence from the coat protein cistron of R 17 bacteriophage RNA. Nature 223, 1009-1014.

Adams, M. D., Celniker, S. E., ...(193 further coauthors).., Venter, J.C. (2000): The genome sequence of *Drosophila melanogaster*. Science 287, 2185-2195.

Adams, M. D., Dubnick, M., Kerlavage, A. R., Moreno, R., Kelley, M., Utterback, T. R., Nagle, J. W., Fields, C. & Venter, C. (1992): Sequence identification of 2375 human brain genes. Nature 355, 632-634.

Alberts, B. & Sternglanz, R. (1977): Recent excitement in the DNA replication problem. Nature 269, 655-661.

Alcock, J. (2001): The triumph of sociobiology. Oxford University Press: Oxford, U.K.

Aldhous, P. (1991): Who needs a genome ethics treaty? Nature 351, 507.

Alloway, J. L. (1932): The transformatjon *in vitro* of R pneumococci into S forms of different specific types by the use of filtered pneumococcus extracts. J. Exp. Med. 55, 91-99.

Alpher, R. A., Bethe, H. & Gamow, G. (1948): The origin of chemical elements. Physical Rev. 73, 803-804.

Altmann, R. (1889): Über Nucleinsäuren. Cited in: A Century of DNA, F. H. Portugal & J. S. Cohen, 1977 MIT Press, Cambridge, Mass.

Ammermann, D. (1979): Die Zellkerne des Ciliaten Stylonychia mytilus. Biol. in uns. Zeit 9 (2), 40-44.

Anderson, C. (1992): New French genome centre aims to prove that bigger really is better. Nature 357, 526-527.

Anderson, C. (1993): Genome project goes commercial. Science 259, 300-302.

Anderson, T. F. (1950): Destruction of bacterial viruses by osmotic shock. J.Appl.Phys.21, 70.

Anderson, T. F. (1953): The morphology and osmotic properties of bacteriophage systems. Cold Spring Harbor Symp. Ouant. Biol. 18, 197-203.

Anderson, T. F., Delbrück, M. & Demerec, M. (1945): Types of morphology found in bacterial viruses. J. Appl. Phys. 16, 264.

Anderson, T. F. & Doermann, A. H. (1952): The intracellular growth of bacteriophages. II. The growth of T3 studied by sonic disintegration and by T6-cyanide lysis of infected cells. J. Gen. Physiol. 35, 657-667.

Anfinsen, C. B., Haber, E., Sela, M. & White, F. H. jr. (1961): The kinetics of formation of native ribonuclease during oxidation of the reduced polypeptide chain. Proc. Natl. Acad. Sci. USA 47, 1309-1314.

Anonymus (1971 ): Lifting replication out of the rut. Nature New Biol. 233, 97.

Arber, W. (1965a): Host-controlled modification of bacteriophage. Ann. Rev. Microbiol. 19, 365-378.

Arber, W. (1965b): Host specificity of DNA produced by *Escherichia coli*. V. The role of methionine in the production of host specificity. J. Mol. Biol. 11, 247-256.

Arber, W. (1968): Host-controlled restriction and modification of bacteriophage. In: The molecular biology of viruses; 18th Symp. Soc. Gen. Microbiol. Cambridge Univ. Press: Cambridge, pp. 295-314.

Arber, W. & Dussoix, D. (1962): Host specificity of DNA produced by *Escherichia coli*. J. Mol. Biol. 5, 18-36.

Arber, W. & Kühnlein, U. (1967): Mutationeller Verlust B-spezifischer Restriktion des Bakteriophagen fd. Path. Microbiol. 30, 946-952.

Arber, W. & Linn, S. (1969): DNA modification and restriction. Ann. Rev. Biochem. 38, 469-500.

Armitage, P. (1953): Statistical concepts in the theory of bacterial mutations. J. Hygiene 51, 162-184.

Arnheim, N. & Erlich, H. (1992): Polymerase chain reaction strategy. Ann. Rev. Biochem. 61, 131-156.

Astbury, W. T. & Bell, F. O. (1938): Some recent developments in the X-ray study of proteins and related structures. Cold Spring Harbor Symp. Quant. Biol. 6, 109-119.

Astbury, W. T. (1947): X-ray studies of nucleic acids. Symp. Soc. Exp. Biol. 1, 66- 76.

Avery, O. T., MacLeod, M. & McCarty, M. (1944): Studies on the chemical nature of the substance inducing transformation of pneumococcal types. Induction of transformation by a desoxyribonucleic acid fraction isolated from pneumococcus Type III. J. Exp. Med. 79, 137-158.

Bachman, B. J. (1987): Linkage map of *Escherichia coli* K-12, Edition 7; in: *Escherichia coli* and *Salmonella typhimurium*. Cellular and molecular biology; Vol. 2, pp. 807-876; Neidhardt, F. C. (ed.): American Society for Microbiology: Washington DC.

Baker, T. A. & Wickner, S. H. (1992): Genetics and enzymology of DNA replication in *Escherichia coli*. Ann. Rev. Genet. 26, 447-477.

Baltimore, D. (1970): Viral RNA-dependent DNA polymerase. Nature 226, 1209-1211.

Bateson, P. & Dawkins, R. (1985): Sociobiology: the debate continues. New Scientist 105; Nr. 1440, 58-60.

Bawden, F. C. & Pirie, N. W. (1937): The isolation and some properties of liquid crystalline substances from solanaceous plants infected with three strains of tobacco mosaic virus. Proc. Roy. Soc. (London) [B] 123, 274-320.

Beadle, G. W. (1946): Genes and the chemistry of the organism. American Scientist 34, 31-53.

Beadle, G. W. & Ephrussi, B. (1936): The differentiation of eye pigments in *Drosophila* as studied by transplantation. Genetics 21, 225-247.

Beadle, G. W. & Tatum, E. L. (1941): Genetic control of biochemical reaction in *Neurospora*. Proc. Natl. Acad. Sci. USA 27, 499-506.

Beckwith, J. (1987): The operon: an historical account. In: *Escherichia coli* and *Salmonella typhimurium*; cellular and molecular biology. F. C. Neidhardt (ed.): American Soc. Microbiol: Washington, D. C., pp. 1439-1443.

Belozersky, A. N. & Spirin, A. S. (1958): A correlation between the composition of deoxyribonucleic and ribonucleic acids. Nature 182, 111-112.

Benzer, S. (1957): The elementary units of heredity. In: The chemical basis of heredity (W. D. McElroy & B. Glass, eds.). John Hopkins Press: Baltimore, pp. 70-93.

Benzer, S. (1961): On the topology of the genetic fine structure. Proc. Natl. Acad. Sci. USA 47, 403-415.

Berg, P., Baltimore, D., Boyer, H. W., Cohen, S. N., Davis, R. W., Hogness, D. S., Nathans, D., Robbin, R., Watson, J. D., Weissman, S. & Zinder, N. (1974): Potential biohazards of recombinant DNA molecules. Proc. Natl. Acad. Sci. USA 71, 2593-2594.

Berger, S. M., Berk, A. J., Harrison, T. & Sharp, P.A. (1977): Spliced segments at the 5' termini of adenovirus-2 late mRNA: A role for heterogeneous nuclear RNA in mammalian cells. Cold Spring Harbor Symp. Quant. Biol. 42, 523-529.

Bernfield, M. R. & Nirenberg, M. W. (1965): RNA codewords and protein synthesis. The nucleotide sequences of multiple codewords for phenylalanine, serine, leucine, and proline. Science 147, 479-484.

Bernstein, J. (1993): Cranks, quarks, and the cosmos. Basic Books: New York.

Bickle, T. A. & Krüger, D. H. (1993): Biology of DNA restriction. Microbiol. Rev. 57, 434-450.

Biessmann, H. & Mason, J. M. (1992): Genetics and molecular biology of telomeres. Adv. Genet. 30, 185-249.

Blackburn, E. H. (1992): Telomerases. Ann. Rev. Biochem. 61, 113-129.

Blanchetot, A., Wilson, V., Wood, D. & Jeffreys, A. J. (1983): The seal myoglobin gene: an unusually long globin gene. Nature 301, 732-734.

Blattner, F. R., Plunkett, G., 3rd Bloch, C.A., Perna, N.T., Burland, V., et al. (1997): The comlete genome sequence of *Escherichia coli* K-12. Science 277, 1453-1474.

Boivin, A. & Vendrely, R. (1947): Sur le role possible des deux acides nucleïques dans la cellule vivate. Experientia 3, 32-34.

Bonner, D. (1946): Biochemical mutations in Neurospora. Cold Spring Harbor Symp. Quant. Biol. 11, 14-24.

Bonner, D. M. (1951 ): Gene-enzyme relationships in Neurospora. Cold Spring Harbor Symp. Quant. Biol. 16, 143-157.

Booth, W. (1989): Oh, I thought you were a man. Science 243, 475.

Bordier, C. & Dubochet, J. (1974): Electron microscopic localization of the binding sites of *Escherichia coli* RNA polymerase in the early promoter region of T7 DNA. Eur. J. Biochem. 44, 617-624.

Borek, E. (1963): The methylation of transfer RNA: Mechanism and function. Cold Spring Harbor Symp. Quant. Biol. 28, 139-148.

Borsook, H. & Huffman, H. (1938): In: Chemistry of amino acids and proteins. C. L. A. Schmidt (ed.): Thomas: Baltimore [cited in Zamecnik (1979)].

Brachet, J. (1947): Nucleic acid in the cell and the embryo. Cambridge Univ. Press: Cambridge, pp. 207-224.

Braunitzer, G., Gehring-Müller, R., Hilschmann, N., Hilse, K., Hobom, G., Rudloff, J. & Wittmann-Liebold, B. (1961): Die Konstitution des normalen adulten Humanhämoglobins. Z. Physiol. Chem. 325, 283-284.

Brenner, S. (1957): On the impossibility of all overlapping triplet codes in information transfer from nucleic acid to proteins. Proc. Natl. Acad. Sci. USA 43, 687-693.

Brenner, S. (1983): Cited in "Thirty years of DNA" (Anonymus). Nature 302, 651-654.

Brenner, S., Jacob, F. & Meselson, M. (1961): An unstable intermediate carrying information from genes to ribosomes for protein synthesis. Nature 190, 576-585.

Bresch, C. & Hausmann, R. (1972): Klassische und molekulare Genetik. Springer: Heidelberg.

Brethnach, R., Mandel, J. L. & Chambon, P. (1977): Ovalbumin gene is split in chicken DNA. Nature 270, 314-319.

Bretscher, M. S. (1966): Polypeptide chain initiation and the characterization of ribosomal binding sites in *E. coli*. Cold Spring Harbor Symp. Quant. Biol. 31, 289-295.

Bridges, C. B. (1935): Salivary chromosome maps. J. Hered. 26, 60.

Britten, R. J. & Davidson, E. H. (1971): Repetitive and non-repetitive DNA sequences and a speculation on the origins of evolutionary novelty. Quart. Rev. Biol. 46, 111-133.

Britten, R. J. & Kohne, D. E. (1968): Repeated sequences in DNA. Science 161, 529-540.

Broker, T. R., Chow, L. T., Dunn, A. R., Gelinas, R. E., Hassell, J. A., Klessig, D. F ., Lewis, J. B., Roberts, R. J. & Zain, B. S. (1977): Adenovirus-2 messengers – an example of baroque molecular architecture. Cold Spring Harbor Symp. Quant. Biol. 42, 531-553.

Brown, D. M. & Todd, A. R. (1952): Nucleotides. Part X. Some observations on the structure and chemical behavior of the nucleic acids. J. Chem. Soc., 52-58.

Brown, D. M. & Todd, A. R. (1955): Nucleic acids. Ann. Rev. Biochem. 24, 311-338.

Brownlee, G. G., Sanger, F. & Barrell, B. G. (1968): The sequence of 5s ribosomal ribonucleic acid. J. Mol. Biol. 34, 379-412.

Brush, S. G. (1974): Should the history of science be rated X? Science 183, 1164-1172.

Büchel, D. E., Gronenborn, B. & Müller-Hill, B. (1980): Sequence of the lactose permease gene. Nature 283, 541-545.

Bukhari, A, I., Shapiro, J. A. & Adhya, S. L. (eds.) (1977): DNA insertion elements, plasmids and episomes. Cold Spring Harbor Lab. Press: New York.

Burgers, P. M. J. (1989): Eukaryotic DNA polymerases $\alpha$ and $\delta$: Conserved properties and interactions, from yeast to mammalia cells. Progr. Nucl. Acids Res. Molec. Biol. 37, 235-280.

Burgess, R. R., Travers, A. A., Dunn, J. J. & Bautz, E. K. F. (1969): Factor stimulating transcription by RNA polymerase. Nature 221, 43-46.

Bussard, A., Naono, S., Gros, F. & Monod, J. (1960): Effets d'un analogue de I'uracil sur les propriétés d'une protéine enzymatique synthétisée en sa présence. Compt. Rend. Acad. Sci. (Paris) 250, 4049-4051.

Butenandt, A., Weidel, W. & Becker, E. (1942): Kynurenin als Augenpigmentbildung auslösendes Agens bei Insekten. Naturwissenschaften 28, 63-64.

Butler, D. (1995): Genetic diversity proposal fails to impress international ethics panel. Nature 377, 373.

Cairns, J., Stent, G.S. & Watson, J. D. (1966): Phage and the origins of molecular biology. Cold Spring Harbor Lab. Press: New York.

Caldwell, P. C. & Hinshelwood, Sir. C. (1950): Some considerations on autosynthesis in bacteria. J. Chem. Soc. 4, 3156-3159.

Campbell, A. M. (1962): Episomes. Adv. Genet. 11, 101-145.

Campbell, A. M. (1993): Thirty years ago in GENETICS: Prophage insertion into bacterial chromosomes. Genetics 133, 433-438.

Campbell, J. W., Duée, E., Hodgson, G;, Mercer, W. D., Stammers, D. K., Wendell, P. L., Muirhead, H. & Watson, H. C. (1971): X-ray diffraction studies on enzymes in the glycolytic pathway. Cold Spring Harbor Symp. Quant. Biol. 36, 165-170.

Canfield, R. E. & Anfinsen, C. B. (1963): Non-uniform labeling of egg white lysozyme. Biochemistry 2, 1073-1078.

Cantacuzene, J. & Bonciu, O. (1926): Modifications subies par des streptocoques d'origine non scarlatineuse au contact de produits scarlatineux filtrés. Compt. Rend. Acad. Sci. (Paris) 182, 1185-1187.

Cantor, C.R. (1990): Orchestrating the human genome project. Science 248, 49-51.

Capaldi, R.A. & Aggeler, R. (2002): Mechanism of the $F_1F_0$-type ATP synthase, a biological rotary motor. Trends Biochem. Sci. 27, 154-160.

Carlson, E. A. (1966) The Gene: A critical history. Saunder Co.: Philadelphia & London, p. 159.

Caspari, E. (1933): Über die Wirkung eines pleiotropen Gens bei der Mehlmotte *Ephestia Kühniella*. Arch. Entwicklungsmech. Organ. 130, 352-381.

Cavalli-Sforza, L. L, & Lederberg, J. (1956): Isolation of pre-adaptive mutants in bacteria by sib selection. Genetics 41, 367-381.

Cech, T. R., Herschlag, D., Piccirilli, J. A. & Pyle A. M. (1992): RNA catalysis by a group I ribozyme. J. Biol. Chem. 267, 17479-17482.

Cerdá-Olmedo, E. & Lipson, E. D. (eds.) (1987): Phycomyces. Cold Spring Harbor Lab. Press: New York.

Chamberlin, M. (1970): Transcription 1970: a summary. Cold Spring Harbor Symp. Quant. Biol. 35, 851-873.

Chamberlin, M. J. (1974): The selectivity of transcription. Ann. Rev. Biochem. 43, 721-775.

Chamberlin, M. & Berg, P. (1962): Deoxyribonucleic acid-directed synthesis of ribonucleic acid by an enzyme from *Escherichia coli*. Proc. Natl. Acad. Sci. USA 48, 81-94.

Charbonneau, H. & Tonks, N. K. (1992): 1002 Protein phosphatases? Ann. Rev. Cell Biol. 8, 463-493.

Chargaff, E. (1950): Chemical specificity of nucleic acids and mechanisms of their enzymatic degradation. Experientia 6, 201-240.

Chargaff, E. (1963): Essays on nucleic acids. Elsevier Press: Amsterdam.

Chargaff, E. (1974): Building the tower of babble. Nature 248, 776-779.

Chargaff, E. (1978): Heraclitean Fire. Rockefeller University Press.

Chargaff, E. (1998): Die Aussicht vom 13. Stock. Neun Essays. Klett-Cotta, Stuttgart.

Chung, C. W., Mahler, H. R. & Enrione, M. (1960): Incorporation of adenine nucleotide into ribonucleic acid by cytoplasmic enzyme preparations of chick embryos. J. Biol. Chem. 235, 1448-1461.

Clark, A. J. & Adelberg, E. A. (1962): Bacterial conjugation. Ann. Rev. Microbiol. 16, 289-319.

Clark, B. F. C. & Marker, K. A. (1966): The role of N-formyl-methionyl-sRNA in protein biosynthesis. J. Mol. Biol. 17, 394-406.

Cochran, W., Crick, F. H. C. & Vand, V. (1952): The structure of synthetic polypeptides. I. The transform of atoms on a helix. Acta Crystallogr. 5, 581-586.

Cohen, G. & Jacob, F. (1959): Sur la répression de la synthèse des enzymes intervenant dans la formation du tryptophane chez *Escherichia coli*. Compt. Rend. Acad. Sci. (Paris) 248, 3490-3492.

Cohen, G. N. & Monod, J. (1957): Bacterial permeases. Bacteriol. Rev. 21, 169-194.

Cohen, J. J. & Duke, R. C. (1992): Apoptosis and programmed cell death in immunity. Ann. Rev. Immunol. 10, 267-293.

Cohen, N. D., Beegen, H., Utter, M. F. & Wrigley, N. G. (1979): A re-examination of the electron microscopic appearance of pyruvate carboxylase from chicken liver. J. Biol. Chem. 254, 1740-1747.

Cohen, S. N., Chang, A. C. Y., Boyer, H. W. & Helling, R. B. (1973): Construction of biologically functional bacterial plasmids in vitro. Proc. Natl. Acad. Sci. USA 70, 3240-3244.

Cohen, S. S. (1947): The synthesis of bacterial viruses in infected cells. Cold Spring Harbor Symp. Quant. Biol. 12, 35-49.

Cohen, S. S. (1948): The synthesis of bacterial viruses. I. The synthesis of nucleic acid and protein in *Escherichia coli* B infected with T2r$^+$ bacteriophage. J. Biol. Chem. 174, 281-294.

Cohen, S. S. (1968): Virus-induced enzymes. Columbia Univ. Press: New York.

Cohen, S. S. & Anderson, T. F. (1946): Chemical studies on host-virus interactions. I. The effect of bacteriophage adsorption on the multiplication of its host, *Escherichia coli* B. With an appendix giving some data on the composition of the bacteriophage, T2. J. Exp. Med. 84, 511-523.

Cohen, S. S. & Stanley, W. M. (1942): The molecular size and shape of the nucleic acid of tobacco mosaic virus. J. Biol. Chem. 144, 589-598.

Cohn, M. (1957): Contributions of studies on the β-galactosidase of *Escherichia coli* to our understanding of enzyme synthesis. Bacteriol. Rev. 21, 140-168.

Cohn, M., Monod, J., Pollock, M. R., Spiegelman, S. & Stanier, R. Y. (1953): Terminology of enzyme formation. Nature 172, 1096.

Comfort, N. C. (2001): The tangled field: Barbara McClintock's search for the patterns of genetic control. Harvard Univ. Press, Cambridge, MA.

Cook, P. J., (1993): The matter of tobacco use. Science262, 1750-1751.

Coulson, A., Kozono, Y., Lutterbach, B., Shownkeen, R., Sulston, J. & Waterston, R. (1991): YACs and the *C. elegans* genome. Bioessays 13, 413-417.

Crea, R., Kraszewski, A., Hirose, T. & Itakura, K. (1978): Chemical synthesis of genes for human insulin. Proc. Natl. Acad. Sci. USA 75, 5765-5769.

Crick, F. (1994): The astonishing hypothesis: The scientific search for the soul. Simon and Schuster.

Crick, F. & Jones, E. (1993): Backwardness of human neuroanatomy. Nature 361, 109-110.

Crick, F. H. C. (1955): On degenerate templates and the adaptor hypothesis. [unpublished communication to the "RNA Tie Club", cited in Crick (1963).]

Crick, F. H. C. (1958): On protein synthesis. Symp. Soc. Exp. Biol. 12, 138-163.

Crick, F. H. C. (1959): The present position of the coding problem. Brookhaven Symp. Biol. 12, 35-39.

Crick, F. H. C. (1963): The recent excitement in the coding problem. Progr. Nucl. Acids Res. Molec. Biol. 1, 163-217.

Crick, F. H. C. (1966a): Codon-anticodon pairing: the wobble hypothesis. J. Mol. Biol. 19, 548-555.

Crick, F. H. C. (1966b): The genetic code – yesterday, today, and tomorrow. Cold Spring Harbor Symp. Quant. Biol. 31, 3-9.

Crick, F. H. C. (1988): What mad pursuit. Basic Books: New York.

Crick, F. H. C., Barnett, L., Brenner, S. & Watts-Tobin, R. J. (1961): General nature of the genetic code for proteins. Nature 192, 1227-1232.

Crick, F. H. C., Griffith, J. S. & Orgel, L. E. (1957): Codes without commas. Proc. Natl. Acad. Sci. USA 43, 416-421.

Crick, F. H. C. & Kendrew, J. C. (1957): X-ray analysis and protein structure. Advances in Protein Chem. 12, 133-214.

Crow, J. F. (1988): A diamond anniversary: The first chromosome map. Genetics 118, 1-3.

Crowley, T. E., Hoey, T., Liu, J. K., Jan, Y. N., Jan, L. Y. & Tjian, R. (1993): A new factor related to TATA-binding protein has highly restricted expression patterns in *Drosophila*. Nature 361, 557-561.

Cyranoski, D. (2001): Japan's ape sequencing effort set to unravel the brain's secrets. Nature 409, 651.

Dalgliesh, C. E. (1953): The template theory and the role of transpeptidation in protein biosynthesis. Nature 171, 1027 -1028.

Danna, K. & Nathans, D. (1971): Specific cleavage of Simian Virus 40 DNA by restriction endonuclease of *Hemophilus influenzae*. Proc. Natl. Acad. Sci. USA 68, 2913-2917.

Danna, K. J., Sack, G. H. Jr. & Nathans, D. (1973): Studies of Simian Virus 40 DNA. VII. A cleavage map of the SV40 genome. J. Mol. Biol. 78, 363-376.

Davern, C. I. & Meselson, M. (1960): The molecular conservation of ribonucleic acid during bacterial growth. J. Mol. Biol. 2, 153-160.

Davis, B. D. (1948): Isolation of biochemically deficient mutants of bacteria by penicillin. J. Am. Chem. Soc. 70, 4267.

Davis, B. D. (1950a): Studies on nutritionally deficient bacterial mutants isolated by means of penicillin. Experientia 6, 41-50.

Davis, B. D. (1950b): Nonfiltrability of the agents of genetic recombination in *Escherichia coli*. J. Bacteriol. 60, 507-508.

Davis, B. D. (1992): Science and politics: tensions between the head and the heart. Ann. Rev. Microbiol. 46. 1-33.

Davis, R. L. (1993): Mushroom bodies and *Drosophila* learning. Neuron 11, 1-14.

Davis, R. W. & Hyman, R. W. (1971): A study in evolution: the DNA base sequence homology between coliphages T7 and T3. J. Mol. Biol. 62, 287-301.

Dawkins, R. (1976): The selfish gene. Oxford Univ. Press: Oxford.

Dawson, M. H. & Sia, R. H. P. (1931): *In vitro* transformation of pneumococcal types, parts I & II. J. Exp. Med. 54, 701-710.

DeBoer, J., Anderson, J.O., de Wit, J., Huijmans, J., et al. (2002): Premature aging in mice deficient in DNA repair and transcription. Science 296, 1276-1279.

DeLucia, P. & Cairns, J. (1969): Isolation of an *E. coli* strain with a mutation affecting DNA polymerase. Nature 224, 1164-1166.

Delbrück, M. (1949): A physicist looks at biology [reprod. in Cairns et. al. (eds.), 1966]. Trans. Connect. Acad. Arts Sci. 38, 173-190.

Delbrück, M. & Bailey, W. T. Jr. (1946): Induced mutations in bacterial viruses. Cold Spring Harbor Symp. Quant. Biol. 11, 33-37.

Delbrück, M. & Stent, G. S, (1957): On the mechanism of DNA replication. In: The chemical basis of heredity (W. D. McElroy & B. Glass, eds.). Johns Hopkins Press: Baltimore, pp. 736-699.

Demerec, M. (1955): What is a gene – twenty years later. American Naturalist 89, 5-20.

Demerec, M., Blomstrand, I. & Demerec, Z. E. (1955): Evidence of complex loci in Salmonella. Proc. Natl. Acad. Sci. USA 41, 359-364.

Demerec, M. & Fano, U. (1945): Bacteriophage-resistant mutants in *Escherichia coli*. Genetics 30, 119-136.

Deutscher, M.P. (ed.) (1990): Methods in enzymatology 182; Guide to protein purification.

Dickson, R. C., Abelson, J., Barnes, W. M. & Reznikoff, W. S. (1975): Genetic regulation: The lac control region. Science 187, 27-35.

Dintzis, H. M. (1961): Assembly of the peptide chains of hemoglobin. Proc. Natl. Acad. Sci. USA 47, 247-261.

Doermann, A. H. (1952): The intracellular growth of bacteriophages. I. Liberation of intracellular bacteriophage T4 by premature lysis with another phage or with cyanide. J. Gen. Physiol. 35, 645-656.

Doermann, A. H. (1953): The vegetative state in the life cycle of bacteriophage: Evidence for its occurrence and its genetic characterisation. Cold Spring Harbor Symp. Ouant. Biol. 18, 3-11.

Doherty, E. A. & Doudna, J.A. (2000): Ribozyme structures and mechanisms. Ann. Rev. Biochem. 69, 597-615.

Doolittle, W. F. (1980): Revolutionary concepts in evolutionary cell biology. Trends Biochem. Sci. 5, 146-149.

Doolittle, W. F. & Sapienza, C. (1980): Selfish genes, the phenotype paradigm and genome evolution. Nature 284, 601-603.

Doolittle, W. F. & Stoltzfus, A. (1993): Genes-in-pieces revisited. Nature 361, 403.

Dressler, D. & Potter, H. (1991): Discovering enzymes. Scientific American Library: New York.

Dubos, R. J. (1976): The professor, the institute, and DNA. Rockefeller Univ. Press: New York.

Dugaiczyk, A., Woo, S. L. C., Colbert, D. A., Lai, E. C., Mace, M. L. Jr. & O'Malley, B. W. (1979): The ovalbumin gene: Cloning and molecular organization of the entire natural gene. Proc. Natl. Acad. Sci. USA 76, 2253-2257.

Dunn, D. B. & Smith, J. D. (1958): The occurrence of 6-methylamino-purine in deoxyribonucleic acids. Biochem. J. 68, 627-636.

Dunn, L. C. (ed.) (1951): Genetics in the 20th century. Macmillan: New York.

Dunn, L. C. (1965): A short history of genetics. McGraw-Hill: New York.

Dussoix, D. & Arber, W. (1962): Host specificity of DNA produced by *Escherichia coli*. II. Control over acceptance of DNA from infecting phage λ. J. Mol. Biol. 5, 37-49.

Ellis, E. L. & Delbrück, M. (1939): The growth of bacteriophage. J. Gen. Physiol. 22, 365-384.

Ellis, R. E., Yuan, J. Y. & Horvitz, H. R. (1991): Mechanisms and functions of cell death. Ann. Rev. Cell Biol. 7, 663-698.

Elthon, T. E. & Stewart, C. R. (1983): A chemiosmotic model for plant mitochondria. BioScience 33, 687-692.

Englesberg, E., Irr, J., Power, J. & Lee, N. (1965): Positive control of enzyme synthesis by gene C in the L-arabinose system. J. Bacteriol. 90, 946-957.

Englesberg, E., Sheppard, D., Squires, C. & Meronk, F. jr. (1969): An analysis of "revertants" of a deletion mutation in the C gene of the L-arabinose gene complex in *Escherichia coli* B/r: Isolation of initiator constitutive mutants ($I^C$). J. Mol. Biol. 43, 281-298.

Enserink, M. (2000a): Start-up claims piece of Iceland's gene pie. Science 287, 951.

Enserink, M. (2000b): Patent office may raise the bar on gene claims. Science 287, 1196-1197.

Epstein, C. J., Goldberger, R. F. & Anfinsen (1963): The genetic control of tertiary protein structure: Studies with model systems. Cold Spring Harbor Symp. Quant. Biol. 28, 439-449.

Felleman, D. J. & Essen, D. van (1991): Cerebral Cortex 1, 1-48 (cited in Crick & Jones 1993).

Fernández-Morán, H., van Bruggen, E. F. J. & Ohtsuki, M. (1966): Macromolecular organization of hemocyanins and apohemocyanins as revealed by electron microscopy. J. Mol. Biol. 16, 191-207.

Feughelman, M., Langridge, R., Seeds, W. E., Stokes, A. R., Wilson, H. R., Hooper, C. W., Wilkins, M. H. F., Barclay, R. K. & Hamilton, L. D. (1955): Molecular structure of deoxyribose nucleic acid and nucleoprotein. Nature 175, 834-838.

Feulgen, R., Behrens, M. & Mahdihassan, S. (1937): Darstellung und Identifizierung der in den pflanzlichen Zellkernen vorkommenden Nucleinsäuren. Z. Physiol. Chem. 246, 203-211.

Feulgen, R. & Rossenbeck, H. (1924): Mikroskopisch-chemischer Nachweis einer Nucleinsäure vom Typus der Thymonucleinsäure und die darauf beruhende selektive Färbung von Zellkernen in mikroskopischen Präparaten. Hoppe-Seyler's Z. physiol. Chem. 135, 203-248.

Fields, S. (2001) Proteomics in genome land. Science 291, 1221-1224.

Fiers, W., Contreras, R., De Wachter, R., Haegemen, W. & Vandenberghe, A. (1971 ): Recent progress in the sequence determination of bacteriophage MS2 RNA. Biochimie 53, 495-506.

Flaks, I. G. & Cohen, S. S. (1959): Virus-induced acquisition of metabolic function. I. Enzymatic formation of 5-hydroxymethyldeoxycytidylate. J. Biol. Chem. 234, 1501-1506.

Fleischmann, R. D., ... (38 coauthors) ... & Venter, J.C. (1995): Whole-genome random sequencing and assembly of Haemophilus influenzae Rd. Science 269, 496-512.

Fletcher, L. (2000): Efforts to commercialize structural genomics may be limited. Nature Biotech. 18, 1036.

Franklin, R. E. & Gosling, R. G. (1953): Molecular configuration in sodium thymonucleate. Nature 171, 740-741.

Frantz, I. D. jr., Zamecnik, P. C., Reese, J. W. & Stephenson, M. L. (1948): The effect of dinitrophenol on the incorporation of alanine labeled with radioactive carbon into the proteins of slices of normal and malignant rat liver. J. Biol. Chem. 174, 773-774.

Freese, E. (1959a): On the molecular explanation of spontaneous and induced mutations. Brookhaven Symp. Biol.12, 63-75.

Freese, E. (1959b): The difference between spontaneous and nucleic acid base-analogue induced mutations of phage T4. Proc. Natl. Acad. Sci. USA 45, 622-633.

Freese, E. (1959c): The specific mutagenic effect of base analogues on phageT4. J. Mol. Biol.1, 87-105.

Fruton, J. S. (1990): Contrasts in scientific style. Research groups in the chemical and biochemical sciences. American Philos. Soc.: Philadelphia.

Fry, C. J. & Peterson, C.L. (2001): Chromatin remodeling enzymes: who's on first? Currents Biol. 11, R185-R197.

Galas, D. J. & Schmitz, A. (1978): DNase footprinting: a simple method for the detection of protein-DNA binding specificity. Nucl. Acids Res. 5, 3157-3170.

Gale, E. F. & Folkes, J. P. (1953): Amino acid incorporation by fragmented staphylococcal cells. Biochem. J. 55, p. xi.

Gamow, G. (1954): Possible relation between deoxyribonucleic acid and protein structures. Nature 173, 318.

Gamow, G., Rich, A. & Yčas, M. (1955): The problem of information transfer from the nucleic acids to proteins. Adv. Biol. & Medical Physics 4, 23-68.

Gaster, B. (1990): The slow diffusion of the DNA paradigm into biology textbooks. Trends Biochem. Sci. 15, 325-327.

Gefter, M. L., Hirota, Y., Kornberg, T., Wechsler, J. A. & Barnoux, C. (1971): Analysis of DNA polymerases II and III in mutants of *Escherichia coli* thermosensitive for DNA synthesis. Proc. Natl. Acad. Sci. USA 68, 3150-3153.

Geiduschek, E. P.,Tocchini-Valentini, G. P. & Sarnat, M. T. (1964): Asymmetric synthesis of RNA *in vitro*: Dependence on DNA continuity and conformation. Proc. Natl. Acad. Sci. USA 52, 486-493.

Georgopoulos, C. (1992): The emergence of the chaperone machines. Trends Biochem. Sci. 17, 295-299.

Gerasimova, T. I. & Corces, V.G. (2001): Chromatin insulators and boundaries: effects on transcription and nuclear organization. Ann. Rev. Genet. 35, 193-208.

Germond, J. E., Hirt, B., Oudet, P., Gross-Bellard, M. & Chambon, P. (1975): Folding of the DNA double helix in chromatin-like structures from simian virus 40. Proc. Natl. Acad. Sci. USA 72, 1843-1847.

Giacomoni, D. (1993): The origin of DNA: RNA hybridization. J. Hist. Biol. 26, 89-107.

Giacomoni, D. & Spiegelman, S. (1962): Origin and biological individuality of the genetic dictionary. Science 138, 1328-1331.

Gibbons, A. (1990): Our chimp cousins get that much closer. Science 250, 376.

Gilbert, W. (1978): Why genes in pieces? Nature 271, 501.

Gilbert, W. (1981): DNA sequencing and gene structure. Science 214, 1305-1312.

Gilbert, W. (1986) cited in: Lewin (1986).

Gilbert, W. & Maxam, A. (1973): The nucleotide sequence of the lac operator. Proc. Natl. Acad. Sci. USA 70, 3581-3584.

Gilbert, W. & Müller-Hill, B. (1967): The lac operator is DNA. Proc. Natl. Acad. Sci. USA 58, 2415-2421.

Gilham, P. T. & Khorana, H. G. (1958): Studies on polynucleotides. I. A new and general method for the chemical synthesis of the C5'-C3' internucleotide linkage. Synthesis of deoxyribo-dinucleotides. J. Am. Chem. Soc. 80, 6212-6222.

Goedell, D. V., Kleid, D. G., Bolivar, F., Heyneker, H. L., Yansura, D. G., Hirose, T., Kraszewski, A., Itakura, K. & Riggs, A. D. (1979): Expression in *Escherichia coli* of chemically synthesized genes for human insulin. Proc. Natl. Acad. Sci. USA 76, 106-110.

Goff, S. A. et al. (54 coauthors) (2002): A draft sequence of the rice genome (*Ozyra sativa* L. ssp. *japonica*). Science 296, 92-100.

Goffeau, A., Barrell, B. G., Bussey, H., Davis, R. W., Dujon, B., Feldmann, H., Galibert, F., Hoheisel, J. D., Jacq, C., Johnston, M., Louis, E. I., Mewes, H. W., Mutakami, Y., Philippsen, P., Tettelin, H. & Oliver, S. G. (1996): Life with 6000 genes. Science 274, 546-567.

Gold, M., Gefter, M., Hausmann, R. & Hurwitz, J. (1966): Methylation of DNA. J. Gen. Physiol. 49, 5-28.

Gold, M. & Hurwitz, J. (1964): The enzymatic methylation of RNA and DNA. VI. Further studies on the properties of the DNA methylation reaction. J. Biol. Chem. 239, 3866-3874.

Golomb, S. W., Welch, L. R. & Delbrück, M. (1958): Construction and properties of comma-free codes. Biol. Medd. Dan. Vid. Selsk. 23, 2-34.

Gottschalk, G. & Schlegel, H. G. (1982): Das Institut für Mikrobiologie der Universität Göttingen. Forum Mikrobiol. 5, 5-8.

Gray, M. W. (1989): The evolutionary origins of organelles. Trends Genet. 5, 294-299.

Green, M. H. (1964): Strand selective transcription of T4 DNA *in vitro*. Proc. Natl. Acad. Sci. USA 52, 1388-1395.

Griffith, F. (1928): The significance of pneumococcal types. J. Hygiene 27, 113-159.

Gros, F., Gilbert, W., Hiatt, H., Kurland, C., Risebrough, R. W. & Watson, J. D. (1961 ): Unstable ribonucleic acid revealed by pulse labeling of *E. coli*. Nature 190, 581-585.

Grunberg-Manago, M. (1963): Polynucleotide phosphorylase. Progr. Nucl. Acids Res. Molec. Biol.1, 93-133.

Grunberg-Manago, M. & Ochoa, S. (1955): Enzymatic synthesis and breakdown of polynucleotides; polynucleotide phosphorylase. J. Am. Chem. Soc. 77, 3165-3166.

Grunberg-Manago, M., Oritz, P. J. & Ochoa, S. (1955): Enzymatic synthesis of nucleic acid-like polynucleotides. Science 122, 907 -910.

Guainville, G. (1992): Le génome à portée de main. La Recherche 23, 1438-1439.

Gullen, D. (1993): Carping on. New Scientist 139, 48.

Hakem, R. & Mak, T. W. (2001): Animal models of tumor-suppressor genes. Ann. Rev. Genet. 35, 209-241.

Hall, B. D. & Spiegelman, S. (1961 ): Sequence complementarity of T2 DNA and T2 specific RNA. Proc. Natl. Acad. Sci. USA 47, 137-146.

Harris, T. (2000): Clear vision for a structure-seeking business. Nature Biotech. 18, 1017.

Harrison, E. (1987): Whigs, prigs and historians of science. Nature 329, 213-214.

Hartl, F. U. & Hayer-Hartl, M. (2002): Molecular chaperones in the cytosol: from nascent chain to folded protein. Science 295, 1852-1858.

Hartl, D. L. & Orel, V. (1992): What did Gregor Mendel think he discovered? Genetics 131, 245-253.

Hausmann, R. & Bresch, C. (1960): Zum Problem der genetischen Rekombination von Bakteriophagen. II. Versuch einer experimentellen Unterscheidung von paarweiser und kompletter Kooperation. Zeitschr. Vererbl. 91, 266-276.

Hayashi, M. & Spiegelman, S. (1961): The selective synthesis of informational RNA in bacteria. Proc. Natl. Acad. Sci. USA 47, 1564-1580.

Hayes, W. (1952): Recombination in Bact. coli K-12: unidirectional transfer of genetic material. Nature 169, 118-119.

Hayes, W. (1953a): Observations on a transmissible agent determining sexual differentiation in *Bacterium coli*. J. Gen. Microbiol. 8, 72-88.

Hayes, W. (1953b): The mechanism of genetic recombination in *Escherichia coli*. Cold Spring Harbor Symp. Quant. Biol. 18, 75-93.

Heidelberger, M. & Avery, O.T. (1923): The soluble specific substance of pneumococcus. J. Exp. Med. 38,73-79.

Hershey, A. D. (1946): Spontaneous mutations in bacterial viruses. Cold Spring Harbor Symp. Quant. Biol. 11, 67-77.

Hershey, A. D. & Chase, M. (1951): Genetic recombination and heterozygosis in bacteriophage. Cold Spring Harbor Symp. Quant. Biol. 16, 471-479.

Hershey, A. D. & Chase, M. (1952): Independent functions of viral protein and nucleic acid in growth of bacteriophage. J. Gen. Physiol. 36, 39-56.

Hershey, A. D. & Rotman, R. (1949): Genetic recombination between host range and plaque-type mutants of bacteriophage in single bacterial cells. Genetics 34, 44- 71.

Hershey, J. W. B. (1991):Translational control in mammalian cells. Ann. Rev. Biochem. 60, 717-755.

Hewish, D. R. & Burgoyne, L. A. (1973): Chromatin sub-structure. The digestion of chromatin DNA at regularly spaced sites by a nuclear deoxyribonuclease. Biochem. Biophys. Res. Commun. 52, 504-510.

Hirs, C. H. W., Moore, S. & Stein, W. H. (1960): The sequence of the amino acid residues in performic acid-oxided ribonuclease. J. Biol. Chem. 235, 633-647.

Hoagland, M. B. (1959): The present status of the adaptor hypothesis. Brookhaven Symp. Biol. 12, 40-46.

Hoagland, M. B., Stephenson, M. L., Scott, J. F., Hecht, L. I. & Zamecnik, P. C. (1958): A soluble ribonucleic acid intermediate in protein synthesis. J. Biol. Chem. 231, 241-257.

Hoagland, M. B., Zamecnik, P. C. & Stephenson, M. L. (1959): A hypothesis concerning the roles of particulate and soluble ribonucleic acids in protein synthesis. Symp. Mol. Biol. Univ. Chicago Press (R. E. Zirkle, ed.): Chicago, pp. 105-114.

Hoffmann-Berling, H., Marvin, D. A. & Dürwald, H. (1963): Ein fädiger DNS-Phage (fd) und ein sphärischer RNS-Phage (fr), wirtspezifisch für männliche Stämme von *E. coli*. 1. Präparation und chemische Eigenschaften von fd und fr. Z. Naturforsch. 18b, 876-883.

Hogness, D. S., Cohn, M. & Monod, J. (1955): Studies on the induced synthesis of β-galactosidase in *Escherichia coli*: the kinetics and mechanism of sulfur incorporation. Biochim. Biophys. Acta 16, 99-116.

Holley, R. W., Apgar, J., Everett, G. A., Madison, J. T., Marquisee, M., Merrill, S. H., Penswick, J. R. & Zamir, A. (1965): Structure of a ribonucleic acid. Science 147, 1462-1465.

Holmes, F. L. (2001): Meselson, Stahl, and the replication of DNA: A history of "The most beautiful experiment in biology". Yale University Press.

Hoppe-Seyler, F. (1871): Über die chemische Zusammensetzung des Eiters. Hoppe-Seyler's Mediz.-Chem. Untersuch. 4, 486-501.

Horgan, J. (1998): The end of science. Abacus: London.

Horgan, J. (1999): The undiscovered mind. Free Press: New York.

Horowitz, N. H. (1948): The one gene-one enzyme hypothesis. Genetics 33, 612-613.

Horowitz, N. H. & Leupold U. (1951): Some recent studies bearing on the one gene-one enzyme hypothesis. Cold Spring Harbor Symp. Quant. Biol. 16, 65-74.

Hotchkiss, R. D. (1951): Transfer of penicillin resistance in pneumococci by the deoxyribonucleate derived from resistant cultures. Cold Spring Harbor Symp. Quant. Biol. 16, 457-461.

Hotchkiss, R. D. (1957): Criteria for quantitative genetic transformation of bacteria. In: The chemical basis of heredity, W. D. Mc Elroy & B. Glass (eds.), Johns Hopkins Press Baltimore, 321-335.

Hovanitz, W. (1953): Textbook of genetics. Elsevier Press: New York.

Huberman, J. A. & Riggs, A. D. (1968): On the mechanism of DNA replication in mammalian chromosomes. J. Mol. Biol. 32, 327-341.

Hug, H., Hausmann, R., Liebeschuetz, J. & Ritchie, D. A. (1986): *In vitro* packaging of foreign DNA into heads of bacteriophage T1. J. Gen. Virol. 67, 333-343.

Hurwitz, J., Bresler, A. & Diringer, R. (1960): The enzymic incorporation of ribonucleotides into polyribonucleotides and the effect of DNA. Biochem. Biophys. Res. Commun. 3, 15-19.

Hurwitz, J. & Bresler, A. E. (1961): The incorporation of ribonucleotides into ribonucleic acids. J. Biol. Chem. 236, 542-548.

Hurwitz, J., Furth, J. J., Anders, M., Ortiz, P. J. & August, J.T. (1961): J. Chim. Phys. 58, 934.

Hurwitz, J., Furth, J. J., Anders, M. & Evans, A. H. (1962): The role of deoxyribonucleic acid in ribonucleic acid synthesis. J. Biol. Chem. 237, 3752-3759.

Huxley, A. (1932): Brave new world. Chatto & Windus Ltd.: London; (1955) Penguin, Harmondsworth, U. K.

Ingman, M., Kaessmann, H., Pääbo, S. & Gyllensten, U. (2000): Mitochondrial genome variation and the origin of modern humans. Nature 408, 708-713.

Ingram, V. M. (1956): A specific chemical difference between the globins of normal human and sickle-cell anaemia haemoglobin. Nature 178, 792-794.

Ingram, V. M. (1957): Gene mutations in human haemoglobin: The chemical difference between normal and sickle cell haemoglobin. Nature 180, 326-328.

Ingram, V. M. (1958): Abnormal human haemoglobins, I. The comparison of normal human and sickle-cell haemoglobins by "fingerprinting" . Biochim. Biophys. Acta 28, 539-545.

Ingram, V. M. (1962): A biochemical approach to genetics. In: Horizons in biochemistry (M. Kasha & B. Pullman, eds.). Academic Press: New York, pp. 145-151.

International Human Genome Sequencing Consortium (2001):Initial sequencing and analysis of the human genome. Nature 409, 860-921.

Ippen, K., Miller, J. H., Scaife, J. & Beckwith, J. (1968): New controlling element in the lac operon of E. coli. Nature 217, 825-827.

Itakura, K., Hirose, T., Crea, R., Riggs, A. D., Heyneker, H. L., Bolivar, F. & Boyer, H. W. (1977): Expression in Escherichia coli of a chemically synthesized gene for the hormone somatostatin. Science 198, 1056-1063.

Jacob, F. & Adelberg, E. A. (1959): Transfert de caractères génétiques par incorporation au facteur sexuell d'Escherichia coli. Compt. Rend. Acad. Sci. (Paris) 249, 189-191.

Jacob, F. & Monod, J. (1961): Genetic regulatory mechanisms in the synthesis of proteins. J. Mol. Biol. 3, 318-356.

Jacob, F., Perrin, D., Sanchez, C. & Monod, J. (1960): L'opéron: groupe de gènes à expression coordonée par un opérateur. Compt. Rend. Acad. Sci. (Paris) 250, 1727-1729.

Jacob, F., Ullman, A. & Monod, J. (1964): Le promoteur: élément génétique nécessaire a I'expression d'un opéron. Compt. Rend. Acad. Sci. (Paris) 258, 3125-3128.

Jacob, F. & Wollman, E. L. (1954): Induction spontane du developpement du bacteriophage λ au cours de la recombinaison génétique chez E. coli K12. Compt. Rend. Acad. Sci. (Paris) 239, 455-456.

Jacob, F. & Wollman, E. L. (1956): Recombinaison génétique et mutants de fertilité. Compt. Rend. Acad. Sci. (Paris) 242, 303-306.

Jacob, F. & Wollman, E. L. (1957): Analyse des groupes de liaison génétique de differentes souches donatrices d'Escherichia coli K12. Compt. Rend. Acad. Sci. (Paris) 245, 1840-1850.

Jacob, F. & Wollman, E. L. (1958a): Genetic and physical determinations of chromosomal segments in *E. coli*. Symp. Soc. Exp. Biol. 12, 75-92.

Jacob, F. & Wollman, E. L. (1958b): Les épisomes, éléments génétiques ajoutes. Compt. Rend. Acad. Sci. (Paris) 247, 154-156.

Jacob, F. & Wollman, E. L. (1961 ): Sexuality and the genetics of bacteria. Academic Press: New York.

Jeffords, J. M. & Daschle, T. (2001): Political issues in the genome era. Science 291, 1249-1251.

Jeffreys, A. J. & Flavell, R. A. (1977): The rabbit ß-globin gene contains a large insert in the coding sequence. Cell 12, 1097-1108.

Jeppesen, P. G. N., Barrell, B. G.; Sanger, F. & Coulson, A. R. (1972): Nucleotide sequence of two fragments from the coat-protein cistron of bacteriophage R17 ribonucleic acid. Biochem. J. 128, 993-1006.

Jordan, E., Saedler, H. & Starlinger, P. (1967): Strong-polar mutations in the transferase gene of the galactose operon in *E. coli*. Mol. Gen. Genet. 100, 296-306.

Jordan, E., Saedler, H. & Starlinger, P. (1968): 0° and strong-polar mutations in the gal operon are insertions. Mol. Gen. Genet. 102, 353-363.

Jordan, P. (1938): The specific attraction between gene molecules. Phys. Z. 39, 711-714.

Jordan, P. (1970): Schöpfung und Geheimnis. Gerhard Stalling: Hamburg.

Josse, J., Kaiser, A. D. & Kornberg, A. (1961): Enzymatic synthesis of deoxyribonucleic acid, VIII. Frequencies of nearest neighbor base sequences in deoxyribonucleic acid. J. Biol. Chem. 236, 864-875.

Jozza, N., Kroemer, G. & Penninger, J.M. (2002): Genetic analysis of the mammalian cell death machinery. Trends Genet. 18, 142-149.

Judson, H. F. (1980): The eigth day of creation. Makers of the revolution in biology. Simon & Schuster: New York (revised edition, 1994).

Jürgensmeier, J. M. (1993): Die Induktion von Apoptose in transformierten Fibroblasten durch TFG-ß-behandelte normale Zellen. Dissertation: Univ. Freiburg.

Kaiser, A. D. & Wu, R. (1968): Structure and function of DNA cohesive ends. Cold Spring Harbor Symp. Quant. Biol. 33, 729-734.

Karlin, S., Bergman, A. & Gentles, A.J. (2001): Annotation of the *Drosophila* genome. Nature 411, 259-260.

Kay, L. E. (2000): Who wrote the book of life? Stanford University Press, Stanford, California.

Keller, E. F. (1983): A feeling for the organism. The life and work of Barbara McClintock. Freeman & Co.: San Francisco.

Kelly, T. J. & Brown, G.W. (2000): Regulation of chromosome replication. Ann. Rev. Biochem. 69, 829-880.

Kendrew, J. C. (1961): The three-dimensional structure of a protein molecule. Scient. American 205 (6), 96-110.

Kennedy, E. P. (1970): The lactose permease system of *Escherichia coli*. In: The lactose operon; Beckwith, J. R. & Zipser, D. (eds.). Cold Spring Harbor Lab. Press: New York, pp. 49-92.

Kerr, J. F. R., Wyllie, A. H. & Currie, A. R. (1972): Apoptosis: A basic biological phenomenon with wide-ranging implications in tissue kinetics. Br. J. Cancer 26, 239-257.

Khorana, H. G., Büchi, H., Ghosh, H., Gupta, N., Jacob,T. M., Kössel, H., Morgan, R., Narang, S. A., Ohtsuka, E. & Wells, R. D. (1966): Polynucleotide synthesis and the genetic code. Cold Spring Harbor Symp. Quant. Biol. 31, 39-49.

Kirschner, R. H., Rusli, M. & Martin, T.E. (1977): Characterization of the nuclear envelope, pore complex, and dense lamina of mouse liver nuclei by high resolution scanning electron microscopy. J. Cell Biol. 72, 118-132.

Kohler, R. (1971): The background to Eduard Buchner's discovery of cell-free fermentation. J. Hist. Biol. 4, 35-61.

Kolb, A., Busby, S., Buc, H., Garges, S. & Adhya, S. (1993): Transcriptional regulation by cAMP and its receptor protein. Ann. Rev. Biochem. 62, 749-795.

Kolter, R. & Yankosfky, C. (1982): Attenuation in amino acid biosynthetic operons. Ann. Rev. Genet. 16, 113-134.

Komeili, A. & O'Shea, E.K. (2001): New perspectives on nuclear transport. Ann. Rev. Genet. 35, 341-364.

Kornberg, A. (1960): Biological synthesis of deoxyribonucleic acid. Science 131, 1503-1508.

Kornberg, A. (1976): For the love of enzymes. In: A. Kornberg, B. L. Horecker, L. Cornudella, J. Oro (eds.): Reflections on biochemistry. In honour of Severo Ochoa. Pergamon Press, Oxford, pp. 243-251.

Kornberg, A. (1989): Never a dull enzyme. Ann. Rev. Biochem. 58, 1-30.

Kornberg, A., Zimmerman, S. B., Kornberg, S. R. & Josse, J. (1959): Enzymatic synthesis of deoxyribonucleic acid. VI. Influence of bacteriophage T2 on the synthesis pathway in host cells. Proc. Natl. Acad. Sci. USA 45, 772-785.

Kornberg, R. D. & Lorch, Y. (1992): Chromatin structure and transcription. Ann. Rev. Cell Biol. 8, 563-587.

Kornberg, S. R., Zimmerman, S. B. & Kornberg, A. (1961): Glucosylation of deoxyribonucleic acid by enzymes from bacteriophage-infected Escherichia *coli*. J. Biol. Chem. 236, 1487-1493.

Kornberg, T. & Gefter, M. L. (1971): Purification and DNA synthesis in cell-free extracts: properties of DNA polymerase II. Proc. Natl. Acad. Sci. USA 68, 761-764.

Kosko, B. & Isaka, S. (Juli 1993): Fuzzy logic. Scient. American 269, 62-67.

Kössel, H., Edwards, K., Fritzsche, E., Koch, W. & Schwarz, Z. (1983): Phylogenetic significance of nucleotide sequence analysis. In: Proteins and nucleic acis in plant systematics; U. Jensen & D. E. Fairbrothers, eds. Springer: Heidelberg.

Kozak, M. (1992): Regulation of translation in eukaryotic systems. Ann. Rev. Cell Biol. 8, 197-225.

Kühnlein, U. & Arber, W. (1972): Host specificity of DNA produced by *Escherichia coli*. XV. The role of nucleotide methylation in *in vitro* B-specific modification. J. Mol. Biol. 63, 9-19.

Lamborg, M. R. & Zamecnik, P. C. (1960): Amino acid incorporation into protein by extracts of *E. coli*. Biochim. Biophys. Acta 42, 206-211.

Langridge, R., Marvin, D. A., Seeds, W. E., Wilson, H. R., Hooper, C. W., Wilkins, M. H. F. & Hamilton, L. D. (1960b): The molecular configuration of deoxyribonucleic acid. II. Molecular models and their Fourier transforms. J. Mol. Biol. 2, 38-64.

Langridge, R., Wilson, H. R., Hooper, C. W., Wilkins, M. H. F. & Hamilton, L. D. (1960a): The molecular configuration of deoxyribonucleic acid. I. X-ray diffraction study of a crystalline form of the lithium salt. J. Mol. Biol. 2, 19-37.

Lea, D. E. & Coulson, C. A. (1949): The distribution of the numbers of mutants in bacterial populations. J. Genet. 49, 264-285.

Leder, P. & Nirenberg, M. (1964): RNA codewords and protein synthesis, II. Nucleotide sequence of a valine RNA codeword. Proc. Natl. Acad. Sci. USA 52, 420-427.

Leder, P., Tilghman, S. M., Tiemeier, D. C., Polsky, F. I., Seidman, J. G., Edgell, M. H., Enquist, L. W., Leder, A. & Norman, B. (1977): The cloning of mouse globin and surrounding gene sequences in bacteriophage λ. Cold Spring Harbor Symp. Quant. Biol. 42, 915-920.

Lederberg, E. M. (1951a): Lysogenicity in E. coli. Genetics 36, 560.

Lederberg, E. M. & Lederberg, J. (1953): Genetic studies of lysogenicity in Escherichia coli. Genetics 38, 51-64.

Lederberg, J. (1947): Gene recombination and linked segregation in Escherichia coli. Genetics 32, 505-525.

Lederberg, J. (1951b): Genetic studies with bacteria, pp. 263-289; in: Dunn, L. C. (ed.), Genetics in the 20th century. Macmillan: New York.

Lederberg, J. (1955): Genetic recombination in bacteria. Science 122, 920-921.

Lederberg, J. (1959): Bacterial reproduction. Harvey Lectures 53, 69-82.

Lederberg, J. (1986): A fortieth anniversary reminiscence. Nature 324, 627-628.

Lederberg, J. (1987): Genetic recombination in bacteria: a discovery account. Ann. Rev. Genet. 21, 23-46.

Lederberg, J., Cavalli, L. L. & Lederberg, E. M. (1952): Sex compatibility in Escherichia coli. Genetics 37, 720-730.

Lederberg, J. & Lederberg, E. M. (1952): Replica plating and indirect selection of bacterial mutants. J. Bacteriol. 63, 399-406.

Lederberg, J., Lederberg, E. M., Zinder, N. D. & Lively, E. R. (1951): Recombination analysis of bacterial heredity .Cold Spring Harbor Symp. Quant. Biol. 16, 413-443.

Lederberg, J. & Tatum, E. L. (1946): Novel genotypes in mixed cultures of biochemical mutants of bacteria. Cold Spring Harbor Symp. Quant. Biol. 11, 113-114.

Lee, K. Y., Wahli, R. & Barbu, E. (1956): Contenu en bases puriques et pyrimidiques des acides deoxyribonucléiques des bactéries. Ann. Inst. Pasteur 91, 212-224.

Leloir, L. F. (1983): Far away and long ago. Ann. Rev. Biochem. 52, 1-15.

Lengyel, J. A., Goldstein, R. N., Marsh, M. & Calendar, R. (1974): Structure of the bacteriophage P2 tail. Virology 62, 161-174.

Lengyel, P. (1976): Ten years in protein synthesis. In: Reflections on Biochemistry, A. Kornberg, B. L. Horecker, L. Cornudella & J. Oró, eds. Pergamon Press: N. York, pp. 309-316.

Lerman, L. (1961): Structural considerations in the interaction of DNA and acridines. J. Mol. Biol. 3, 18-30, see also: Brenner, Crick & Orgel, J. Mol. Biol. 3, 121 (1961).

Levene, P. A. (1921): On the structure of thymus nucleic acid and on its possible bearing on the structure of plant nucleic acid. J. Biol. Chem. 48, 119-125.

Levinthal, C. & Fisher, H. W. (1953): Maturation of phage and the evidence of phage precursors. Cold Spring Harbor Symp. Quant. Biol. 18, 29-34.

Lewin, R. (1986): Proposal to sequence the human genome stirs debate. Science 232, 1596-1600.

Li, J. J. & Deshaies, R. J. (1993): Exercising self-restraint: discouraging illicit acts of S and M in eukaryotes. Cell 74, 223-226.

Lipmann, F. (1963): Messenger ribonucleic acid. Progr. Nucl. Acids Res. Molec. Biol. 1, 135-161.

Lipmann, F. (1969): Polypeptide chain elongation in protein biosynthesis. A protein grows by single unit addition on the ribosome-reactor with messenger RNA as conveyor belt. Science 164, 1024-1031.

Lipmann, F., Hülsmann, W. C., Hartmann, G., Boman, H. G. & Acs, G. (1959): Amino acid activation and protein synthesis. J. Cell. Comp. Physiol. 54, 75-88.

Loeb, T. & Zinder, N. D. (1961): A bacteriophage containing RNA. Proc. Natl. Acad. Sci. USA 47, 282-289.

Loftfield, R. B., Grover, J. W. & Stephenson, M. L. (1953): Possible role of proteolytic enzymes in protein synthesis. Nature 171, 1024-1025.

Lucas-Lenard, J. & Lipmann, F. (1971 ): Protein biosynthesis. Ann. Rev. Biochem. 40, 409-448.

Luria, S. E. (1945): Mutations of bacterial viruses affecting their host range. Genetics 30, 84.

Luria, S. E. & Anderson, F. F. (1942): The identification and characterization of bacteriophages with the electron microscope. Proc. Natl. Acad. Sci. USA 28, 127-130.

Luria, S. E. & Delbrück, M. (1943): Mutations of bacteria from virus sensitivity to virus resistance. Genetics 28, 491-511.

Luria, S. E., Delbrück, M. & Anderson, T. F. (1943): Electron microscope studies of bacterial viruses. J. Bacteriol. 46, 57-77.

Luria, S. E. & Human, M. L. (1952): A non-hereditary host-induced variation of bacterial viruses. J. Bacteriol. 64, 557-569.

Lwoff, A. & Gutmann, A. (1950): la libération de bacteriophages par la lyse d'une bacterie lysogène. Compt. Rend. Acad. Sci. (Paris) 230, 154-156.

Lwoff, A., Siminovitch, L. & Kjeldgaard, N. (1950): Induction de la production de bactériophages chez une bactérie lysogène. Ann. Inst. Pasteur 79, 815-858.

Maaløe, O. & Symonds, N. (1953): Radioactive sulfur tracer studies on the reproduction of T4 bacteriophage. J. Bacteriol. 65, 177-182.

Marcker, K. (1965): The formation of N-formyl-methionyl-sRNA. J. Mol. Biol. 14, 63-70.

Marcker, K. & Sanger, F. (1964): N-Formyl-methionyl-S-RNA. J. Mol. Biol. 8, 835-840.

Marians, K. J. (1992): Prokaryotic DNA replication. Ann. Rev. Biochem. 61, 673-719.

Marsh, R. E., Corey, R. B. & Pauling, l. (1955): An investigation on the structure of silk fibroin. Biochim. Biophys. Acta 16, 1-16.

Marshall, E. (2001): A plan to release data within six months. Science 292, 188.

Marx, J. (1993): Cell death studies yield cancer clues. Science 259, 760-761.

Mathis, D. J. & Chambon, P. (1981): The SV40 early region TATA box is required for accurate in vitro initiation of transcription. Nature 290, 310-315.

Matson, S. W. (1991): DNA helicases of Escherichia coli. Progr. Nucl. Acids Res. Molec. Biol. 40, 289-326.

Matsutani, S. & Ohtsubo, E. (1993): Distribution of the Shigella sonnei insertion elements in Enterobacteriaceae. Gene 127, 111-115.

Matthaei, J. H., Jones, O. W., Martin, R. G. & Nirenberg, M. S. (1962): Characteristics and composition of RNA coding units. Proc. Natl. Acad. Sci. USA 48, 666-677.

Matthaei, J. H. & Nirenberg, M. W. (1961): Characterization of DNase-sensitive protein synthesis in E. coli extracts. Proc. Natl. Acad. Sci. USA 47, 1580-1588.

Maxam, A. M. & Gilbert, W. (1977): A new method for sequencing DNA. Proc. Natl. Acad. Sci. USA 74, 560-564.

Maynard Smith, J. (1982): Evolution and the theory of games. Cambridge Univ. Press: Cambridge, U. K.

Maynard Smith, J., Smith, N. H., O'Rourke, M. & Spratt, B. G. (1993): How clonal are bacteria? Proc. Natl. Acad. Sci. USA 90, 4384-4388.

McCarty, M., Taylor, H. E. & Avery, O. T. (1946): Biochemical studies of environmental factors essential in transformation of pneumococcal types. Cold Spring Harbor Symp. Quant. Biol. 11, 177-183.

McClintock, B. (1951): Chromosome organisation and genic expression: Cold Spring Harbor Symp. Quant. Biol. 16, 13-47.

McClintock, B. (1956): Controlling elements and the gene. Cold Spring Harbor Symp. Quant. Biol. 21, 197-216.

McClintock, B. (1965): The control of gene action in maize.Brookhaven Symp. Biol. 18, 162-184.

McEachern, M. J., Krauskopf, A. & Blackburn, E.H. (2000): Telomeres and their control. Ann. Rev. Genet. 34, 331-358.

McKeown, M. (1992): Alternative mRNA splicing. Ann. Rev. Cell Biol. 8, 133-155.

Medawar, P. (1978): Philosophy of ignorance. Nature 272, 772-774.

Meselson, M. & Stahl, F. W. (1958): The replication of DNA in Escherichia coli. Proc. Natl. Acad. Sci. USA 44, 671-682.

Meselson, M. & Yuan, R. (1968): DNA restriction enzyme from E. coli. Nature 217, 1110-1114.

Meyer, J. & Arber, W. (1986): 100 Jahre Escherichia coli. Bescheidene Anfänge – unerwartete Folgen. Naturwissensch. Rundschau 39, 467-473.

Miescher, F. (1871): Über die chemische Zusammensetzung der Eiterzellen. Hoppe-Seyler's Mediz.-Chem. Untersuch. 4, 441-460.

Miller, R. V. & Kokjohn, T. A. (1990): General microbiology of recA: environmental and evolutionary significance. Ann. Rev. Microbiol. 44, 365-394.

Mirsky, A. E. (1951): Some chemical aspects of the cell nucleus; p. 127-153. In: Genetics in the 20th century, L. C. Dunn, ed. Macmillan: N. York.

Mirsky, A. E. & Pauling, L. (1936): Hydrogen bonds involved in maintaining native protein structure. Proc. Natl. Acad. Sci. USA 22, 439-447.

Mitchell, P. (1961): Coupling of phosphorylation to electron and hydrogen transfer by a chemiosmotic type of mechanism. Nature 191, 144-148.

Mitchell, P. (1967): Translocation through natural membranes. Adv. Enzymol. Relat. Areas 29, 33-87.

Mitchell, P. (1979): Keilin's respiratory chain concept and its chemiosmotic consequences. Science 206, 1148-1159.

Mitchell, P. & Moyle, J. (1965): Stochiometry of proton translocation through the respiratory chain and adenosine triphosphatase systems of rat liver mitochondria. Nature 208, 147-151.

Mitchell,P. & Moyle, J. (1968): Proton translocation coupled to ATP hydrolysis in rat liver mitochondria. Eur. J. Biochem. 4, 530-539.

Monk, R. (1993): Invisible Johnny. Nature 326, 668-669.

Monod, J., Changeux, J.-P. & Jacob, F. (1963): Allosteric proteins and cellular control systems. J. Mol. Biol. 6, 306-329.

Morgan, A. R. (1993): Base mismatches and mutagenesis: how important is tautomerism? Trends Biochem. Sci 18, 160-163.

Morgan, T. H. (1910): Sex limited inheritance in *Drosophila*. Science 32, 120-122.

Morgan, T. H., Bridges, C. B. & Sturtevant, A. M (1925): The genetics of *Drosophila*. Bibliogr. Genetica 2, 1-262.

Mosig, G. (1983): Appendix: T4-genes and gene products, In: Mathews, C. K., Kutter, E. M., Mosig, G. & Berget, P. B (eds.): Bacteriophage T4. American Society for Microbiology: Washington DC., pp. 362-374.

Mouriquand, C., Gilly, C. & Wolff, C. (1972): Ultrastructure du chromosome: données fournies par l'observation du chromosome entier. Ann. Génét. 15, 249-256.

Muller, H. J. (1927): Artificial transmutation of the gene. Science 66, 84-87.

Näär, A. M., Lemon, B. D. & Tjian, R. (2001): Transcriptional coactivator complexes. Ann. Rev. Biochem. 70, 475-501.

Naughton, M. A. & Dintzis, H. M. (1962): Sequential biosynthesis of the peptide chains of hemoglobin. Proc. Natl. Acad. Sci. USA 48, 1822-1830.

Nelson, K. E., Paulson, I. T., Heidelberg, J. F. & Fraser, C. M. (2000): Status of genome projects for nonpathogenic bacteria and archaea. Nature Biotechnol. 18, 1049-1054.

Neumann, J. (1966): Theory of self-reproducing automata. Univ. Illinois Press: Urbana, Ill.

Newmark, P. (1985): Is megasequencing madness? Nature 323, 291.

Nigg, E. A. (2001): Mitotic kinases as regulators of cell division and its check points. Nature Rev. 2, 21-32.

Nirenberg, M. W., Jones, O. W., Leder, P., Clark, B. F. C., Sly, W. S. & Pestka, S. (1963): On the coding of genetic information. Cold Spring Harbor Symp. Quant. Biol. 28, 549-557.

Nirenberg, M. & Leder, P. (1964): RNA codewords and protein synthesis. Science 145, 1399-1407.

Nirenberg, M., Leder, P., Bernfield, M., Brimacombe, R., Trupin, J., Rottman, F. & O'Neal, C. (1965): RNA codewords and protein synthesis, VII. On the general nature of the RNA code. Proc. Natl. Acad. Sci. USA 53, 1161-1167.

Nirenberg, M.W. & Matthaei, J. H. (1961): The dependence of cell-free protein synthesis in *E. coli* upon naturally occurring or synthetic polyribonucleotides. Proc. Natl. Acad. Sci. USA 47, 1588-1602.

Noll, M. (1974): Subunit structure of chromatin. Nature 251, 249-251.

Noller, H. F. (1991): Ribosomal RNA and translation. Ann. Rev. Biochem. 60, 191-227.

Nomura, M., Hall, B. D. & Spiegelman, S. (1960): Characterization of RNA synthesized in *Escherichia coli* after bacteriophage T2 infection. J. Mol. Biol. 2, 306-326.

Novick, A. & Szilard, L. (1951): Genetic mechanisms in bacteria and bacterial viruses. I. Experiments on spontaneous and chemically induced mutations of bacteria growing in the chemostat. Cold Spring Harbor Symp. Quant. Biol. 16, 337-343.

Nüsslein, V., Otto, B., Bonhoeffer, F. & Schaller, H. (1971): Function of DNA polymerase III in DNA replication. Nature New Biol. 234, 285-286.

Ochman, H. & Selander, R. K. (1984): Evidence for clonal population structure in *Escherichia coli*. Proc. Natl. Acad. Sci. USA 81, 198-201.

Oeppen, J. & Vaupel, J. W. (2002): Broken limits to life expectancy. Science 296, 1029-1030.

Okazaki, R., Arisawa, M. & Sugino, A. (1971): Slow joining of newly replicated DNA chains in DNA polymerase I-deficient *Escherichia coli* mutants. Proc. Natl. Acad. Sci. USA 68, 2954-2957.

Okazaki, R., Okazaki, T., Sakabe, K., Sugimoto, K., Kainuma, R., Sugino, A. & Iwatsuki, N. (1968): *In vivo* mechanism of DNA chain growth. Cold Spring Harbor Symp. Quant. Biol. 33, 129-143.

Okazaki, T. & Okazaki, R. (1969): Mechanism of DNA chain growth, IV. Direction of synthesis of T4 short DNA chains as revealed by exonucleolytic degradation. Proc. Natl. Acad. Sci. USA 64, 1242-1248.

Olby, R. (1970): Francis Crick and the central dogma. Daedalus 99, 939-987.

Olby, R. (1974): The path to the double helix. Macmillan: London. (Revised edition, 1994, Dover Publ.: New York.)

Olby, R, (1979): Mendel, no Mendelian? J. Hist. Sci.17, 53-72.

Olins, A. L. & Olins, D. E. (1974): Spheroid chromatin units (V bodies). Science 183, 330-332.

Olins, D. E. & Olins, A. L. (1978): Nucleosomes: The structural quantum in chromosomes. American Scientist 66, 704-711.

Oliver, S. G., van der Aart, Q. I. M. ...143 names follow in alphabetical order, then... Zimmermann, F. K., Sgouros, I. G. (1992): The complete DNA sequence of yeast chromosome IIID. Nature 357, 38-46.

Olshansky, S. J., Carnes, B. A. & Désesquelles, A. (2001): Prospects for human longevity. Science 291, 1491-1492.

Orgel, L. E. & Crick, F. H. C. (1980): Selfish DNA: the ultimate parasite. Nature 284, 604-607.

Orphanides, G. & Reinbarg, D. (2002): A unified theory of gene expression. Cell 108, 439-451.

Pace, N. R. & Marsh, T. L. (1985): RNA catalysis and the origin of life. Origins of Life 16, 97-116.

Painter, T. S. (1935): The morphology of the third chromosome in the salivary gland of *Drosophila melanogaster* and a new cytological map of this element. Genetics 20, 301-326.

Pardee, A. B., Jacob, F. & Monod, J. (1958): Sur I'expression et le rôle des alléles « inductible » et « constitutif » dans la synthèse de la ß-galactosidase chez les zygotes d' *Escherichia coli*. Compt. Rend. Acad. Sci. (Paris) 246, 3125-3127.

Pardee, A. B., Jacob, F. & Monod, J. (1959): The genetic control and cytoplasmic expression of "inducibility" in the synthesis of ß-galactosidase by *E. coli*. J. Mol. Biol. 1, 165-178.

Parodi, A. J. (2000): Protein glucosylation and its role in protein folding. Ann. Rev. Biochem. 69, 69-93.

Pauling, C. & Hamm, L. (1969): Properties of a temperature-sensitive, radiation-sensitive mutant of *Escherichia coli*, II. DNA replication. Proc. Natl. Acad. Sci. USA 64, 1195-1202.

Pauling, L. (1974): Molecular basis of biological specificity. Nature 248, 769-771.

Pauling, L. & Corey, R. B. (1950): Two hydrogen-bonded spiral configurations of the polypeptide chain. J. Am. Chem. Soc. 72, 5349.

Pauling, L. & Corey, R. B. (1951): The pleated sheet, a new layer configuration of polypeptide chain. Proc. Natl. Acad. Sci. USA 37, 251-256.

Pauling, L. & Corey, R. B. (1953a): A proposed structure for the nucleic acids. Proc. Natl. Acad. Sci. USA 39, 84-97.

Pauling, L. & Corey, R. B. (1953b): Structure of the nucleic acids. Nature 171, 346.

Pauling, L. & Delbrück, M. (1940): The nature of the intermolecular forces operative in biological processes. Science 92, 77-79.

Pauling, L., et. al. (14 authors) (1985): Effect of dietary ascorbic acid on the incidence of spontaneous mammary tumors in R III mice. Proc. Natl. Acad. Sci. USA 82, 5185-5189.

Pauling, L., Itano, H. A., Singer, S. J. & Wells, I. C. (1949): Sickle cell anemia, a molecular disease. Science 110, 543-548.

Paulson, J. R. & Laemmli, U. K. (1977): The structure of histone-depleted metaphase chromosomes. Cell 12, 817-828.

Perutz, M. F. (1987): Physics and the riddle of life. Nature 326, 555-558.

Peterson, E. A. & Greenberg, D. M. (1952): Characteristics of the amino acid-incorporation system of liver homogenates. J. Biol. Chem. 194, 359-375.

Phillips, D. H. (1983): Fifty years of benzo(a)pyrene. Nature 303, 468-472.

Phillips, D. M. (1974): Nuclear shaping in the absence of microtubules in scorpion spermatids. J. Cell Biol. 62, 911-917.

Piper, A. (1998): Light on a dark lady. Trends Biochem. Sci. 23, 151-154.

Plunkett, G. III, Burland, V., Daniels, D. L. & Blattner, F. R. (1993): Analysis of the Escherichia coli genome. III. DNA sequence of the region from 87.2 to 89.2 minutes. Nucl. Acids Res. 21, 3391-3398.

Portugal, F. H. & Cohen, J. S. (1977): A century of DNA. MIT Press: Cambridge (Mass.).

Preiss, J., Berg, P., Ofengand, E. J., Bergmann, F.H. & Dieckmann, M. (1959): The chemical nature of the RNA-amino acid compound formed by amino acid-activating enzymes. Proc. Natl. Acad. Sci. USA 45, 319-328.

Pribnow, D. (1975): Bacteriophage T7 early promoters: nucleotide sequences of two RNA polymerase binding sites. J. Mol. Biol. 99, 419-443.

Pryer, N. K., Wuestehube, L. J. & Schekman, R.(1992): Vesicle-mediated protein sorting. Ann. Rev. Biochem. 61, 471-516.

Ptashne, M. & Gann, A. (2002): Genes and signals. Cold Spring Harbor Laboratory Press, Cold Spring Harbor, N.Y.

Reznikoff, W. S., Siegele, D. A., Cowing, D. W. & Gross, C. A. (1985): The regulation of transcription initiation in bacteria. Ann. Rev. Genet. 19, 355-387.

Riley, M. & Serres, M.H. (2000): Interim report on genomics of Escherichia coli. Ann. Rev. Microbiol. 54, 341-411.

Rich, A. (1960): A hybrid helix containing both deoxyribose and ribose polynucleotides and its relation to the transfer of information between the nucleic acids. Proc. Natl. Acad. Sci. USA 46, 1044-1053.

Richardson, J. P. (1993): Transcription termination. Crit. Rev. Biochem. Mol. Biol. 28, 1-30.

Rickenberg, H. V., Cohen, G. N., Buttin, G. & Monod, J. (1956): La galactoside permease d'Escherichia coli. Ann. Inst. Pasteur 91, 829-857.

Ris, H. & Kubai, F. (1970): Chromosome structure. Ann. Rev. Genet. 4, 263-294.

Roberts, L. (1988): Watson may head genome office. Science 240, 878-879.

Roberts, L. (1992): Why Watson quit as project head. Science 256, 301-302.

Robertson, D. (2001): Racially defined haplotype project debated. Nature Biotechnol. 19, 795-796.

Ross, J., Aviv, H., Scolnick, E. & Leder, P. (1972): In vitro synthesis of DNA complementary to purified rabbit globin mRNA. Proc. Natl. Acad. Sci. USA 69, 264-268.

Rougeon, F., Kourilsky, P. & Mach, B. (1975): Insertion of a rabbit ß-globin gene sequence into an *E. coli* plasmid. Nucl. Acids Res. 2, 2365-2378.

Rout, M.P., Aitchison, J.D., Suprapto, A., Hjertaas, K., Zhao, Y. & Chait, B.T. (2000): The yeast nuclear pore complex: composition, architecture, and transport mechanism. J. Cell Biol. 148, 635-651.

Rowen, L. & Kornberg, A. (1978): Primase, the dnaG protein of *Escherichia coli*. J. Biol. Chem. 253, 758-764.

Royal, A., Garapin, A., Cami, B., Perrin, F., Mandel, J. L., Lemeur, M., Brégégègre, F., Gannon, F., LePennec, J. P., Chambon, P. & Kourilsky, P. (1979): The ovalbumin gene region: common features in the organisation of three genes expressed in chicken oviduct under hormonal control. Nature 279, 125-132.

Ruvolo, M., Disotell, T. R., Allard, M. W., Brown, W. M. & Honeycutt, R. L. (1991): Resolution of the african hominoid trichotomy by use of mitochondrial gene sequences. Proc. Natl. Acad. Sci. USA 88, 1570-1574.

Ryle, A. P., Sanger, F., Smith, L. F. & Kitai, R. (1955): The disulphide bonds of insulin. Biochem. J. 60, 541-556.

Saedler, H. & Starlinger, P. (1967): 0° Mutations in the galactose operon in *E. coli*. I. Genetic characterization. Mol. Gen. Genet. 100, 178-189.

Saedler, H. & Starlinger, P. (1992): Twenty-five years of transposable element research in Köln. In: Fedoroff, N. & Botstein, D. (eds.): The dynamic genome. Cold Spring Harbor Lab. Press: New York.

Saiki, R., Scharf, S., Faloona, F., Mullis, K. B., Horn, G. T., Erlich, H. A. & Arnheim, N. (1985): Enzymatic amplification of ß-globin genomic sequences and restriction site analysis for diagnosis of sickle cell anemia. Science 230, 1350-1354.

Sakai, M., Morinaga, T., Urano, Y., Watanabe, K., Wegmann,T. G. & Tamaoki, T. (1985): The human α-fetoprotein gene. Sequence organization and the 5'flanking region. J. Biol. Chem. 260, 5055-5060.

Salpeter, M. M. & Farquhar, M. G. (1981): High resolution analysis of the secretory pathway in mammotrophs of the rat anterior pituitary. J. Cell Biol. 91, 240-246.

Sangalli, A. (1992): Fuzzy logic goes to market. New Scientist 133, 36-39.

Sanger, F. (1959): Chemistry of insulin. Determination of the structure of insulin opens the way to greater understanding of life processes. Science 129, 1340-1344.

Sanger, F. (1988): Sequences, sequences, and sequences. Ann. Rev. Biochem. 57, 1-28.

Sanger, F., Air, G. M., Barrell, B. G., Brown, N. L., Coulson, A. R., Fiddes, J. C., Hutchison III, C. A., Slocombe, P. M. & Smith, M. (1977): Nucleotide sequence of bacteriophage ΦX174 DNA. Nature 265, 687-695.

Sanger, F., Brownlee, G. G., Barrell, B. G. (1965): A two-dimensional fractionation procedure for radioactive nucleotides. J. Mol. Biol. 13, 373-398.

Sanger, F. & Coulson, A. (1975): A rapid method for determining sequences in DNA by primed synthesis with DNA polymerase. J. Mol. Biol. 94, 441-448.

Sanger, F., Donelson, J. E., Coulson, A. R., Kössel, H. & Fischer, D. (1974): Determination of a nucleotide sequence in bacteriophage f1 DNA by primed synthesis with DNA polymerase. J. Mol. Biol. 90,315-333.

Sanger, F., Nicklen, S. & Coulson, A. R. (1977): DNA sequencing with chain-terminating inhibitors. Proc. Natl. Acad. Sci. USA 74, 5463-5467.

Sayre, A. (1975): Rosalind Franklin and DNA. W.W. Norton: New York.

Schildkraut, C. L., Wierzchowski, K. L., Marmur, J., Green, D. M. & Doty, P. (1962): A study of the base sequence homology among the T series of bacteriophages. Virology 18, 43-55.

Schlesinger, M. (1934): Zur Frage der chemischen Zusammensetzung des Bakteriophagen. Biochem. Z. 273, 306-311.

Schmidt, F., Besemer, J. & Starlinger, P. (1976): The isolation of IS1 and IS2 DNA. Mol. Gen. Genet. 145, 145-154.

Schoenheimer, R. (1942): The dynamic state of body constituents. Harvard Univ. Press: Cambridge [zitiert in Zamecnik, P. C. (1979)].

Schramm, G. & Dannenberg, H. (1944): Über die Ultraviolettabsorption des Tabakmosaikvirus. Ber. Chem. Ges. 77, 53-60.

Schrödinger, E. (1944): What is life? Cambridge University Press, Cambridge, U.K.

Schwartz, R. M. & Dayhoff, M. 0. (1978): Origins of prokaryotes, eukaryotes, mitochondria and chloroplasts. Science 199, 395-403.

Schwemmle, M., Schickinger, J., Bader, M., Sarre, T. F. & Hilse, K. (1991): A 60-kDa protein from rabbit reticulocytes specifically recognizes the capped 5'end of ß-globin mRNA. Eur. J. Biochem. 201, 139-145.

Segerstråle; U. (2000): Defenders of the truth: The battle for science in the sociobiology debate and beyond. Oxford University Press: Oxford, U.K.

Shapiro, J. A. (1969): Mutations caused by the insertion of genetic material into the galactose operon of Escherichia coli. J. Mol. Biol. 40, 93-105.

Sibley, C. G. & Ahlquist, J. E. (1984): The phylogeny of the hominoid primates, as indicated by DNA-DNA hybridisation. J. Mol. Evol. 20, 2-15.

Siebenlist, U. & Gilbert, W. (1980): Contacts between Escherichia coli RNA polymerase and an early promoter of phage T7. Proc. Natl. Acad. Sci. USA 77, 122-126.

Siebenlist, U., Simpson, R. B. & Gilbert, W. (1980): E. coli RNA polymerase interacts homologously with two different promoters. Cell 20, 269-281.

Simpson, A. J. G. et al. (115 coauthors) (2000): The genome sequence of the plant pathogen Xylella fastidiosa. Nature 406, 151-157.

Simpson, R. T. (1991): Nucleosome positioning: occurrence, mechanisms, and functional consequences, Progr. Nucl. Acids Res. Molec. Biol. 40, 143-184.

Singer, H. F., Jones, O. W.& Nirenberg, M. W. (1963): The effect of secondary structure on the template activity of polyribonucleotides. Proc. Natl. Acad. Sci. USA 49, 392-399.

Sinsheimer, R. L. (1959): Is the nucleic acid message in a two-symbol code? J. Mol. Biol. 1, 218-220.

Smellie, R. M. S. (1963): The biosynthesis of ribonucleic acid in animal systems. Progr. Nucl. Acids Res. Molec. Biol. 1, 27-58.

Smith, H. 0. & Wilcox, K. W. (1970): A restriction enzyme from Hemophilus influenzae. J. Mol. Biol. 51, 379-391.

Smith, J. D., Arber, W. & Kühnlein, U. (1972): Host specificity of DNA produced by Escherichia coli. XIV. The role of nucleotide methylation in in vivo B-specific modification. J. Mol. Biol. 63, 1-8.

So, S. G. & Downey, K. M. (1992): Eukaryotic DNA replication. Crit. Rev. Biochem. Mol. Biol. 27, 129-155.

Southern, E. M. (1975): Detection of specific sequences among DNA fragments separated by gel electrophoresis. J. Mol. Biol. 98, 503-517.

Spackman, D. H., Stein, W. H. & Moore, S. (1960): The disulfide bonds of ribonuclease. J. Biol. Chem. 235, 648-659.

Speyer, J. F., Lengyel, P., Basilio, C. & Ochoa, S. (1962): Synthetic polynucleotides and the amino acid code, IV. Proc. Natl. Acad. Sci. USA 48, 441-448.

Speyer, J. F., Lengyel, P., Basilio, C., Wahba, A. J., Gardner, R. S. & Ochoa, S. (1963): Synthetic polynucleotides and the amino acid code. Cold Spring Harbor Symp. Quant. Biol. 28, 559-567.

Spiegelman, S. (1961): The relation of informational RNA to DNA. Cold Spring Harbor Symp. Quant. Biol. 26, 75-90.

Srb, A. M. & Horowitz, N. H. (1944): The ornithine cycle in *Neurospora* and its genetic control. J. Biol. Chem. 154, 129-139.

Stanley, W. M. (1935): Isolation of a crystalline protein possessing the properties of tobacco-mosaic virus. Science 81, 644-645.

Starlinger, P. (1980): IS elements and transposons. Plasmid 3, 241-259.

Starlinger, P. (1993): What do we still need to know about transposable element Ac? Gene 135, 251-255.

Stent, G. S. (1968): That was the molecular biology that was. Science 160, 390-395.

Stent, G. S. (1978): Paradoxes of progress. Freeman & Co.: San Francisco.

Stent, G. S. & Weisblat, D. A. (März 1982): Die Entwicklung eines einfachen Nervensystems. Spektrum der Wissenschaft: Heidelberg.

Stevens, A. (1960): Incorporation of the adenine ribonucleotide into RNA by cell fractions from *E. coli* B. Biochem. Biophys. Res. Commun. 3, 92-96.

Stillman, B. (1989): Initiation of eukarotic DNA replication *in vitro*. Ann. Rev. Cell Biol. 5, 197-245.

Stone, E. M., Rothblum, K. N. & Schwartz, R. J. (1985): Intron-dependent evolution of chicken glyceraldehyde phosphate dehydrogenase gene. Nature 313, 498-500.

Strasser, A., O'Connor, L. & Dixit, V.M. (2000): Apoptosis signaling. Ann. Rev. Biochem. 69, 217-245.

Straus, D. B. & Goldwasser, E. (1961): Uridine nucleotide incorporation into pigeon liver microsome ribonucleic acid. J. Biol. Chem. 236, 849-856.

Strohman, R. C. (2001): Genomics and human life span – what is left to extend? Nature Biotechnology 19, 195.

Sturtevant, A. H. (1913): The linear arrangement of six sex-linked factors in *Drosophila*, as shown by their mode of association. J. Exp. Zool. 14, 43-59.

Sueoka, N. (1961): Compositional correlation between deoxyribonucleic acid and protein. Cold Spring Harbor Symp. Quant. Biol. 26, 35-43.

Sugino, A. & Okazaki, R. (1973): RNA-linked DNA fragments *in vitro*. Proc. Natl. Acad. Sci. USA 70, 88-92.

Sumner, J. (1926): The isolation and crystallization of the enzyme urease. J. Biol. Chem. 69, 436-439.

Sweitzer, T. D., Love, D. C. & Hanover, J. A. (2000): Regulation of nuclear import and export. Current Top. Cell Regul. 36, 77-94.

Szilard, L. (1960): The control of the formation of specific proteins in bacteria and in animal cells. Proc. Natl. Acad. Sci. USA 46, 277-302.

Tatum, E. L. & Bonner, D. (1944): Indole and serine in the biosynthesis and breakdown of tryptophane. Proc. Natl. Acad. Sci. USA 30, 30-37.

Taylor, A. L. & Thoman, M. S. (1964): The genetic map of *Escherichia coli* K-12. Genetics 50, 659-677.

Temin, H. M. & Mizutani, S. (1970): RNA-dependent DNA polymerase in virions of Rous Sarcoma Virus. Nature 226, 1211-1213.

Tener, G. M., Khorana, H. G., Markham, R. & Pol, E. H. (1958): Studies on polynucleotides. II. The synthesis and characterization of linear and cyclic thymidine oligonucleotides. J. Am. Chem. Soc. 80, 6223-6230.

The Arabidopsis Genome Initiative (2000): Analysis of the genome sequence of the flowering plant *Arabidopsis thaliana*. Nature 4 08, 796-815.

Tilghman, S. M., Tiemeier, D. C., Polsky, F., Edgell, M. H., Seidman, J. G., Leder, A., Enquist, L. W., Norman, B. & Leder, P. (1977): Cloning specific segments of the mamalian genome: bacteriophage λ containing mouse globin and surrounding gene sequences. Proc. Natl. Acad. Sci. USA 74, 4406-4410.

Tilghman, S. M., Tiemeier, D. C., Seidman, J. G., Peterlin, J. V. & Leder, P. (1978): Intervening sequence of DNA identified in the structural portion of a mouse ß-globin gene. Proc. Natl. Acad. Sci. USA 75, 725-729.

Timoféeff-Ressovsky, N. W., Zimmer, K. G. & Delbrück, M. (1935): Über die Natur der Genmutation und der Genstruktur. Nachr. Gesellsch. Wissensch. Göttingen, Neue Folge Fachgr. VI, 190-245.

Tissières, A. & Watson, J. D. (1958): Ribonucleoprotein particles from *Escherichia coli*. Nature 182, 778-780.

Travassos, J. (1930): Nota sobre o phenomeno da agglutinabilidade transmissivel de Cantacuzene e Bonciu. Brazil Médico 44, 699-702.

Travassos, L. R. (1979): Bacterial transformation revisited: familiar and unfamiliar results of the 1920-1930 decade. ASM News 45, 420-422.

Tsugita, A., Gish, D. T., Young, I., Fraenkel-Conrat, H., Knight, C. A. & Stanley, W. M. (1960): The complete amino acid sequence of the protein of tobacco mosaic virus. Proc. Natl. Acad. Sci. USA 46, 1463-1469.

Tullius, T. D. (1989): Physical studies of protein DNA complexes. Ann. Rev. Biophys. Biophys. Chem. 18, 213-237.

Unwin, P. N. T. & Milligan, R. A. (1982): A large particle associated with the perimeter of the nuclear pore complex. J. Cell Biol. 93, 63-75.

Vaux, D. L., Weissman, I. L. & Kim, S. K. (1992): Prevention of programmed cell death in *Caenorhabditis elegans* by human bcl-2. Science 258, p. 1955-1957.

Veenstra, G. J. C. & Wolffe, A. P. (2001): Gene-selective developmental roles of general transcription factors. Trends Biochem. Sci. 26, 665-671.

Venter, J. C. et al. (274 coauthors) (2001): The sequence of the human genome. Science 291, 1304-1351.

Visconti, N.&Delbrück, M. (1953):The mechanism of genetic recombination in phage. Genetics 38, 5-33.

Vogt, V. (1969): Breaks in DNA stimulate transcription by core RNA polymerase. Nature 223, 854-855.

Vogel, G. (2001): A parakeet genome project? Science 291, 1187.

Volkin, E. & Astrachan, L. (1956): Intracellular distribution of labelled RNA after phage infection of *E. coli*. Virology 2, 433-437.

Volkin, E. & Astrachan, L. (1957): RNA metabolism in T2-infected *Escherichia coli*. In: The chemical basis of heredity. Johns Hopkins Press: Baltimore, 686-695.

Volkin, E., Astrachan, L. & Countryman, J. L. (1958): Metabolism of RNA phosphorus in *E. coli* infected with bacteriophage T7. Virology 6, 545-555.

Wade, N. (1973): Microbiology: hazardous profession faces new uncertainties. Science 182, 566-567.

Waller, J. P. (1963): The NH$_2$-terminal residues of the proteins from cell-free extracts of *E. coli*. J. Mol. Biol. 7, 483-496.

Watson, J. D. (1968): The double helix. Atheneum: New York.

Watson, J. D. (1990): The human genome project: past, present and future. Science 248, 44-49.

Watson, J. D. & Crick, F. H. C. (1953a): A structure for deoxyribose nucleic acid. Nature 171, 737-738.

Watson, J. D. & Crick, F. H. C. (1953b): Genetic implications of the structure of deoxyribonucleic acid. Nature 171, 964-969.

Watson J. D. & Crick, F. H. C. (1953c): The structure of DNA. Cold Spring Harbor Symp. Quant. Biol. 18, 123-134.

Watson, J. D. & Hayes, W. (1953): Genetic exchange in *Escherichia coli* K-12: evidence for three linkage groups. Proc. Natl. Acad. Sci. USA 39, 416-426.

Weidel, W. & Kellenberger, E. (1955): The *E. coli* B-receptor for phage T5. II. Electron microscopic studies. Biochem. Biophys. Acta 17, 1-9.

Weinberg, S. (1992): Dreams of a final theory. Pantheon Books, New York.

Weintraub, S. B. & Frankel, F. R. (1972): Identification of the T4rIIB gene product as a membrane protein. J. Mol. Biol. 70, 589-615.

Weinzierl, R. O. J., Dynlacht, B. D. & Tjian, R. (1993): Largest subunit of *Drosophila* transcription factor II D directs assembly of a complex containing TBP and a coactivator. Nature 362, 511-517.

Weiss, R. A. (1993): Dorian Gray mice. Nature 362, 411.

Weiss, S. B. (1960): Enzymatic incorporation of ribonucleoside triphosphates into the interpolynucleotide linkages of ribonucleic acid. Proc. Natl. Acad. Sci. USA 46, 1020-1030.

Weiss, S. B. & Gladstoner L. (1959): A mammalian system for the incorporation of cytidine triphosphate into ribonucleic acid. J. Am. Chem. Soc. 81, 4118-4119.

Weiss, S. B. & Nakamoto,T. (1961a): Net synthesis of ribonucleic acid with a microbial enzyme requiring deoxyribonucleic acid and four ribonucleoside triphosphates. J. Biol. Chem. 236, PC 18-20.

Weiss, S. B. & Nakamoto, T. (1961b): The enzymatic synthesis of RNA: nearest-neighbour base frequencies. Proc. Natl. Acad. Sci. USA 47, 1400-1405.

Whatley, J. M., John, P. & Whatley, F. R. (1979): From extracellular to intracellular: the establishment of mitochondria and chloroplasts. Proc. Roy. Soc. (London) [B] 204, 165-187.

Wilkins, M. H. F. (1956): Physical studies of the molecular structure of deoxyribose nucleic acid and nucleoprotein. Cold Spring Harbor Symp. Quant. Biol. 21, 75-90.

Wilkins, M. H. F., Stokes, A. R. & Wilson, H. R. (1953): Molecular structure of deoxypentose nucleic acid. Nature 171, 738- 740.

Williams, G. T. (1991): Programmed cell death: apoptosis and oncogenesis. Cell 65, 1097-1098.

Wilson, E. (1975): Sociobiology: The new Synthesis. Harvard Univ. Press: Cambridge (Mass.).

Wimberly, B. T., Brodersen, D. E., Clemons, W. M. jr., Morgan-Warren, R. J., Carter, A. P., Vonrhein, C., Hatsch, T. & Ramakrishna, V. (2000): Structure of the 30S ribosomal subunit. Nature 407, 327-339.

Wittmann, H. G. (1962): Proteinuntersuchungen an Mutanten des Tabakmosaikvirus als Beitrag zum Problem des genetischen Codes. Zeitschr. Vererbl. 93, 491-530.

Woldringh, C. L. (1976): Morphological analysis of nuclear separation and cell division during the life cycle of *Escherichia coli*. J. Bacteriol. 125, 248-257.

Wollman, E. L. (1966): Bacterial conjugation. In: Phage and the origins of molecular biology. J. Cairns, G. S. Stent & J. D. Watson, eds. Cold Spring Harbor Lab. Press: New York, pp. 216-225.

Wollman, E. L. & Jacob, F. (1955): Sur le mécanisme du transfert de matériel génétique au cours de la recombinaison chez *Escherichia coli* K12. Compt. Rend. Acad. Sci. (Paris) 240, 2449-2451.

Wollman, E. L., Jacob, F. & Hayes, W. (1956): Conjugation and genetic recombination in *E. coli*. Cold Spring Harbor Symp. Quant. Biol. 21, 141-162.

Wu, H. (1931): Chinese J. Physiol. 5, 321-344 [cited in Ann. N. Y. Acad. Sci. 325, Smith, E. L., 107-120, p. 117].

Wu, R. & Kaiser, A. D. (1968): Structure and base sequence in the cohesive ends of bacteriophage λ. DNA. J. Mol. Biol. 35, 523-537.

Wu, R. & Taylor, E. (1971): Nucleotide sequence analysis of DNA. II. Complete nucleotide sequence of the cohesive ends of bacteriophage λ. DNA. J. Mol. Biol. 57, 491-511.

Wyatt, G. R. (1951): Recognition and estimation of 5-methylcytosine in nucleic acids. Biochem. J. 48, 581-584.

Wyatt, G. R. & Cohen, S. S. (1953): The bases of the nucleic acids of some bacterial and animal viruses: The occurence of 5-hydroxymethyl-cytosine. Biochem. J. 55, 774-782.

Yankeelov, J. A. jr. & Coggins, J. R. (1971): Construction of space-filling models of protein using dihedral angles. Cold Spring Harbor Symp. Quant. Biol. 36, 585-587.

Yankofsky, S. A. & Spiegelman, S. (1962a): The identification of the ribosomal RNA cistron by sequence complementarity, I. Specificity of complex formation. Proc. Natl. Acad. Sci. USA 48, 1069-1078.

Yankofsky, S. A. & Spiegelman, S. (1962b): The identification of the ribosomal RNA cistron by sequence complementarity: II. Saturation and competitive interaction at the RNA cistron. Proc. Natl. Acad. Sci. USA 48, 1466-1474.

Yanofsky, C. (1988): Transcriptional attenuation. J. Biol. Chem. 263, 609-612.

Yanofsky, C., Carlton, B. C., Guest, J. R., Helinskyr, D. R. & Henningr U. (1964): On the colinearity of gene structure and protein structure. Proc. Natl. Acad. Sci. USA 51, 266-272.

Yasuzumi, G. (1955): Electron microscopic study on constituent chromofilaments of metabolic chromosomes. Biochem. Biophys. Acta. 16, 322-329.

Yčas, M. (1959): Aspects of ribonucleic acid synthesis. Symp. Mol. Biol. Univ. Chicago Press: Chicagor 115-121.

Yčas, M. & Vincent, W. S. (1960): A ribonucleic acid fraction from yeast related in composition to deoxyribonucleic acid. Proc. Natl. Acad. Sci. USA 46, 804-811.

Yu, J. et al. (97 coauthors) (2002): A draft sequence of the rice genome (*Oryza sativa* L. ssp. *indica*). Science 296, 79-92.

Yusupov, M. M., Yusupova, G. Z., Baucom, A., Lieberman, K., Earnest, T. N., Cate, J. H. D. & Noller, H. F. (2001): Crystal structure of the ribosome at 5.5 Å resolution. Science 292, 883-896.

Zachau, H. G., Dütting, D. & Feldmannr H. (1966): The structure of two serine transfer ribonucleic acids. Hoppe-Seyler's Z. physiol. Chem. 347, 212-235.

Zakian, V. A. (1989): Structure and function of telomeres. Ann. Rev. Genet. 23, 579-604.

Zamecnik, P. (1984): The machinery of protein synthesis. Trends Biochem. Sci. 9, 464-466.

Zamecnik, P. C. (1979): Historical aspects of protein synthesis. Ann. N. Y. Acad. Sci. 325, 269-301.

Zamecnik, P. C. & Keller, E. B. (1954): Relation between phosphate energy donors and incorporation of labelled amino acids into proteins. J. Biol. Chem. 209, 337-353.

Zamecnik, P. C., Stephenson, M. L. & Scott, J. F. (1960): Partial purification of soluble RNA. Proc. Natl. Acad. Sci. USA 46, 811-822.

Zamenhof, S., Alexander, H. E. & Leidyr G. (1953): Studies on the chemistry of the transforming activity. I. Resistance to physical and chemical agents. J. Exp. Med. 98, 373-379.

Zawel, L. & Reinberg, D. (1993): Initiation of transcription by RNA polymerase II: A multistep process. Progr. Nucl. Acids Res. Molec. Biol. 44, 67-108.

Zhang, J. & Xu, M. (2002): Apoptotic DNA fragmentation and tissue homeostasis. Trends Cell Biol. 12, 84-89.

Zinder, N. D. (ed.) (1975): RNA phages. Cold Spring Harbor Lab. Press: New York.

Zinder, N. D. & Lederberg, J. (1952): Genetic exchange in Salmonella. J. Bacteriol. 64, 679-699.

## General Sources and Further Reading

Brock, T.D. ( 1990): The emergence of bacterial genetics. Cold Spring Harbor Lab. Press: New York, 346 pp. (Detailed history of formal bacterial and phage genetics; molecular genetics is considered only marginally.)

Cairns, J., Stent, G.S. & Watson, J.D. (eds.) (1996): Phage and the origins of molecular biology. Cold Spring Harbor Lab. Press: New York, 340 pp. (Festschrift commemorating Max Delbrück's 60th birthday, with scientific and personal reports by his colleagues and friends.)

Carlson, E.A. (1966): The gene: a critical history. Saunders: Philadelphia, 301 pp. (Historical analyses with regard to our changing views about the gene.)

Chadarevian, S. (2002): Designes for life. Cambridge University Press: Cambridge, U.K., 423 pp. (Very detailed, profoundly researched analysis of the developments in molecular biology and science policy after World War II, centered on work done in Cambridge, U.K.)

Crick, F. (1988). What mad pursuit. Basic Books: New York, 182 pp. (Autobiographical, personal view of the problems and historic development of molecular biology.)

Dunn, L.C. (1965): A short history of genetics. McGraw-Hill: New York, 261 pp. (Mendel and classical genetics.)

Echols, H. (2001): Operators and promoters. University of California Press: Berkeley and Los Angeles, California, 466 pp. (Detailed historical account, with emphasis on the problems of gene regulation in bacteria and bacteriophages, especially phage λ.)

Feyerabend, P. (1978): Science in a free society. NLB: London, 221 pp. (Intelligently chaotic, liberal, alternative view of the scientific establishment.)

Judson, H.F. (1980): The eighth day of creation. Makers of the revolution in biology. Simon & Schuster: New York, 686 pp.; revised edition, 1994. (Detailed historical appaisal of molecular biology; a journalistic approach with many personal interviews of the scientists directly involved; a masterly accomplishment with a wealth of information otherwise not obtainable.)

Kay, L.E. (2000): Who wrote the book of life? A history of the genetic code. Stanford University Press: Stanford, California, 441 pp. (A scholar investigation relating to the historical elaboration of the concept of genetic information, by a professional historian of science.)

Lwoff, A. & Ullmann, A. (eds.) (1979): Origins of molecular biology. A tribute to Jacques Monod. Academic Press: New York, 246 pp. (Reminiscences of colleagues and collaborators in memory of Jacques Monod.)

Olby, R. (1974): The path to the double helix. MacMillan: London, 510 pp.; revised edition in 1994. (Detailed historical analysis of research related to DNA.)

Portugal, F.H. & Cohen, J.S. (1977): A century of DNA. MIT Press: Cambridge, Mass., 384 pp. (Biochemistry and molecular biology of nucleic acids, since the time of their discovery.)

Stent, G.S. (1979): Paradoxes of progress. Freeman: San Francisco, 231 pp. (Epistemological considerations with special emphasis on molecular biology.)

Sturtevant, A.H. (1966): A history of genetics. Harper & Row: New York, 165 pp. (Classical genetics.)

Watson, J.D. (1968): The double helix: Atheneum: New York, 133 pp. (Autobiographical, self-ironic account of the events leading to the discovery of the structure of DNA.)

# INDEX

# COPYRIGHT REGISTER

The following copyright holders have given their kind permission to reproduce the corresponding materials listed below. In the case of copyrighted reproductions not mentioned in this list, the copyright holders are cited in the text. For English versions of this book, Kluiver Academic Publishers Dordrecht is the copyright holder for all figures whose authorship is not expressly mentioned in the text ; for editions in other languages, copyrights are held by the Wissenschaftliche Buchgesellschaft Darmstadt; of these figures, Figs. 5.8, 10.1, 11.3, and 14.2, are by C. Baumann, Fig. 22.2 (right), by R. Freyer, and the others by the author. Photographic artwork is by M. Messerschmid.

Fig. 1.7: ©Donald E. Nicholson, Univ. Leeds, U.K.; Fig. 1.9: ©Linus Pauling Institute of Science and Medicine, Palo Alto, California; Fig. 1.11: Sir John C. Kendrew, Cambridge, U.K.; Fig. 1.12: © Academic Press, London; Fig. 1.13: © 1971 Cold Spring Harbor Laboratory, Cold Spring Harbor, New York; Fig. 2.2: © New Scientist, IPC Magazines, London; Fig. 2.7: © American Society for Microbiology, Washington DC; Fig. 2.8a: Reprinted by permission from Nature 191, 144-148, Copyright © 1961 Macmillan Magazines Ltd; Fig. 2.8b: Copyright © 1983 American Institute of Biological Sciences, Washington DC; Fig. 2.15a: © Genetics Society of America, Bethesda, Maryland; Figs. 3.3, 3.4, 3.5: © Sigma Xi, The Scientific Research Society, Research Triangle Park, NC; Fig. 4.3: Reproduced from J. Gen. Physiol. (1993) 22, 365-384, by copyright permission of Rockefeller Univ. Press, New York; Fig. 4.6: © 1951 Cold Spring Harbor Laboratory , Cold Spring Harbor, New York; Fig. 4.9a: © Genetics Society of America, Bethesda, Maryland; Fig. 4.9b: © American Society for Microbiology, Washington DC; Figs. 4.11 & 4.12: © Cold Spring Harbor Labaratory, Cold Spring Harbor, New York; Fig. 5.1: © Elsevier Science Publ. BV, Amsterdam, NL; Fig. 5.2a: Reprinted by permission from Nature 171, 738-741, copyright © 1971 Macmillan Magazines Ltd; Fig. 5.2c: © Academic Press, London; Fig. 5.3: © Birkhäuser Verlag, Basel, CH; Fig. 5.5a: Reprinted by permission from Nature 175, 834-838, copyright © 1955 Macmillan Magazines Ltd; Fig. 6.1: Reprinted by permission from Nature 175, 737-738, copyright © 1953 Macmillan Magazines Ltd; Fig. 6.3: © Elsevier Science Publishers BV, Amsterdam, NL; Figs. 6.4, 6.5, 6.6: © The Johns Hopkins University Press, Baltimore, MD; Fig. 6.8: © Academic Press, Inc.; Figs. 7.1, 7.4: Reproduced from J. Gen. Physiol (1974) 62, 911-917 by copyright permission of the Rockefeller Univ. Press, New York; Figs. 8.1, 8.2, 8.3, 8.5, 8.6: Reprinted by permission from Nature 173, 318; 171, 1027-28; 182,111-12; 192, 1227-32; 178, 792-94; copyright © 1954, 1953, 1958, 1961, and 1956, Macmillan Magazines Ltd; Fig. 8.7: © Johns Hopkins University Press, Baltimore, MD; Fig. 9.1: © American Society for Biochemistry

& Molecular Biology, Baltimore, MD; Figs. 10.4, 10.7: ©1966 Cold Spring Harbor Laboratory , Cold Spring Harbor, New York; Fig. 12.1: © Editions GAUTHIER-VILLARS, Paris; Fig. 12.2: © Academic Press, Orlando, Florida; Fig. 12.3: © Genetics Society of America, Bethesda, Maryland; Fig. 12.4a: © Academic Press, Orlando, Florida; Fig. 13.1: © Academic Press, London, U.K.; Figs. 13.2, 13.3: © Editions GAUTHIER-VILLARS, Paris; Fig. 15.1: © Academic Press, London, U.K.; Fig. 17.1 top: Reprinted by permission from Nature 269, 655-61 , copyright © 1977 Macmillan Magazines Ltd; Fig. 17.1 bottom: © Annual Reviews, Inc.; Fig. 18.1: © GIT Verlag GmbH, Darmstadt; Fig. 18.2: Reproduced by permission from Nature 361, 109-110, copyright © 1993 Macmillan Magazines Ltd., London, U.K.; Fig. 19.1: © Academic Press Ltd, London, U.K.; Fig. 19.3: Reproduced from The Journal of General Physiology (1966) 49(6) Part 2, p.11 by copyright permission of the Rockefeller University Press, New York; Fig. 19.6: © Academic Press Ltd., London, U.K.; Fig. 20.2: © Firma Boehringer Mannheim; Fig. 21.1: © 1971 University of Chicago Press, Chicago, IL; Fig. 21.2: Reproduced from The Journal of General Physiology (1981) 91, 240-246 by copyright permission of the Rockefeller University Press, New York; Fig. 21.3: Reproduced from the Journal of Cell Biology (1977) 72, 118-132, and (1982) 93, 63-75, by copyright permission of the Rockefeller University Press, New York; Fig. 21.4: © Academic Press, Orlando, Florida; Fig. 21.5 left: © 1956 Cold Spring Harbor Laboratory, Cold Spring Harbor, New York; Fig. 21.5 right: Reprinted by permission from Nature 175, 834-838, copyright © 1955 Macmillan Magazines Ltd, London, U.K.; Fig. 21.6, top: © Sigma Xi, The Scientific Research Society, Research Triangle Park, NC; Fig. 21.6, bottom: © Elsevier Science Publishers B.V. Amsterdam, NL; Fig. 21.8: © Cell Press, Cambridge, Massachusetts; Fig. 21.11a: Reprinted by permission from Nature 313, 498-500; copyright © 1985 Macmillan Magazines Ltd., London, U.K.; Fig. 21.11 b: © The American Society for Biochemistry & Molecular Biology; Fig. 21.12: Permission from Ann. Rev. Cell Biol. 8, © 1992 by Ann. Reviews Inc.; Fig. 22.4: by permission of Oxford University Press, Oxford, U.K.; Fig. 23.1: © 1988 Elsevier Publications, Cambridge, U.K.; Fig. 24.1: Reprinted by permission from Nature 362, 511-517, copyright © 1993 Macmillan Magazines Ltd., London, U.K.

# ACKNOWLEDGEMENTS

This book could not have been completed without the help of many dedicated collaborators. The active interest of many students in the lectures on historical aspects of biology, held by my colleagues at the University of Freiburg, C. Bresch, J. Campos-Ortega, R. Hertel, H. Kössel, and myself, was the original motivation for writing the present text.

If I here thank all the critics who read versions of the manuscript, that does not mean that I always followed their advise or that they agree with all my points of view; nevertheless, the same thanks to all: W. Arber, S. Bahn, J. Bandel, C. Baumann, C. Bresch, F. Brombacher, A. Doerries, R. Freyer, M. Fuchs, E. Härle, C. Hausmann, J. Hausmann, R. Hausmann (my special thanks to my wife and children!), R. Hertel, K. Hilse, D. Horn, G. Igloi, H. Jäger, H. Kleinig, C. Korsten, H. Kössel, B. Kromer, A. Luckenbach, S. Lutz, S. Müller, M. Neumann, M. Reck, J. Schliebitz, P. Starlinger, A. Technau, R. Wehler, S. Weist, M. Willemsen.

C. Baumann I would like to thank for her drawings and M. Messerschmid for the photographic artwork, B. Probst and D. Becherer for their patient literature searches, C. Tajeddine for endless hours of typing and preparing the layout, and M. Willemsen for helping to prepare the glossary.

AUG 0 3 2004 DATE DUE

APR 1 7 2007

DEMCO INC 38-2971